Advances in temperate agroforestry

It is widely recognised that agriculture is a significant contributor to global warming and climate change. Agriculture needs to reduce its environmental impact and adapt to current climate change whilst still feeding a growing population, i.e. become more 'climate-smart'. Burleigh Dodds Science Publishing is playing its part in achieving this by bringing together key research on making the production of the world's most important crops and livestock products more sustainable.

Based on extensive research, our publications specifically target the challenge of climate-smart agriculture. In this way we are using 'smart publishing' to help achieve climate-smart agriculture.

Burleigh Dodds Science Publishing is an independent and innovative publisher delivering high quality customer-focused agricultural science content in both print and online formats for the academic and research communities. Our aim is to build a foundation of knowledge on which researchers can build to meet the challenge of climate-smart agriculture.

For more information about Burleigh Dodds Science Publishing simply call us on +44 (0) 1223 839365, email info@bdspublishing.com or alternatively please visit our website at www.bdspublishing.com.

Related titles:

Advances in Conservation Agriculture Volume 1
Print (ISBN 978-1-78676-264-1); Online (ISBN 978-1-78676-266-5, 978-1-78676-267-2)

Agroforestry for sustainable agriculture
Print (ISBN 978-1-78676-220-7); Online (ISBN 978-1-78676-222-1, 978-1-78676-223-8)

Promoting pollination and pollinators in farming
Print (ISBN 978-1-80146-098-9); Online (ISBN 978-1-80146-100-9, 978-1-80146-101-6)

Protecting natural capital and biodiversity in the agri-food sector
Print (ISBN 978-1-80146-351-5); Online (ISBN 978-1-80146-352-2, 978-1-80146-353-9)

Reconciling agricultural production with biodiversity conservation
Print (ISBN 978-1-78676-348-8); Online (ISBN 978-1-78676-350-1, 978-1-78676-351-8)

Understanding and fostering soil carbon sequestration
Print (ISBN 978-1-78676-969-5); Online (ISBN 978-1-78676-971-8, 978-1-78676-972-5)

Understanding and preventing soil erosion
Print (ISBN 978-1-80146-379-9); Online (ISBN 978-1-80146-380-5, 978-1-80146-381-2)

Chapters are available individually from our online bookshop:
https://shop.bdspublishing.com

climate•SMART•publishing
IN AGRICULTURAL SCIENCE

BURLEIGH DODDS SERIES IN AGRICULTURAL SCIENCE
NUMBER 156

Advances in temperate agroforestry

Edited by Professor María Rosa Mosquera-Losada, Universidade de Santiago de Compostela, Spain; Dr Ladislau Martin, Embrapa, Brazil; Professor Anastasia Pantera, Agricultural University of Athens, Greece; and Dr Allison Chatrchyan, Cornell University, USA

burleigh dodds
SCIENCE PUBLISHING

Published by Burleigh Dodds Science Publishing Limited
82 High Street, Sawston, Cambridge CB22 3HJ, UK
www.bdspublishing.com

Burleigh Dodds Science Publishing, 1518 Walnut Street, Suite 900, Philadelphia, PA 19102-3406, USA

First published 2025 by Burleigh Dodds Science Publishing Limited

Library of Congress Control Number: 2025931724

British Library Cataloguing in Publication Data
A catalogue record for this book is available from the British Library

ISBN 978-1-80146-719-3 (print)
ISBN 978-1-80146-720-9 (ePub)
ISBN 978-1-80146-721-6 (PDF)
ISSN 2059-6936 (print)
ISSN 2059-6944 (online)

DOI: 10.19103/AS.2024.0141

Typeset by Deanta Global Publishing Services, Dublin, Ireland

Contents

Series list

Title	Series number
Achieving sustainable cultivation of maize - Vol 1 From improved varieties to local applications Edited by: Dr Dave Watson, CGIAR Maize Research Program Manager, CIMMYT, Mexico	001
Achieving sustainable cultivation of maize - Vol 2 Cultivation techniques, pest and disease control Edited by: Dr Dave Watson, CGIAR Maize Research Program Manager, CIMMYT, Mexico	002
Achieving sustainable cultivation of rice - Vol 1 Breeding for higher yield and quality Edited by: Prof. Takuji Sasaki, Tokyo University of Agriculture, Japan	003
Achieving sustainable cultivation of rice - Vol 2 Cultivation, pest and disease management Edited by: Prof. Takuji Sasaki, Tokyo University of Agriculture, Japan	004
Achieving sustainable cultivation of wheat - Vol 1 Breeding, quality traits, pests and diseases Edited by: Prof. Peter Langridge, The University of Adelaide, Australia	005
Achieving sustainable cultivation of wheat - Vol 2 Cultivation techniques Edited by: Prof. Peter Langridge, The University of Adelaide, Australia	006
Achieving sustainable cultivation of tomatoes Edited by: Dr Autar Mattoo, USDA-ARS, USA and Prof. Avtar Handa, Purdue University, USA	007
Achieving sustainable production of milk - Vol 1 Milk composition, genetics and breeding Edited by: Dr Nico van Belzen, International Dairy Federation (IDF), Belgium	008
Achieving sustainable production of milk - Vol 2 Safety, quality and sustainability Edited by: Dr Nico van Belzen, International Dairy Federation (IDF), Belgium	009
Achieving sustainable production of milk - Vol 3 Dairy herd management and welfare Edited by: Prof. John Webster, University of Bristol, UK	010
Ensuring safety and quality in the production of beef - Vol 1 Safety Edited by: Prof. Gary Acuff, Texas A&M University, USA and Prof. James Dickson, Iowa State University, USA	011
Ensuring safety and quality in the production of beef - Vol 2 Quality Edited by: Prof. Michael Dikeman, Kansas State University, USA	012
Achieving sustainable production of poultry meat - Vol 1 Safety, quality and sustainability Edited by: Prof. Steven C. Ricke, University of Arkansas, USA	013
Achieving sustainable production of poultry meat - Vol 2 Breeding and nutrition Edited by: Prof. Todd Applegate, University of Georgia, USA	014

Acknowledgements

We wish to acknowledge the following for their help in reviewing particular chapters:

- Chapter 2: Romane Mettauer, SAS-Institut Agro-INRAE, France and Emeritus Prof. Andy Gordon, University of Guelph, Canada
- Chapter 5: Dr Naresh Thevathasan, University of Guelph, Canada
- Chapter 11: Dr Robert Borek, Institute of Soil Science and Plant Cultivation, Poland and Dr Raphael Paut, INRAE, France
- Chapter 16: Prof. Srinivasa Mentreddy, Alabama A&M University, USA and Dr Susan Leopold, United Plant Savers, USA

Introduction

The USDA defines agroforestry as: 'the intentional mixing of trees and shrubs into crop and animal production systems to create environmental, economic and social benefits.' With growing concern about climate change and increasing interest in agroecological approaches to achieve a more sustainable agriculture, agroforestry has become a major focus of research. This collection reviews key advances in the subject such as ways of assessing ecosystem service benefits from agroforestry and developments in different types of silvopastoral and silvoarable practice. The book builds on an earlier 2019 volume published by Burleigh Dodds Science Publishing: *Agroforestry for sustainable agriculture*.

The book is split into three parts. Part 1 chapters discuss the value of agroforestry ecosystem services, including the benefits of temperate agroforestry in promoting soil health and enhancing carbon sequestration, as well as the application of natural capital accounting to valuation of ecosystem services. Part 2 chapters cover advances in silvopastoral systems, highlighting different types of system such as forest/timber planting, orchards/vineyards and adding trees to pasture. Finally, Part 3 chapters explores advances in silvorable systems, looking at different types such as shelterbelts and alley cropping. It also discusses development in forest farming and urban agroforestry.

Part 1 Measuring and valuing agroforestry ecosystem services

The first group of chapters point out the key benefits of temperate agroforestry and the application of natural capital accounting in valuing ecosystem services. Chapter 1 focuses on the impact of climate change and unsustainable management practices on soil health. It explores how temperate cropland agroforestry systems can promote soil biological health. By altering community composition, agroforestry systems result in enhanced population densities and activities of soil microorganisms and an increase in diversity of soil fauna. The chapter highlights the particular contribution of alley cropping and shelterbelt systems to promoting soil health and providing many environmental benefits.

Chapter 2 continues to point out the key benefits of temperate agroforestry, with a focus on its application to mitigate climate change through enhancing carbon sequestration in soil and plant biomass. The chapter evaluates the potential of carbon sequestration in temperate agroforestry systems by analysing existing data from temperate agroforestry practices in different regions worldwide. It is projected that agroforestry practices have the potential to annually offset 20%, 18%, 12%, and 8% of total CO2 emissions in the UK, Europe, Canada and the USA, respectively. However, there is a need for further research to more accurately estimate reliable values for carbon sequestration rates.

Chapter 3 discusses natural capital: the stock of renewable and non-renewable resources that combine to provide flows of services and benefits for humanity. Studies suggest that over half of the global gross domestic product (GDP) is highly or moderately dependant on nature. Agroforestry has been identified as a nature-based solution for addressing on-going issues such as climate change and loss of biodiversity. However, quantitative information on the natural capital benefits associated with agroforestry remains relatively sparse. The chapter demonstrates how individual agroforestry projects could be recognized as assets on the balance sheet of an enterprise. The chapter explores how to achieve transparent record keeping of natural ecosystem assets, their extent and condition. This will support decision-making, planning, risk analysis, investment and monitoring over time to support practices such as agroforestry.

Part 2 Advances in silvopastoral systems

The next group of chapters focus on different types of silvopastoral systems. Chapter 4 provides an overview of silvopastoral systems where trees are grown in grazed or mown pasture in a regular or varied pattern. There are many ways of achieving silvopastoral systems, such as re-spacing an established woodland or planted forest by selective thinning and establishing pasture, or by planting trees into already established pasture. Silvopasture plays an important role in delivering ecosystem services like more sustainable food production, boosting climate resilience, improving air and water quality, and promoting biodiversity. Agroforestry clearly aligns will with objectives of agro-environment measures and is likely to be supported by conservation bodies and community development groups.

Chapter 5 study focuses on the Mediterranean region with a climate conducive to the growth of evergreen and sclerophyllous species with limited wood productivity compared to their counterparts in Northern and Central Europe. This limitation has historically led to Mediterranean forests with a rich tradition of harvesting animal and plant products. In the Iberian Peninsula e.g. it has been estimated that silvopastoral systems cover about 6.5 million hectares, representing approximately 30% of the total agricultural area. The chapter explores the complex interaction between the key oak species in the region (holm oak, cork oak and pyrenean oak) and monogastric animals. The chapter discuses the key role of Mediterranean silvopastoral systems as a resilient and sustainable model of agroforestry, incorporating ecological conservation with socio-economic development and cultural preservation. To the face the challenges that comes with the integration of such a model, emphasis must be laid upon the application of innovative approaches, focusing on the need for targeted conservation strategies, policy support, and the promotion of sustainable practices.

With the rise in the global population and per capita consumption of livestock products, Chapter 6 highlights the resulting risk of deforestation to clear land for grazing. A possible alternative are integrated crop-livestock systems (ICLS). The chapter discusses the integration of livestock into herbaceous crops and trees resulting in a system known as integrated crop-livestock-forestry systems (ICLFS) or agro-silvopastoral systems. Grazing in fruit tree crops is referred to as orchard grazing, a fruit-tree-based agroforestry system (FTB-AFS) involving orchard intercropping. The focuses on systems based on fruit trees and vines. Advantages of integration include increasing land productivity, weed control, improved aeration and soil fertilization with reduced inputs and environmental impact. The chapter also addresses challenges such as competition for water, nutrients and light, damage to trees and/or to the green cover, potential contamination risks and management challenges related to the complexity of livestock.

Chapter 7 discusses associating trees and shrubs with pasture as a type of silvopastoral systems (SPS). The chapter comprises four case studies of SPS from Argentina, categorised into two or three-strata systems involving beef cattle. Two-strata systems are composed of pastures or natural grassland with a combination of timber and fruit tree species. Triple-strata systems also incorpórate shrubs as an intermediate layer. The upper layer consists of a diverse variety of multipurpose trees. The mid layer contains shrubs species of high protein value with the third layer of pasture. As the chapter points out, these systems deliver a range of benefits

The chapter further goes on to explain the various environmentl services delivered by these SPS. They increase biomass and soil organic carbon (SOC) storage while maintaining agricultural Production and profitability. SPS also help in the conservation of biodiversity and water.

Part 3 Advances in silvoarable and other systems

The final group of chapters focuses on the advances in silvorable and other systems of agroforestry. Chapter 8 highlights both the role of biomass production to provide bioenergy and bioproducts and potential trade-offs in losing land for crop cultivation. The chapter provides a fascinating case study in which widespread deployment of riparian buffers and windbreaks is modelled across Europe, in over 81,000 landscapes. The chapter discusses how multifunctional windbreaks and riparian buffers can promote perennial energy crop cultivation alongside cultivation of food crops, as well delivering other benefits such as enhanced soil carbon sequestration. However, it also show the importance of variables such as type of soil and historic land-use in determining impact. It also highlight the need for an integrated approach to planning, implementing and incentivising such schemes (which e.g. cut across energy and environmental programmes and may involve very different groups of stakeholders).

Chapter 9 highlights that understanding and predicting biodiversity in agroecosystems, especially in agroforestry, is challenging due to diverse biological, environmental, and management variables. The chapter describes a modelling approach to compare associated biodiversity in different agroecosystems. The application of ecological models helps to analyse space and time dynamics along with the overall behaviour of complex systems. The chapter focusses on alley cropping as an agroforestry systems that consists of alternating crop alleys which constitutes the arable part of the system with rows of trees and an understory vegetation strip. Benefits include enhancing biodiversity conservation in agricultural landscapes. The highlights how modelling can support environmental managers and practitioners so the best policies can be implemented for biodiversity conservation in agricultural landscapes.

Forest farming, as discussed in Chapter 10, refers to cultivating non-timber crops under a forest canopy. Forest farming is a growing sector of agroforestry in the U.S e.g.. The chapter focuses on temperate forest farming of woodland botanicals. Forest farming aims to both diversify income and biological diversity as well as improve resilience to climate change. The chapter reviews key forest farming practices, including modern tools such as digital site selection software such as PlantShoe as well as those helping with, plant and tree identification. The chapter also discusses regulations and frameworks affecting forest farming concerning how to balance management of natural resources with a successful farming operation.

The final chapter explores developments in urban agroforestry, Chapter 11 defines it as 'a form of nature-based solution that can be applied at the whole city or town scale. The design of an agroforestry subsystem embodies the idea that trees or other woody plants can act as a "mother" to adjacent crops, mushrooms, or animals and in certain circumstances the "children" can return the favour'. The chapter reviews the literature relating to designing resilient urban agroforestry systems using approaches such as PAMASAL: purpose, advice, measures of success, agroecology, starting points, adaptive management and layout. The chapter also discusses two case studies on urban agroforestry in practice in the UK.

Part 1

Measuring and valuing agroforestry ecosystem services

Chapter 1

Assessing the benefits of temperate cropland agroforestry for promoting soil biological health

Lukas Beule, Anna Vaupel, Zita Bednar-Konski and Andrea Krähmer, Julius Kühn Institute (JKI) – Federal Research Centre for Cultivated Plants, Germany

1 Introduction

Soil is one of the most valuable finite natural resources available to humankind. Soils are not only essential for production of food but also for air and water quality, climate mitigation, biodiversity, food security, and consequently human health. Yet, due to climate change and unsustainable management practices, the integrity of soils is at stake across the globe (FAO 2015). Soil 'health' is the most recent attempt to describe the vitality of soil. Soil health emerged from the concept of soil quality, which has been evolving ever since the importance of soil was recognized beyond soil fertility (i.e. suitability of soil to produce food, feed, timber, and fiber) in the 1990s (Bünemann et al. 2018). Since then, in addition to its agricultural potential, soils have been recognized as crucial for maintaining the functioning of terrestrial ecosystems by regulating soil-derived ecosystem functions and services at various scales (e.g. nutrient cycling, climate regulation, biodiversity, and the maintenance of air and water quality). Moreover, the biological component of soils has been acknowledged in addition to physical and chemical soil properties (Pankhurst et al. 1997). In contrast to the concept of soil quality, the scope of soil health extends beyond

http://dx.doi.org/10.19103/AS.2024.0141.02

the anthropocentric notion to the broader sustainability aims that include planetary health (Pankhurst et al. 1997; Lehmann et al. 2020). More than two decades ago, soil health has been defined as 'the capacity of soil to function as a vital living system, within ecosystem and land-use boundaries, to sustain plant and animal productivity, maintain or enhance water and air quality, and promote plant, animal and human health' (Doran and Zeiss 2000). However, the term 'soil health' is often used synonymously to 'soil quality' as the conceptualization and operationalization of soil as a multi-functional system continues to be a matter of an ongoing debate (Bünemann et al. 2018).

A decade ago, the essential role of soil in provision of ecosystem services that are crucial for human health and existence was emphasized by the term 'soil security' (McBratney et al. 2014). Furthermore, analogous to food and water security, soil security has been proposed to be equivalent to other human rights (Koch et al. 2013). As soil security covers the need for access to soil-derived ecosystem services, which strongly depends on soil health, soil health is in a broader sense an important pillar of soil security and therefore a part of a political sphere. On account of recognition of the importance of soil health, research efforts have been undertaken to develop approaches to quantify and monitor soil health. Within this context, identification of suitable indicators of soil health is essential and an ongoing process (Bünemann et al. 2018; Rinot et al. 2019).

Indicators of soil health can be classified into physical, chemical, and biological indicators (Bünemann et al. 2018), although some indicators cannot be assigned exclusively to just one of the three categories (Lehmann et al. 2020). Even though the importance of soil biota for soil health has been recognized over 25 years ago (Pankhurst et al. 1995, 1997), biological indicators usually account for less than 20% of indicators in soil health assessments (Lehmann et al. 2020). Inadequate inclusion of biological indicators can be attributed to several constraints, such as (1) long history of soil fertility assessments that have been predominantly based on chemical and physical soil indicators, (2) complexity of soil biota and biological processes, (3) highly contextual-dependent interpretation of biological indicators, and (4) ongoing elaboration of effective and practical methodological approaches. The most commonly included soil biological indicators are microbial biomass and activity, earthworm abundance, and mineralization rates (Lehmann et al. 2020). However, numerous other biological indicators with strong emphasis on soil microorganisms are being initiated for inclusion in assessment of soil health (see Bünemann et al. 2018; Lehmann et al. 2020; Fierer et al. 2021).

Soil biota are crucial drivers of numerous soil-derived ecosystem services (Kibblewhite et al. 2008; Wall et al. 2012; Schulte et al. 2014) and healthy soils are often associated with extremely complex communities of soil organisms, which range from microscopic microorganisms to soil megafauna. The large diversity in shape and size of soil organisms is also mirrored by their enormous taxonomic and genetic diversity. A recent study estimated that soil is the most biodiverse habitat on our planet as it is home to approximately 59% of all species (Anthony et al. 2023). The high biodiversity of soil biota is also reflected by their community assemblage and the provision of functions across spatial and temporal scales as well as in dependence of environmental conditions and management (Jeffery and Verheijen 2020). The selection of suitable biological indicators and quantitative assessment of their contribution to soil health is thus highly context-specific. Therefore, it has been proposed that research on soil health should rather revolve around identifying optimal management practices for individual farming systems under local environmental conditions rather than on finding a universal set of soil health indicators (Jeffery and Verheijen 2020; Fierer et al. 2021).

In addition to the preservation and restoration of soil health, there are multiple ecological, economical, and socio-cultural expectations placed on agricultural systems. For example, agricultural systems should not just be resistant and resilient to climate change but should also be able to mitigate climate change. They should improve yield stability, be accepted by society, and preserve biodiversity. Overall, agricultural systems are expected to deliver multiple functions simultaneously. In this context, agroforestry systems are proposed as promising multifunctional land-use systems that have the potential to meet the high requirements above while improving soil health, especially its biological components (Marsden et al. 2020; Beule et al. 2022b). In this chapter, we present recent advances in the field of soil biological health in temperate cropland agroforestry systems with a focus on alley cropping and shelterbelt systems.

2 The promotion of soil biological health through agroforestry – from microorganisms to earthworms

Soil organisms have various lifestyles and assemble in complex communities across space and time. Compared to agricultural systems without trees, agroforestry systems introduce additional complexity in space and constancy. Through the incorporation of woody perennials, agroforestry systems introduce spatial heterogeneity as pointed out for soil properties such as soil organic carbon (Bambrick et al. 2010) as well as soil microbial (Beule and Karlovsky 2021a) and faunal communities (Cardinael et al. 2019). Although less studied, the tree component of agroforestry systems also introduces temporal heterogeneity due to differences in phenology and longevity (Stamps and Linit 1997). This spatio-temporal heterogeneity within agroforestry systems is expected to introduce new habitats for soil organisms.

2.1 Promotion of soil microbial communities and processes

Soil microorganisms comprise prokaryotic and eukaryotic organisms such as bacteria, fungi, and archaea. Microbial communities make substantial contributions to soil health, thus, their community characteristics (e.g. biomass, diversity, and community composition) and activity are important indicators in soil health assessments.

Among soil microbial indicators, quantification of the microbial biomass is frequently included in soil health assessments (Lehmann et al. 2020). There are various methodological approaches to quantify the biomass of soil microorganisms ranging from the traditional fumigation–extraction method to molecular techniques that target marker genes of specific microbial groups (Jörgensen et al. 2024). Although each method comes with its own advantages and limitations, across methods, a growing body of literature revealed that trees in temperate agroforestry systems promote the biomass of soil microorganisms (Unger et al. 2013; Beuschel et al. 2019; Beule et al. 2020; Guillot et al. 2021). While some studies showed that microbial populations are exclusively promoted under the trees (Beuschel et al. 2019), others were able to show a gradual decrease with increasing distance from the trees (Luo et al. 2022). These discrepancies are likely related to differences in tree age across sites (Clivot et al. 2020; Beule and Karlovsky 2021b) as well as sampling time and frequency (i.e. single or repeated sampling) within sites (*cf.* Beuschel et al. 2019; Luo et al. 2022). Additionally, studies using molecular techniques (Beule et al. 2019a, 2020, 2022a) or phospholipid fatty acid (PLFA) analysis (Guillot et al. 2021) revealed that spatial gradients from the trees are present for some but not all microbial groups, which may serve as an additional explanation for the absence/presence of spatial gradients across studies.

In addition to the biomass of soil microorganisms, the importance of including their diversity in soil health assessments has been pointed out over two decades ago (Nielsen et al. 2002). Back then, the authors pointed out the urge for methodological development of tools to assess microbial diversity. Since then, methods that investigate microbial diversity developed rapidly, culminating in the development of high-throughput sequencing technologies. As of writing, amplicon-sequencing approaches of marker genes to assess soil microbial diversity are the most commonly used (Schöler et al. 2017).

Microbial diversity is commonly assessed using indices of alpha-, beta-, and gamma-diversity, whereas alpha diversity describes the diversity within a given system (e.g. Shannon diversity and species richness), beta diversity measures the diversity among systems (e.g. community similarity), and gamma diversity measures the diversity at large spatial scales (e.g. diversity at landscape level). To date, only a few studies have used amplicon sequencing to analyze soil microbial diversity in temperate agroforestry systems.

Studies that used amplicon sequencing to investigate microbial alpha-diversity in temperate agroforestry systems showed variable responses of soil bacteria and fungi. While soil bacterial diversity remained largely unchanged in response to the poplar trees in alley cropping systems in Germany (Beule and Karlovsky 2021a), soil fungal richness in the same systems increased with decreasing distance to the trees (Beule et al. 2021). A recent study that investigated the response of the soil fungal community to the establishment of an alley cropping system over one-and-a-half years was able to show that species richness decreased under the trees, whereas Shannon diversity remained unaffected (Beule and Karlovsky 2021b). In contrast, in Canadian windbreak systems, the Shannon diversity of bacteria increased with increasing distance to the trees, whereas the Shannon diversity of fungi showed no consistent response to distance (Rivest et al. 2020). Another study investigated tree-distance and tree-species effects on soil fungal communities in a temperate alley cropping system and found no effects of tree-distance and tree-species on fungal diversity (Vaupel et al. 2023). Therefore, responses of soil microbial alpha-diversity to agroforestry practice remain inconsistent and the reason for the observed discrepancies among studies remains yet unknown. We speculate that these discrepancies are due to differences in the plant species composition and management of the studied agroforestry systems among studies and require further investigation, especially across time given that all studies above determined alpha diversity only at a time point.

In contrast, studies that investigated variations in microbial community composition within alley cropping systems (i.e. beta-diversity) using amplicon sequencing consistently showed that soil bacterial (Beule and Karlovsky 2021a) and fungal communities (Beule et al. 2021; Vaupel et al. 2023) under trees are compositionally different from those in neighboring crops. Effects on community composition were reported as early as 6 months after the establishment of the agroforestry system (Beule and Karlovsky 2021b), which is unsurprising considering the complex interactions in the plant–soil–microbe system. These results confirm earlier studies that used PLFA analysis and were able to show that temperate agroforestry can alter microbial community composition (Lacombe et al. 2009; Unger et al. 2013).

Several studies that investigated changes in community composition in response to agroforestry identified the members of the soil microbial communities that were associated with the compositional shifts. A recent study from a Mediterranean walnut–pea alley cropping system used PLFA analysis and found that the trees promote Gram-positive and Gram-negative bacteria as well as Actinobacteria (Guillot et al. 2021). The promotion of gram-negative bacteria and Actinobacteria agrees with the findings of other studies conducted in other temperate agroforestry systems (Unger et al. 2013; Beule et al. 2020). Another study that sequenced soil bacterial communities in three poplar-based alley cropping systems in Germany found similar shifts in community composition in response to the trees across sites (Beule and Karlovsky 2021a). These shifts included, among other things, a greater relative abundance of members of the genera *Actinomycetospora*, *Bradyrhizobium*, and *Mesorhizobium* and lower relative abundance of *Pseudarthrobacter* spp. under the trees than in the crop rows and an adjacent open cropland. At the same study sites, the authors sequenced fungal communities and showed that *Basidiomycota* were strongly promoted by the trees (Beule and Karlovsky 2021a). Furthermore, the trees increased the community share of ectomycorrhizal fungi (EMF) such as *Cortinarius*, *Geopora*, *Inocybe*, *Laccaria*, and *Preussia* spp. (Beule et al. 2021; Vaupel et al. 2023). Likewise, studies investigating communities of arbuscular mycorrhizal fungi (AMF) found compositional differences between tree and crop rows of alley cropping systems (Chifflot et al. 2009; Battie-Laclau et al. 2019; van Tuinen et al. 2020) as well as between crop rows of alley cropping systems and open croplands (Bainard et al. 2012). We expect these compositional changes to translate into functional changes with consequences for soil and plant health. For example, changes in microbial community composition can affect the abundance of beneficial and pathogenic microorganisms in soil with consequences for agricultural production. Following the one health concept, soil microorganisms cannot just have indirect and direct impacts on plant health but also on the health of animals and humans (Banerjee and van der Heijden 2023). The importance of the soil microbiome for plant health has been recognized for decades and the ability of soils to suppress plant diseases in the presence of pathogens (soil suppressiveness) has become an important field of research. Given their importance, soil suppressiveness and plant health in temperate agroforestry systems are vastly understudied. Although a recent study conducted in an alley cropping system in Germany found the first evidence of enhanced soil suppressiveness through agroforestry (Vaupel et al. 2023), validation is yet lacking. The same holds true for measures of plant health. The first and yet only study on plant health and crop diseases in temperate alley cropping systems showed that crops cultivated in temperate alley cropping systems are at least as healthy as those cultivated in neighboring open croplands (Beule et al. 2019b). There is an urgent need for further studies on soil suppressiveness and plant health in temperate agroforestry systems.

Microorganisms further regulate biogeochemical cycles and a number of studies thus explored functional genes involved in the nitrogen cycle covering N_2 fixation, nitrification, and denitrification. The majority of studies showed that genes involved in denitrification are promoted through agroforestry (Beule et al. 2020, 2022a). A recent study demonstrated that denitrifier gene abundance may not be a reliable predictor of N_2O fluxes under field conditions (Luo et al. 2022). Despite functional gene abundance, numerous studies report changes in soil enzyme activities in response to agroforestry. Studies that determined the activities of a range of common soil enzymes (e.g. β-glucosidase, N-acetyl-β-glucosaminidase, and dehydrogenase) in agroforestry systems generally reported higher activity under the trees as compared to the neighbouring crops (Mungai et al. 2005; Udawatta et al. 2008, 2009; Unger et al. 2013; Beuschel et al. 2019). These findings are congruent with studies that investigated substrate-induced respiration rates. For example, using a wide range of substrates, Mungai et al. (2005) and Beuschel et al. (2019) were able to show that substrate-induced respiration was increased through trees. Using a narrower range of substrates, Guillot et al. (2021) recently confirmed these findings. Furthermore, Mungai et al. (2005) and Beuschel et al. (2019) reported greater diversity of substrate utilization ability in the tree rows than in neighbouring rows. Since soil microorganisms are key drivers of biogeochemical cycles, we propose that composition and functional changes of the microbiome induced through agroforestry affect soil nutrient cycling in these systems. For example, Beule et al. (2022b) recently hypothesized that root-associated microorganisms support the 'safety-net'-role of tree roots for leached nutrients by enhancing nutrient uptake of tree roots and thereby contributing to nutrient pumping. Yet, the potential contributions of soil microorganisms to improved soil nutrient cycling in temperate agroforestry systems remain largely unexplored.

Overall, temperate agroforestry (1) promotes the population size of soil microorganisms, especially the population size of soil fungi and (2) alters the composition and functions of the soil microbiome, which may result in improved soil suppressiveness and altered soil nutrient cycling. For alpha diversity, however, variable effects are observed among studies with yet unknown underlying reasons. We suggest that future studies should aim to explore the contribution of the soil microbiome to soil functions in agroforestry systems.

2.2 Promotion of soil faunal communities

Faunal communities in soil are highly diverse and build complex food webs that highlight their interconnectedness. Since soil fauna is a key driver of several ecosystem functions, the global loss of soil biodiversity poses a serious threat to ecosystem functioning (Wagg et al. 2014). Soil faunal communities comprise a large variety of organisms that are considered indicators of soil health including earthworms (Paoletti 1999), nematodes (Neher 2001), and numerous arthropod groups (Menta and Remelli 2020). Among these, earthworms are a major component of soil faunal communities and are routinely used as biological indicators in soil health assessments (Lehmann et al. 2020). They contribute to numerous ecological functions (Blouin et al. 2013) such as enhanced water infiltration, improved nutrient availability, and suppression of phytopathogens. Because of these beneficial functions, the presence of earthworms has been shown to have a positive impact on crop yield (van Groenigen et al. 2014). Therefore, earthworms play a crucial role in global food production (Fonte et al. 2023).

Soil health indicators related to earthworms include population size, biomass, community composition, activity (e.g. number of casts on the soil surface), and functional parameters (e.g. water infiltration and aggregate formation). Methods suitable for the assessment of earthworm population size, biomass, and community composition include, among other things, hand sorting, chemical extraction (e.g. with formaldehyde, allyl isothiocyanate, or mustard), electrical octet method, Kempson extraction, or combinations of some of these methods. In contrast, the functions of earthworms are usually assessed indirectly by inferring them from the population parameters listed previously. Especially information on the abundance of ecological groups (anecic, endogeic, and epigeic as per Bouché 1972) is widely used to estimate earthworm functions. Recently, however, it has been argued that the use of ecological groups of earthworms can be misleading as ecological groups are not necessarily congruent with functional groups (Bottinelli and Capowiez 2021). Consequently, it has been proposed that future research should combine different methodologies and include advanced technologies such as stable isotopic techniques, X-ray tomography, and earthworm tagging in order to improve our understanding of the role earthworms in soil processes (Bartlett et al. 2010; Blouin et al. 2013). Additionally, the impact of earthworm community shifts on certain ecological functions can be investigated directly by, for example, measuring macroporosity and water infiltration rates (Capowiez et al. 2009). However, even though earthworms are among the most studied faunal groups in temperate cropland agroforestry systems, direct measurements of earthworm functions in these systems are yet missing.

A quarter of a century ago, Seiter et al. (1999) investigated bacterial and fungal biomass in an alley cropping system in the USA and made a noteworthy field observation on earthworm activity. Although the authors did not show their data, they observed that the majority of tree litter was pulled into earthworm burrows and that earthworm casts accumulated in close proximity to the trees (Seiter et al. 1999). Since then, numerous studies investigated earthworms in temperate cropland agroforestry systems and found predominantly positive effects (Price and Gordon 1998; Cardinael et al. 2019; D'Hervilly et al. 2020, 2021, 2022, 2024; Vaupel et al. 2023). For example, Cardinael et al. (2019) investigated earthworms in 13 alley cropping systems in France and were able to show greater population densities of earthworms in the tree rows of the agroforestry systems than in adjacent open croplands without trees. More than 20 years earlier, Price and Gordon (1998) found a similar promotion of earthworms in the poplar tree rows as compared to the crop rows of an alley cropping system in Canada. In the same study, however, the authors observed a decrease of earthworm population densities in the tree rows with ash trees as compared to the crop rows, suggesting an influence of tree species (Price and Gordon 1998). In 2023, Vaupel et al. published a study that investigated earthworms in an alley cropping system in Germany comprised of blocks of three different tree species (two different poplar clones and black locust). The authors found a strong promotion of earthworm population densities in the tree rows as compared to the crop rows and an adjacent open cropland, irrespective of the tree species. Furthermore, the authors were able to show a gradual decline of the population density of endogeic earthworms from the tree rows into the crop rows of the agroforestry system, which they attributed to the increased input of tree litter in close proximity to the trees (Vaupel et al. 2023). Additionally, they observed that anecic earthworms were strongly promoted in the tree rows, leading to a community shift towards anecics. In contrast to endogeic species, the positive impact of the tree rows on anecic earthworms did not extend into the crop rows, which was attributed to the presence of tillage in the crop rows (Vaupel et al. 2023). Similar results were obtained by D'Hervilly and co-authors (2020) who reported increased relative abundance of anecic earthworms within the earthworm communities in the tree rows as compared to the crop rows of two alley cropping systems in France. One year later, these results were confirmed in another alley cropping system in France, highlighting the positive impact of trees on this ecological group of earthworms (D'Hervilly et al. 2021). Overall, the existing literature reports predominantly positive effects of temperate cropland agroforestry systems on earthworms.

However, a number of knowledge gaps remain regarding the influence of tree row design and management on earthworm communities (e.g. tree age, tree species, tree harvesting cycles, and tree row orientation). Furthermore, the magnitude of spillover effects of earthworms from the trees into the neighboring crops is highly understudied. The quantification of spillover effects requires the investigation of migration patterns of earthworms from the trees into the crops and vice versa. In a recent study, D'Hervilly et al. (2024) aimed to quantify earthworm migration patterns in an alley cropping system in France by investigating earthworm fluxes from the tree rows into the crop rows and vice versa. The authors observed greater earthworm fluxes from the crop rows into the tree rows than from the tree rows into the crop rows. However, fluxes were largely based on the presence of a single endogeic earthworm species (*Allolobophora chlorotica*) and the authors suspect different migration patterns for other species that could not be successfully quantified using their traps (e.g. epigeic species). Finally, the authors found a strong temporal variation of earthworm densities within the agroforestry system between March and May, suggesting that migration patterns might differ throughout the year (D'Hervilly et al. 2024).

Similarly, D'Hervilly et al. (2022) observed seasonal variations of the spatial distribution of several soil faunal groups within a walnut-based alley cropping agroforestry system in France. The authors found seasonal variations across spring, summer, and autumn for Chilopoda, adult Coleoptera, Isopoda, and Diplopoda. Furthermore, the authors classified their soil faunal individuals into soil-, litter-, animal-, and plant-feeding organisms. While population densities of soil-feeding organisms were greater in the tree rows than in the crop rows throughout all seasons, tree rows promoted litter-feeding organisms only during spring and summer (D'Hervilly et al. 2022). While animal-feeding organisms were promoted through the tree rows in spring, plant-feeding organisms showed lower population densities in the tree rows than in the crop rows during this season (D'Hervilly et al. 2022).

The effect of temperate agroforestry practice on soil invertebrates has been investigated by Park et al. (1994). The authors obtained invertebrates from soil cores (collected at 0–15 and 15–30 cm soil depth) using Tullgren funnels in a poplar-based alley cropping system in the United Kingdom. Their results showed greater invertebrate populations and biomass in the tree rows and at 1 m distance into the crop rows than at 3 and 6 m into the crop rows as well as in an adjacent open cropland (Park et al. 1994). Almost 30 years later, Szigeti et al. (2022) applied Tullgren funnels to investigate microarthropods at 0–10 cm soil depth in two shelterbelts and an adjacent open croplands in Hungary. The shelterbelts harboured greater population size and species richness of *Collembola* as compared to the neighbouring crops (Szigeti et al. 2022). Likewise, the population densities of Acarina were greater in the shelterbelts than in the crops (Szigeti et al. 2022). Similar results were obtained by Olejniczak (2007) who used the same methodological approach as Szigeti et al. (2022) in three shelterbelt systems of different ages (6, 7, and 170 years) in Poland. The authors found mostly decreasing population densities of Acarina with increasing distance from the trees and reported that these trends were depending on shelterbelt age and sampling time (Olejniczak 2007). In contrast to Acarina, the effect of shelterbelts on the population density of Collembola remains inconclusive and was strongly affected by sampling season (Olejniczak 2007).

In 1998, Griffiths et al. published a study in which they investigated the impacts of alley cropping agroforestry on slugs in the United Kingdom. By utilizing pipe, refuge, and pitfall traps, they found greater diversity and activity density of slugs in the tree rows as compared to the crop rows and open croplands. Additionally, the authors assessed damage to pea plants cultivated at their study site and reported increased plant damage and reduced counts of emerged pea seedlings in close proximity to the trees (Griffiths et al. 1998). In contrast, Boinot et al. (2019b) investigated overwintering invertebrates with emergence traps in seven alley cropping systems in France and found that population densities of economically relevant pests (*Aphidae*, *Elateridae*, and slugs) were lower in the tree rows than in the crop rows. At their study sites, the majority of investigated overwintering invertebrate groups was promoted by the tree rows. For example, they found greater numbers of Heteroptera, Hymenoptera, Lepidoptera, Neuroptera, Psocoptera, Blattoptera, Collembola, Isopoda, Acarina, Araneae, Opiliones, snails, and Pseudoscorpionida under the trees than in the crop rows. Coleoptera, Homoptera, and Myriapoda, however, showed the opposing trend (Boinot et al. 2019b). Furthermore, tree rows had a positive effect on the overwintering of beneficial invertebrates, such as *Coccinellidae*, *Chrysopidae*, and *Formicidae*, whereas numbers of overwintering *Carabidae*, *Staphylinidae*, and *Syrphidae* declined under the trees as compared to the crop rows (Boinot et al. 2019b). In the same year, Pardon et al. (2019) published a study in which they investigated ground-dwelling arthropods in six alley cropping agroforestry systems in Belgium using pitfall traps. They observed greater activity densities of woodlice and millipedes in the tree rows than in the crop rows as well as decreasing activity densities in the crop rows with increasing distance from the trees. For Carabids and rove beetles, however, the authors reported the opposing trend (Pardon et al. 2019). Another study from 2019 that used pitfall traps in a walnut-based alley cropping agroforestry system in France found similar results on the activity density of ground beetles (Martin-Chave et al. 2019b). In the same system, the authors conducted one of the few studies that directly investigated the functions of soil fauna by determining the predation of *Cydia pomonella* larvae (Martin-Chave et al. 2019a). Predation was mainly driven by ground beetles and Opiliones and was greater in the agroforestry system than in the reference system without trees. Predation of *Cydia pomonella* larvae, however, showed temporal variations between June and August (Martin-Chave et al. 2019a). In 2024, Matevski and colleagues installed pitfall and emergence traps at five paired alley cropping and open cropland systems in Germany and assessed spider abundance, taxonomic and functional diversity. The authors found that agroforestry systems generally promoted the abundance as well as taxonomic and functional diversity of the overall spider communities as compared to open croplands. Interestingly, they found that the promotion of forest specialists and eurytopic spiders in the agroforestry systems without negatively affecting open-habitat specialists. Furthermore, they found that spider abundance and taxonomic diversity in the crop rows peaked when crop pests arrived, which likely resulted in increased pest control potential through agroforestry (Matevski et al. 2024).

Overall, the existing literature on soil fauna in temperate cropland agroforestry systems reports predominantly positive effects on their population density and diversity. In a recent review article, Marsden et al. (2020) pointed out that so far only a few studies conducted direct measurements of functions of soil fauna in agroforestry systems and stressed that future studies should aim to include such measurements. We strongly agree with Marsden et al. (2020)

and argue that direct measurements of functions are needed to understand the advantages and disadvantages of agroforestry systems for essential ecosystem services driven by soil fauna.

2.3 Mechanisms by which agroforestry promotes soil biological health

There are several different agricultural management practices that can promote soil health. For example, in a recent meta-analysis, Nunes et al. (2020) showed that reduced tillage intensity and crop residual removal and permanent plant cover enhance indicators of soil biological health. The authors further pointed out that biological indicators of soil health are most effectively promoted through perennial cropping systems with zero soil disturbance. The introduction of woody perennials into agricultural systems through the implementation of agroforestry creates areas with zero soil disturbance under the trees and many authors argue that this is a key factor for the promotion of soil life under the trees (Beuschel et al. 2019, Beule et al. 2020, D'Hervilly et al. 2021). Furthermore, the absence of soil disturbance under the trees leads to the emergence of a permanent herbaceous understory vegetation (Boinot et al. 2019a), which has been shown to be important in maintaining microbial communities (Battie-Laclau et al. 2019). Additionally, trees provide permanent plant cover and their aboveground tree leaf litter is not removed, serving as soil cover as well as a resource for soil organisms. In alley cropping systems, the amount of tree leaf litter generally decreases exponentially with increasing distance to the trees (Swieter et al. 2018, 2022), which may explain gradients of indicators of soil health from the trees into the neighbouring crops. In addition to tree litter quantity and its spatial distribution, the quality of the tree litter is yet another important driver of soil microbial activity in agroforestry systems (Clivot et al. 2020). The decomposition of fine root litter of trees is expected to further contribute to the maintenance of soil organisms, especially at deep soil depths (Beule et al. 2022a).

3 Measuring soil biological health in agroforestry systems

3.1 Spatial heterogeneity in agroforestry systems and sampling design

As reported by numerous studies, trees in temperate agroforestry systems introduce spatial heterogeneity of soil health indicators. The vast majority of studies that investigated spatial heterogeneity in these systems determined horizontal heterogeneity (e.g. spatial gradients from the trees into the neighboring crops at a defined soil depth). In alley cropping systems, sampling at multiple distances along gradients spanning from the tree rows into the crop rows is routinely done to capture spatial heterogeneity. Such a study design further allows calculating area-weighted averages in order to obtain aggregated values for the crop rows (Beule et al. 2019b; Schmidt et al. 2021) or the agroforestry system as a whole (Luo et al. 2022; Shao et al. 2023). This further enables comparisons of agroforestry systems with other land-use systems such as open fields (Veldkamp et al. 2023). The distances to trees, however, must be selected carefully as pointed out by Shao et al. (2023) who observed an overlap of fertilizer application in the middle of the crop rows due to the working width of the operating farmer.

Although soils are three-dimensional systems and their variation in physical and chemical composition can span up orders of magnitude – even along a single soil profile – most studies of soil organisms in temperate agroforestry systems do not account for soil depth. Since variations in soil properties with depth are among the main characteristics of soil (Mulla and McBratney 2001), it is unsurprising that soil microbial communities adapt to this and change in their community composition with depth (Fierer et al. 2003b). Furthermore, it is well established that soil microbial communities generally decrease in biomass and activity with soil depth, which is mainly attributed to resource limitations at deeper depths (Fierer et al. 2003a).

The introduction of trees into agricultural systems introduces both horizontal and vertical spatial variations in soil. For example, one of the most prominent benefits of agroforestry systems is the concept of the 'safety-net'-role of deep tree roots, which is based on differences in the distribution of crop and tree roots across soil depths that enables a complementary rather than competitive use of resources (Allen et al. 2004). While a large body of literature explored horizontal variations of soil biological health indicators in temperate agroforestry systems by collecting samples at a single soil depth, only very few studies explored both horizontal and vertical variations by additionally collecting samples at multiple soil depths.

In 2005, Mungai et al. investigated the activities of β-glucosidase and fluorescein diacetate (FDA) hydrolase at multiple depths (0–10, 10–20, and 20–30 cm) in two alley cropping systems in Missouri, USA. Although the differences in enzyme activity between the tree and the crop rows were mostly higher in the tree rows, they varied between sampling year, and soil depth. Ten years later, Udawatta et al. (2014) collected soil samples at 0–10, 10–20, and 20–30 cm soil depth at one of the two alley cropping systems of Mungai et al. (2005). The authors analyzed the activities of β-glucosidase, N-acetyl-β-glucosaminidase, FDA hydrolase, and dehydrogenase and found no differences between the tree row and the middle of the crop alley at any soil depth. In 2019, Beuschel and co-workers reported that patterns of spatial heterogeneity in microbial population size, enzyme actives, and substrate-induced respiration rates in three alley cropping systems were evident at 0–5 but not at 5–20 cm topsoil. The authors also accounted for horizontal heterogeneity and collected soil samples at multiple distances from the tree rows into the crop rows. For years following their soil sampling, Beule et al. (2022a) collected soil samples at two of the three study sites of Beuschel et al. (2019) in topsoil (0–15 and 15–30 cm) and subsoil (30–45 and 45–60 cm). They further collected soil samples at the same distances from the trees as Beuschel et al. (2019). Beule et al. (2022a) used real-time polymerase chain reaction to quantify taxonomical and functional groups of soil microorganisms and found that the microbial population size decreases with increasing soil depth and the promotion of microorganisms through trees in subsoil is stronger than in topsoil. Yet, the subsoil microbiome and its functions in temperate cropland agroforestry systems remain largely unexplored.

3.2 Temporal heterogeneity in agroforestry systems

In addition to spatial variation, the examples previously discussed (cf. Mungai et al. 2005; Udawatta et al. 2014; Beuschel et al. 2019; Beule et al. 2022a) highlight the need to further account for temporal variation in soil health assessments in temperate cropland agroforestry systems. As of writing, only a few studies covered temporal dynamics of soil biological health indicators in agroforestry systems by using repeated sampling (Mungai et al. 2005; Beule and Karlovsky 2021b; Luo et al. 2022; D'Hervilly et al. 2022). Studies that investigated soil organisms across seasons reported strong temporal dynamics for communities of both soil micro- (Beule and Karlovsky 2021b) and macroorganisms (D'Hervilly et al. 2022). Temporal dynamics of soil organisms are strongly affected by agricultural management. For example, it is well established that crop rotation and its diversity are important drivers of soil communities (D'Acunto et al. 2016; Torppa and Taylor 2022). However, studies investigating soil organisms across an entire crop rotation within agroforestry systems are absent. Likewise, extensive studies on soil health in relation to tree age and/or tree harvesting cycles are absent. For example, in systems that use short-rotation trees, regular tree harvests will result in repeating temporal changes of, among other things, tree litter input (above- and belowground), root exudation, and shading with potential consequences for soil organisms. Yet, the effects of tree harvesting and tree age on soil life in agroforestry systems remain largely unexplored. We therefore encourage researchers to study soil health across time and under consideration of crop and tree rotation cycles in order to identify optimal combinations as well as management practices of trees and crops that maximize the soil health benefits of agroforestry systems.

4 Future trends

4.1 Consideration of spatio-temporal heterogeneity

While a growing body of literature covers horizontal spatial heterogeneity within agroforestry systems (e.g. sampling at multiple distances from the trees into neighboring crops), only a few studies account for vertical heterogeneity (i.e. different soil depths). We thus encourage researchers to extend their study design to multiple soil depths and to consider subsoil. We further recommend accounting for temporal heterogeneity in agroforestry systems by sampling at multiple points in time under consideration of crop and tree rotation cycles. Such studies may be conducted by repeated sampling across multiple years or by using space-for-time substitution.

Finally, in order to evaluate the changes in soil health associated with the conversion of agricultural systems without trees to agroforestry, we would like to stress the importance of comparing agroforestry systems to land-use systems without trees. Although the importance of including non-agroforestry plots as a reference land use has been pointed out more than 25 years ago (Stamps and Linit 1998), several studies lack this comparison. A good example of a study that successfully implemented a reference land use at multiple study sites is the work of Veldkamp et al. (2023). The authors studied cropland and grassland alley cropping systems and compared those systems to neighboring open cropland and grassland systems without trees, respectively. Since the open cropland or grassland systems were managed identically to the corresponding crop or grass rows of the agroforestry systems, the authors were able to compare the land-use systems (i.e. agroforestry versus open cropland or grassland). The authors further accounted for horizontal spatial heterogeneity within their agroforestry systems. We argue that such study designs are of great value but are aware that they are difficult to realize.

4.2 Consideration of the full diversity of soil organisms

While soil bacterial and fungal communities are by far the most studied group of microorganisms in temperate agroforestry systems, other groups of microorganisms are highly understudied or yet ignored. For example, although soil archaea contribute to important soil functions in biogeochemical cycles (Offre et al. 2013), most prominently their contribution to the oxidation of ammonia in nitrifaction, this group of microorganisms has rarely been quantified in temperate agroforestry systems (Beule et al. 2019a, 2020; Vaupel et al. 2023) and has not yet been investigated using modern sequencing approaches in these systems. Therefore, their diversity and community composition in agroforestry systems remains unexplored. Likewise, soil protists are ubiquitous and contribute to soil nutrient cycling (Geisen et al. 2018; Oliverio et al. 2020) but remain yet unexplored in temperate agroforestry systems. While there are a number of studies on earthworms in temperate agroforestry systems (Price and Gordon 1998; Cardinael et al. 2019; D'Hervilly et al. 2020, 2021, 2022, 2024; Vaupel et al. 2023) and few on Acarina, Araneae, and Collembola (Olejniczak 2007; Boinot et al. 2019b; Szigeti et al. 2022, Matevski et al. 2024), other soil faunal groups are highly understudied. Likewise, studies on burrowing mammals are scarce (Klaa et al. 2005). Finally, only a handful of studies performed direct measurements of the functions of soil fauna in temperate cropland agroforestry systems (Marsden et al. 2020). We thus argue that in addition to inferring potential functions from ecological count data, researchers should aim to directly assess the functions of soil organisms in agroforestry systems.

4.3 Consideration of the full diversity of temperate agroforestry systems

Over the last three decades, the literature investigating soil biological health indicators was largely focused on alley cropping systems with poplar (Park et al. 1994; Chifflot et al. 2009; Beaudette et al. 2010; Sun et al. 2018; Beuschel et al. 2019; Cardinael et al. 2019; Pardon et al. 2019; Beule et al. 2019a, 2020, 2021, 2022a; Beule and Karlovsky 2021a, 2021b; Luo et al. 2022; Shao et al. 2023, Matevski et al. 2024) or walnut trees (Griffiths et al. 1998; Cardinael et al. 2019; Guillot et al. 2019, 2021; Martin-Chave et al. 2019a, 2019b; Pardon et al. 2019; Gao et al. 2019; D'Hervilly et al. 2020, 2021, 2022, 2024; van Tuinen et al. 2020). Other tree species investigated cover, among other things, alder (Seiter et al. 1999), ash (Griffiths et al. 1998; Bainard et al. 2012), black locust (Nii-Annang et al. 2009; Sun et al. 2018; Szigeti et al. 2022; Vaupel et al. 2023, Matevski et al. 2024), maple (Bainard et al. 2012; Bardhan et al. 2013; Udawatta et al. 2014; Szigeti et al. 2022), oak (Olejniczak 2007; Udawatta et al. 2008, 2009; Weerasekara et al. 2016), pecan (Lee and Jose 2003; Mungai et al. 2005), and spruce (Bainard et al. 2012). Yet, there are only a few studies that investigated mixed tree species within the tree rows (Griffiths et al. 1998; Olejniczak 2007; Battie-Laclau et al. 2019; Cardinael et al. 2019; Pardon et al. 2019; Boinot et al. 2019b; D'Hervilly et al. 2020; Szigeti et al. 2022; Mettauer et al. 2023; Vaupel et al. 2025) or evaluated the effects of different tree species at the same study site (Price and Gordon 1998; Bainard et al. 2012; Vaupel et al. 2023). Furthermore, studies comparing the impacts of differences in soil (e.g. tillage vs. no-till), crop (e.g. major vs. minor crops), and farm management (e.g. organic vs. conventional farming) on soil biological health within agroforestry systems are either scarce or absent. Finally, as stated previously, most studies on soil biological health were conducted in alley cropping systems, while other types of cropland agroforestry systems dominant in the temperate zone, such as shelterbelts, remain largely unexplored. Likewise, except for a few studies

Published by Burleigh Dodds Science Publishing Limited, 2025.

(Amorim et al. 2023) soil health in silvopastoral (integration of woody perennials and livestock) and agrosilvopastoral systems (integration of woody perennials, crops, and livestock) remains highly understudied in the temperate zone.

Future research efforts should also focus on urban and peri-urban agroforestry given an ever-increasing urban population density and the need to increase food sovereignty and enhance ecosystem services provision in cities (Clark and Nicholas 2013). In a pioneering study on AMF and EMF colonization of 26 tree species in urban and rural environments in Canada, Bainard et al. (2011) found that urbanization reduces mycorrhizal colonization. The urban factors that led to the observed reduction in colonization, however, remained unclear. The anthropogenic stressors experienced by urban and agricultural soils have intersections but are not identical and urban soils are often strongly degraded through physical (e.g. sealing and compaction) and chemical stressors (e.g. contaminants). In this context, urban and peri-urban agroforestry practices may offer promising solutions to restore the soil health of urbanized soils. We acknowledge, however, that suitable study sites for temperate urban and peri-urban agroforestry may not be widely available. Furthermore, the high complexity of these systems creates large dissimilarities among systems, which may represent hurdles for certain scientific investigations. Despite these hurdles, urban and peri-urban agroforestry systems have tremendous potential to improve soil health as well as the overall sustainability of urbanized areas and thus require more investigations.

Overall, we strongly encourage researchers to consider the full diversity of temperate agroforestry systems as well as their management in future research.

5 Conclusion

Soil is a finite resource. The thin and fragile layer of soil that covers Earth's ice-free land surface sustains an incredibly large number of organisms, including us humans. Therefore, maintaining soil health must be among our top priorities as a species. Furthermore, soil health is recognized as one of 13 principles of agroecology, a concept that presents transition pathways to more sustainable food systems (Wezel et al. 2020). Consequently, the existential need to preserve soil health is mirrored by numerous political agendas and has become the goal of several multinational research funding initiatives. In this chapter, we show that temperate cropland agroforestry systems strongly promote the biological components of soil health. This includes, among other things, enhanced population densities and activities of soil microorganisms and fauna as well as increased diversity of soil fauna under cropland agroforestry as compared to open croplands without trees. Additionally, agroforestry alters the community composition of soil organisms, which is expected to result in altered functions of soil communities. Overall, the changes in soil communities associated with agroforestry practices benefit soil biological health. In addition to the promotion of soil health, multifunctional agroforestry systems offer a variety of environmental benefits. We argue that the land-use change from open cropland to agroforestry provides substantial environmental benefits that outperform those of many sustainable agricultural management practices. However, cropland agroforestry does not exclude additional integration of sustainable management practices but allows their simultaneous inclusion in the system. Therefore, agroforestry can make major contributions to improving and maintaining soil health for future generations. Considering the annual global loss of fertile soil, restoration of degraded soils becomes ever more important and agroforestry offers promising solutions to rehabilitate degraded soils and restore soil health. Overall, agroforestry provides high multifunctionality and represents a key land-use system for the agroecological transition of our food systems. Consequently, we propose agroforestry as a priority land-use system that should receive adequate subsidies and political attention.

6 Where to look for further information

Standard introductions to the subject of improved soil health through agroforestry include:

- Barrios, E., Coe, R., Place, F., Sileshi, G. W., & Sinclair, F. (2023). Nurturing Soil Life through Agroforestry: The Roles of Trees in the Ecological Intensification of Agriculture. In Biological Approaches to Regenerative Soil Systems (pp. 265-278). CRC Press.
- Dollinger, J., & Jose, S. (2018). Agroforestry for soil health. Agroforestry Systems, 92, 213-219.
- Barrios, E., Sileshi, G. W., Shepherd, K., & Sinclair, F. (2012). Agroforestry and soil health: linking trees, soil biota and ecosystem services. In Soil ecology and ecosystem services (pp. 315-330). Oxford Academic.

Centres of expertise to reach out to for information resources include:

- The Center for International Forestry Research and World Agroforestry (CIFOR-ICRAF) (https://www.cifor-icraf.org/)
- The European Agroforestry Federation (EURAF) (https://www.euraf.net/)
- National Research Institute for Agriculture, Food and Environment (https://www.inrae.fr/en)
- USDA National Agroforestry Center (NAC) (https://www.fs.usda.gov/nac/)
- The authors' own consultancy, which have conducted numerous research projects related to soil biological health in temperate agroforestry systems

7 References

Allen SC, Jose S, Nair PKR, et al. (2004) Safety-net role of tree roots: evidence from a pecan (*Carya illinoensis* K. Koch)–cotton (*Gossypium hirsutum* L.) alley cropping system in the southern United States. *Forest Ecology and Management* 192:395–407. https://doi.org/10.1016/j.foreco.2004.02.009

Amorim HCS, Ashworth AJ, O'Brien PL, et al. (2023) Temperate silvopastures provide greater ecosystem services than conventional pasture systems. *Sci Rep* 13:18658. https://doi.org/10.1038/s41598-023-45960-0

Anthony MA, Bender SF, van der Heijden MGA (2023) Enumerating soil biodiversity. *PNAS* 120:e2304663120. https://doi.org/10.1073/pnas.2304663120

Bainard LD, Klironomos JN, Gordon AM (2011) The mycorrhizal status and colonization of 26 tree species growing in urban and rural environments. *Mycorrhiza* 21:91–96. https://doi.org/10.1007/s00572-010-0314-6

Bainard LD, Koch AM, Gordon AM, et al. (2012) Temporal and compositional differences of arbuscular mycorrhizal fungal communities in conventional monocropping and tree-based intercropping systems. *Soil Biol Biochem* 45:172–180. https://doi.org/10.1016/j.soilbio.2011.10.008

Bambrick AD, Whalen JK, Bradley RL, et al. (2010) Spatial heterogeneity of soil organic carbon in tree-based intercropping systems in Quebec and Ontario, Canada. *Agroforest Syst* 79:343–353. https://doi.org/10.1007/s10457-010-9305-z

Banerjee S, van der Heijden MGA (2023) Soil microbiomes and one health. *Nat Rev Microbiol* 21:6–20. https://doi.org/10.1038/s41579-022-00779-w

Bardhan S, Jose S, Udawatta RP, Fritschi F (2013) Microbial community diversity in a 21-year-old temperate alley cropping system. *Agroforest Syst* 87:1031–1041. https://doi.org/10.1007/s10457-013-9617-x

Bartlett MD, Briones MJI, Neilson R, et al. (2010) A critical review of current methods in earthworm ecology: from individuals to populations. *Eur J Soil Biol* 46:67–73. https://doi.org/10.1016/j.ejsobi.2009.11.006

Battie-Laclau P, Taschen E, Plassard C, et al. (2019) Role of trees and herbaceous vegetation beneath trees in maintaining arbuscular mycorrhizal communities in temperate alley cropping systems. *Plant Soil* 453:153–171. https://doi.org/10.1007/s11104-019-04181-z

Beaudette C, Bradley RL, Whalen JK, et al. (2010) Tree-based intercropping does not compromise canola (*Brassica napus* L.) seed oil yield and reduces soil nitrous oxide emissions. *Agric Ecosyst Environ* 139:33–39. https://doi.org/10.1016/j.agee.2010.06.014

Beule L, Lehtsaar E, Corre MD, et al. (2020) Poplar rows in temperate agroforestry croplands promote bacteria, fungi, and denitrification genes in soils. *Front Microbiol* 10:3108. https://doi.org/10.3389/fmicb.2019.03108

Beule L, Arndt M, Karlovsky P (2021) Relative abundances of species or sequence variants can be misleading: soil fungal communities as an example. *Microorganisms* 9:589. https://doi.org/10.3390/microorganisms9030589

Beule L, Corre MD, Schmidt M, et al. (2019a) Conversion of monoculture cropland and open grassland to agroforestry alters the abundance of soil bacteria, fungi and soil-N-cycling genes. *PLoS One* 14:e0218779. https://doi.org/10.1371/journal.pone.0218779

Beule L, Lehtsaar E, Rathgeb A, Karlovsky P (2019b) Crop diseases and mycotoxin accumulation in temperate agroforestry systems. *Sustainability* 11:2925. https://doi.org/10.3390/su11102925

Beule L, Guerra V, Lehtsaar E, Vaupel A (2022a) Digging deeper: microbial communities in subsoil are strongly promoted by trees in temperate agroforestry systems. *Plant Soil* 480:423–437. https://doi.org/10.1007/s11104-022-05591-2

Beule L, Vaupel A, Moran-Rodas VE (2022b) Abundance, diversity, and function of soil microorganisms in temperate alley cropping agroforestry systems: a review. *Microorganisms* 10:616. https://doi.org/10.3390/microorganisms10030616

Beule L, Karlovsky P (2021a) Tree rows in temperate agroforestry croplands alter the composition of soil bacterial communities. *PLoS One* 16:e0246919. https://doi.org/10.1371/journal.pone.0246919

Beule L, Karlovsky P (2021b) Early response of soil fungal communities to the conversion of monoculture cropland to a temperate agroforestry system. *PeerJ* 9:e12236. https://doi.org/10.7717/peerj.12236

Beuschel R, Piepho H-P, Joergensen RG, Wachendorf C (2019) Similar spatial patterns of soil quality indicators in three poplar-based silvo-arable alley cropping systems in Germany. *Biol Fertil Soils* 55:1–14. https://doi.org/10.1007/s00374-018-1324-3

Blouin M, Hodson ME, Delgado EA, et al. (2013) A review of earthworm impact on soil function and ecosystem services. *Eur J Soil Sci* 64:161–182. https://doi.org/10.1111/ejss.12025

Boinot S, Fried G, Storkey J, et al. (2019a) Alley cropping agroforestry systems: reservoirs for weeds or refugia for plant diversity? *Agric Ecosyst Environ* 284:106584. https://doi.org/10.1016/j.agee.2019.106584

Boinot S, Poulmarc'h J, Mézière D, et al. (2019b) Distribution of overwintering invertebrates in temperate agroforestry systems: implications for biodiversity conservation and biological control of crop pests. *Agric Ecosyst Environ* 285:106630. https://doi.org/10.1016/j.agee.2019.106630

Bottinelli N, Capowiez Y (2021) Earthworm ecological categories are not functional groups. *Biol Fertil Soils* 57:329–331. https://doi.org/10.1007/s00374-020-01517-1

Bouché M (1972) *Lombriciens de France. Ecologie et Systématique*. INRA, Paris.

Bünemann EK, Bongiorno G, Bai Z, et al. (2018) Soil quality – A critical review. *Soil Biol Biochem* 120:105–125. https://doi.org/10.1016/j.soilbio.2018.01.030

Capowiez Y, Cadoux S, Bouchant P, et al. (2009) The effect of tillage type and cropping system on earthworm communities, macroporosity and water infiltration. *Soil Till Res* 105:209–216. https://doi.org/10.1016/j.still.2009.09.002

Cardinael R, Hoeffner K, Chenu C, et al. (2019) Spatial variation of earthworm communities and soil organic carbon in temperate agroforestry. *Biol Fertil Soils* 55:171–183. https://doi.org/10.1007/s00374-018-1332-3

Chifflot V, Rivest D, Olivier A, et al. (2009) Molecular analysis of arbuscular mycorrhizal community structure and spores distribution in tree-based intercropping and forest systems. *Agric Ecosyst Environ* 131:32–39. https://doi.org/10.1016/j.agee.2008.11.010

Clark KH, Nicholas KA (2013) Introducing urban food forestry: a multifunctional approach to increase food security and provide ecosystem services. *Landsc Ecol* 28:1649–1669. https://doi.org/10.1007/s10980-013-9903-z

Clivot H, Petitjean C, Marron N, et al. (2020) Early effects of temperate agroforestry practices on soil organic matter and microbial enzyme activity. *Plant Soil* 453:189–207. https://doi.org/10.1007/s11104-019-04320-6

D'Acunto L, Semmartin M, Ghersa CM (2016) Uncultivated margins are source of soil microbial diversity in an agricultural landscape. *Agric Ecosyst Environ* 220:1–7. https://doi.org/10.1016/j.agee.2015.12.032

D'Hervilly C, Marsden C, Capowiez Y, et al. (2021) Trees and herbaceous vegetation strips both contribute to changes in soil fertility and soil organism communities in an agroforestry system. *Plant Soil* 463:537–553. https://doi.org/10.1007/s11104-021-04932-x

D'Hervilly C, Bertrand I, Capowiez Y, et al. (2022) Seasonal variations in macrofauna distribution according to the distance from a herbaceous strip in a Mediterranean alley cropping plot. *Appl Soil Ecol* 170:104309. https://doi.org/10.1016/j.apsoil.2021.104309

D'Hervilly C, Bertrand I, Berlioz L, et al. (2024) Tracking earthworm fluxes at the interface between tree rows and crop habitats in a Mediterranean alley cropping field. *Eur J Soil Biol* 120:103572. https://doi.org/10.1016/j.ejsobi.2023.103572

D'Hervilly C, Marsden C, Hedde M, Bertrand I (2020) Sown understory vegetation strips impact soil chemical fertility, associated microorganisms and macro-invertebrates in two temperate alley cropping systems. *Agrofor Syst*. https://doi.org/10.1007/s10457-020-00501-w

Doran JW, Zeiss MR (2000) Soil health and sustainability: managing the biotic component of soil quality. *Appl Soil Ecol* 15:3–11. https://doi.org/10.1016/S0929-1393(00)00067-6

FAO (2015) *Status of the World's Soil Resources - Technical Summary*. FAO, Rome, Italy.

Fierer N, Allen AS, Schimel JP, Holden PA (2003a) Controls on microbial CO2 production: a comparison of surface and subsurface soil horizons. *Glob Chang Biol* 9:1322–1332. https://doi.org/10.1046/j.1365-2486.2003.00663.x

Fierer N, Schimel JP, Holden PA (2003b) Variations in microbial community composition through two soil depth profiles. *Soil Biol Biochem* 35:167–176. https://doi.org/10.1016/S0038-0717(02)00251-1

Fierer N, Wood SA, Bueno de Mesquita CP (2021) How microbes can, and cannot, be used to assess soil health. *Soil Biol Biochem* 153:108111. https://doi.org/10.1016/j.soilbio.2020.108111

Fonte SJ, Hsieh M, Mueller ND (2023) Earthworms contribute significantly to global food production. *Nat Commun* 14:5713. https://doi.org/10.1038/s41467-023-41286-7

Gao P, Zheng X, Wang L, et al. (2019) Changes in the soil bacterial community in a chronosequence of temperate walnut-based intercropping systems. *Forests* 10:299. https://doi.org/10.3390/f10040299

Geisen S, Mitchell EAD, Adl S, et al. (2018) Soil protists: a fertile frontier in soil biology research. *FEMS Microbiol Rev* 42:293–323. https://doi.org/10.1093/femsre/fuy006

Griffiths J, Phillips DS, Compton SG, et al. (1998) Responses of slug numbers and slug damage to crops in a silvoarable agroforestry landscape. *J Appl Ecol* 35:252–260. https://doi.org/10.1046/j.1365-2664.1998.00291.x

Guillot E, Hinsinger P, Dufour L, et al. (2019) With or without trees: Resistance and resilience of soil microbial communities to drought and heat stress in a Mediterranean agroforestry system. *Soil Biol Biochem* 129:122–135. https://doi.org/10.1016/j.soilbio.2018.11.011

Guillot E, Bertrand I, Rumpel C, et al. (2021) Spatial heterogeneity of soil quality within a Mediterranean alley cropping agroforestry system: comparison with a monocropping system. Eur J Soil Biol 105:103330. https://doi.org/10.1016/j.ejsobi.2021.103330

Jeffery S, Verheijen FGA (2020) A new soil health policy paradigm: pay for practice not performance! *Environ Sci Pol* 112:371–373. https://doi.org/10.1016/j.envsci.2020.07.006

Joergensen RG, Hemkemeyer M, Beule L, et al. (2024) A hitchhiker's guide: estimates of microbial biomass and microbial gene abundance in soil. *Biol Fertil Soils* 60:457–470. https://doi.org/10.1007/s00374-024-01810-3

Kibblewhite MG, Ritz K, Swift MJ (2008) Soil health in agricultural systems. *Philos Trans Royal Soc B* 363:685–701. http://doi.org/10.1098/rstb.2007.2178

Klaa K, Mill PJ, Incoll LD (2005) Distribution of small mammals in a silvoarable agroforestry system in Northern England. *Agroforest Syst* 63:101–110. https://doi.org/10.1007/s10457-004-1110-0

Koch A, McBratney A, Adams M, et al. (2013) Soil security: solving the global soil crisis. *Global Policy* 4:434–441. https://doi.org/10.1111/1758-5899.12096

Lacombe S, Bradley RL, Hamel C, Beaulieu C (2009) Do tree-based intercropping systems increase the diversity and stability of soil microbial communities? *Agric Ecosyst Environ* 131:25–31. https://doi.org/10.1016/j.agee.2008.08.010

Lee K-H, Jose S (2003) Soil respiration and microbial biomass in a pecan – cotton alley cropping system in Southern USA. *Agroforest Syst* 58:45–54. https://doi.org/10.1023/A:1025404019211

Lehmann J, Bossio DA, Kögel-Knabner I, Rillig MC (2020) The concept and future prospects of soil health. *Nat Rev Earth Environ* 1:544–553. https://doi.org/10.1038/s43017-020-0080-8

Luo J, Beule L, Shao G, et al. (2022) Reduced soil gross N2O emission driven by substrates rather than denitrification gene abundance in cropland agroforestry and monoculture. *J Geophys Res Biogeosci* 127:e2021JG006629. https://doi.org/10.1029/2021JG006629

Marsden C, Martin-Chave A, Cortet J, et al. (2020) How agroforestry systems influence soil fauna and their functions: a review. *Plant Soil* 453:29–44. https://doi.org/10.1007/s11104-019-04322-4

Martin-Chave A, Béral C, Capowiez Y (2019a) Agroforestry has an impact on nocturnal predation by ground beetles and Opiliones in a temperate organic alley cropping system. *Bio Control* 129:128–135. https://doi.org/10.1016/j.biocontrol.2018.10.009

Martin-Chave A, Béral C, Mazzia C, Capowiez Y (2019b) Agroforestry impacts the seasonal and diurnal activity of dominant predatory arthropods in organic vegetable crops. *Agroforest Syst* 93:2067–2083. https://doi.org/10.1007/s10457-018-0309-4

Matevski D, Sagolla V, Beule L, et al. (2024) Temperate alley-cropping agroforestry improves pest control potential by promoting spider abundance and functional diversity. *J Appl Ecol.* https://doi.org/10.1111/1365-2664.14797

McBratney A, Field DJ, Koch A (2014) The dimensions of soil security. *Geoderma* 213:203–213. https://doi.org/10.1016/j.geoderma.2013.08.013

Menta C, Remelli S (2020) Soil health and arthropods: from complex system to worthwhile investigation. *Insects* 11:54. https://doi.org/10.3390/insects11010054

Mettauer R, Thoumazeau A, Le Gall S, et al. (2023) Soil health in temperate agroforestry: influence of tree species and position in the field. *Arch Agronom Soil Sci* 69:1781–1800. https://doi.org/10.1080/03650340.2022.2116013

Mulla DJ, McBratney AB (2001) Soil spatial variability. In: Warrick, A.W., eds., *Soil Physics Companion*. CRC Press, Boca Raton, FL, pp. 343–373.

Mungai NW, Motavalli PP, Kremer RJ, Nelson KA (2005) Spatial variation of soil enzyme activities and microbial functional diversity in temperate alley cropping systems. *Biol Fertil Soils* 42:129–136. https://doi.org/10.1007/s00374-005-0005-1

Neher DA (2001) Role of nematodes in soil health and their use as indicators. *J Nematol* 33:161–168.

Nielsen MN, Winding A, Binnerup S, et al. (2002) *Microorganisms as Indicators of Soil Health*. National Environmental Research Institute, Denmark.

Nii-Annang S, Grünewald H, Freese D, et al. (2009) Microbial activity, organic C accumulation and 13C abundance in soils under alley cropping systems after 9 years of recultivation of quaternary deposits. *Biol Fertil Soils* 45:531–538. https://doi.org/10.1007/s00374-009-0360-4

Nunes MR, Karlen DL, Veum KS, et al. (2020) Biological soil health indicators respond to tillage intensity: a US meta-analysis. *Geoderma* 369:114335. https://doi.org/10.1016/j.geoderma.2020.114335

Offre P, Spang A, Schleper C (2013) Archaea in biogeochemical cycles. *Annu Rev Microbiol* 67:437–457. https://doi.org/10.1146/annurev-micro-092412-155614

Olejniczak I (2007) Soil mesofauna (Acarina and Collembola) along transects through shelterbelts of different age and adjacent fields. https://doi.org/10.5281/zenodo.3266986

Oliverio AM, Geisen S, Delgado-Baquerizo M, et al. (2020) The global-scale distributions of soil protists and their contributions to belowground systems. *Sci Adv* 6:eaax8787. https://doi.org/10.1126/sciadv.aax8787

Pankhurst C, Doube BM, Gupta VVSR (1997) *Biological Indicators of Soil Health*, 1st edition. CABI Publishing, Wallingford.

Pankhurst CE, Hawke BG, McDonald HJ, et al. (1995) Evaluation of soil biological properties as potential bioindicators of soil health. *Aust J Exp Agric* 35:1015–1028. https://doi.org/10.1071/ea9951015

Paoletti MG (1999) The role of earthworms for assessment of sustainability and as bioindicators. *Agric Ecosyst Environ* 74:137–155. https://doi.org/10.1016/S0167-8809(99)00034-1

Pardon P, Reheul D, Mertens J, et al. (2019) Gradients in abundance and diversity of ground dwelling arthropods as a function of distance to tree rows in temperate arable agroforestry systems. *Agric Ecosyst Environ* 270–271:114–128. https://doi.org/10.1016/j.agee.2018.10.017

Park J, Newman SM, Cousins SH (1994) The effects of poplar (P. trichocarpa × deltoides) on soil biological properties in a silvoarable system. *Agroforest Syst* 25:111–118. https://doi.org/10.1007/BF00705671

Price GW, Gordon AM (1998) Spatial and temporal distribution of earthworms in a temperate intercropping system in southern Ontario, Canada. *Agroforest Syst* 44:141–149. https://doi.org/10.1023/A:1006213603150

Rinot O, Levy GJ, Steinberger Y, et al. (2019) Soil health assessment: a critical review of current methodologies and a proposed new approach. *Sci Total Environ* 648:1484–1491. https://doi.org/10.1016/j.scitotenv.2018.08.259

Rivest M, Whalen JK, Rivest D (2020) Variation of soil microbial and earthworm communities along an agricultural transect with tree windbreak. *Agroforest Syst* 94:1639–1649. https://doi.org/10.1007/s10457-019-00476-3

Schmidt M, Corre MD, Kim B, et al. (2021) Nutrient saturation of crop monocultures and agroforestry indicated by nutrient response efficiency. *Nutr Cycl Agroecosyst*. https://doi.org/10.1007/s10705-020-10113-6

Schöler A, Jacquiod S, Vestergaard G, et al. (2017) Analysis of soil microbial communities based on amplicon sequencing of marker genes. *Biol Fertil Soils* 53:485–489. https://doi.org/10.1007/s00374-017-1205-1

Schulte RPO, Creamer RE, Donnellan T, et al. (2014) Functional land management: a framework for managing soil-based ecosystem services for the sustainable intensification of agriculture. *Environ Sci Policy* 38:45–58. https://doi.org/10.1016/j.envsci.2013.10.002

Seiter S, Ingham ER, William RD (1999) Dynamics of soil fungal and bacterial biomass in a temperate climate alley cropping system. *Appl Soil Ecol* 12:139–147. https://doi.org/10.1016/S0929-1393(98)00163-2

Shao G, Martinson GO, Corre MD, et al. (2023) Impacts of monoculture cropland to alley cropping agroforestry conversion on soil N2O emissions. *GCB Bioenergy* 15:58–71. https://doi.org/10.1111/gcbb.13007

Stamps WT, Linit MJ (1997) Plant diversity and arthropod communities: implications for temperate agroforestry. *Agroforest Syst* 39:73. https://doi.org/10.1023/A:1005972025089

Stamps WT, Linit MJ (1998) The problem of experimental design in temperate agroforestry. *Agroforest Syst* 44:187–196. https://doi.org/10.1023/A:1006215130892

Sun H, Koal P, Gerl G, et al. (2018) Microbial communities and residues in robinia- and poplar-based alley cropping systems under organic and integrated management. *Agroforest Syst* 92:35–46. https://doi.org/10.1007/s10457-016-0009-x

Swieter A, Langhof M, Lamerre J, Greef JM (2018) Long-term yields of oilseed rape and winter wheat in a short rotation alley cropping agroforestry system. *Agroforest Syst* 93:1853–1864. https://doi.org/10.1007/s10457-018-0288-5

Swieter A, Langhof M, Lamerre J (2022) Competition, stress and benefits: trees and crops in the transition zone of a temperate short rotation alley cropping agroforestry system. *J Agronom Crop Sci* 208:209–224. https://doi.org/10.1111/jac.12553

Szigeti N, Berki I, Vityi A, Winkler D (2022) Soil mesofauna and herbaceous vegetation patterns in an agroforestry landscape. *Agroforest Syst* 96:773–786. https://doi.org/10.1007/s10457-022-00739-6

Torppa KA, Taylor AR (2022) Alternative combinations of tillage practices and crop rotations can foster earthworm density and bioturbation. *Appl Soil Ecol* 175:104460. https://doi.org/10.1016/j.apsoil.2022.104460

Udawatta RP, Kremer RJ, Adamson BW, Anderson SH (2008) Variations in soil aggregate stability and enzyme activities in a temperate agroforestry practice. *Appl Soil Ecol* 39:153–160. https://doi.org/10.1016/j.apsoil.2007.12.002

Udawatta RP, Kremer RJ, Garrett HE, Anderson SH (2009) Soil enzyme activities and physical properties in a watershed managed under agroforestry and row-crop systems. *Agric Ecosyst Environ* 131:98–104. https://doi.org/10.1016/j.agee.2008.06.001

Udawatta RP, Kremer RJ, Nelson KA, et al. (2014) Soil quality of a mature alley cropping agroforestry system in temperate North America. *Commun Soil Sci Plant Anal* 45:2539–2551. https://doi.org/10.1080/00103624.2014.932376

Unger IM, Goyne KW, Kremer RJ, Kennedy AC (2013) Microbial community diversity in agroforestry and grass vegetative filter strips. *Agroforest Syst* 87:395–402. https://doi.org/10.1007/s10457-012-9559-8

van Groenigen JW, Lubbers IM, Vos HMJ, et al. (2014) Earthworms increase plant production: a meta-analysis. *Sci Rep* 4:6365. https://doi.org/10.1038/srep06365

van Tuinen D, Tranchand E, Hirissou F, et al. (2020) Carbon partitioning in a walnut-maize agroforestry system through arbuscular mycorrhizal fungi. *Rhizosphere* 15:100230. https://doi.org/10.1016/j.rhisph.2020.100230

Vaupel A, Bednar Z, Herwig N, et al. (2023) Tree-distance and tree-species effects on soil biota in a temperate agroforestry system. *Plant Soil* https://doi.org/10.1007/s11104-023-05932-9

Vaupel A, Küsters M, Toups J, et al. (2025) Trees shape the soil microbiome of a temperate agrosilvopastoral and syntropic agroforestry system. *Sci Rep* 15:1550. https://doi.org/10.1038/s41598-025-85556-4

Veldkamp E, Schmidt M, Markwitz C, et al. (2023) Multifunctionality of temperate alley cropping agroforestry outperforms open cropland and grassland. *Commun Earth Environ* 4:1–10. https://doi.org/10.1038/s43247-023-00680-1

Wagg C, Bender SF, Widmer F, van der Heijden MGA (2014) Soil biodiversity and soil community composition determine ecosystem multifunctionality. *PNAS* 111:5266–5270. https://doi.org/10.1073/pnas.1320054111

Wall DH, Bardgett RD, Behan-Pelletier V, et al. (2012) *Soil Ecology and Ecosystem Services*. OUP, Oxford.

Wezel A, Herren BG, Kerr RB, et al. (2020) Agroecological principles and elements and their implications for transitioning to sustainable food systems: a review. *Agron Sustain Dev* 40:40. https://doi.org/10.1007/s13593-020-00646-z

Weerasekara C, Udawatta RP, Jose S, et al. (2016) Soil quality differences in a row-crop watershed with agroforestry and grass buffers. *Agroforest Syst* 90:829–838. https://doi.org/10.1007/s10457-016-9903-5

Chapter 2

Assessing the benefits of temperate agroforestry in enhancing carbon sequestration

Augustine K. Osei[1] and Maren Oelbermann, University of Waterloo, Canada

1 Introduction: moving from conservation to carbon - a historical overview

Cultivating agricultural crops in combination with trees dates back to the beginning of plant and animal domestication (King, 1987). Since then, numerous agroforestry practices have been developed throughout tropical and temperate environments. For example, the *Dehesa* or *Montado* system in Europe's Mediterranean region, which integrates cork (*Quercus suber* L.) and holm oak (*Quercus ilex* L.) with livestock grazing and cereal cultivation, has been used for ~4500 years (Nerlich et al., 2013). Although wood pastures, the combination of forests with pastures and field crops, date to Neolithic times in northern Europe, this farming system was still widely used in this region until the early twentieth century (King, 1987; Nerlich et al., 2013). Other common European agroforestry systems included orchard intercropping, hedgerows, and windbreaks (Nerlich et al., 2013). Prior to North American colonization, First Nations were active land managers by using fire to improve wildlife forage, hunt, and travel and boost medicinal plant and berry production (Castleden et al., 2009; Hoffman, 2022). In southwestern USA and northern Mexico, the soil

1 Corresponding author (a2osei@uwaterloo.ca).

http://dx.doi.org/10.19103/AS.2024.0141.05

beneath mesquite (*Prosopis* spp.), a nitrogen-fixing tree, was used to fertilize crops (Bainbridge et al., 1990). However, European settlement of North America introduced new agricultural methods, but some of the European agroforestry practices like tree-based intercropping, orchard intercropping, wood pastures, windbreaks, and home gardens were adopted in the new world (Williams et al., 1997). Furthermore, China has a long history of practicing agroforestry, where Han Dynasty (206 BC to AD 220) administrators recommended to develop forested areas that also included crop and livestock production (You, 1991). Thus, the frequency and distribution of woody vegetation throughout the temperate zone, on a global scale, was tipped toward tree species suitable and useful for human use, which caused trees to become an integral part of the agricultural landscape (Worms, 2021).

While human societies evolved and modernized so did food production systems. This also changed agricultural practices which became increasingly mechanized and intensified, causing trees to disappear from the landscape (Dupraz et al., 2018). In parallel, new policies and economic opportunities through increasing crop yields, combined with the development of agrochemicals, synthetic fertilizes, and crop breeding in the mid-twentieth century, transformed agroforestry practices to large-scale single-crop systems (Dupraz et al., 2018). Trees were considered a nuisance since they obstructed large machinery in the modern agricultural landscape. Furthermore, agroforestry wood pastures did not provide adequate nutritional requirements for modern livestock breeds and were converted to forests (Luick, 2009). This created a separation between agriculture and forestry (Reeg et al., 2008).

In the early twentieth century, intensified agricultural management practices contributed to soil compaction, erosion, loss of soil organic carbon (SOC), and soil fertility. Agricultural techniques and technologies brought by the Europeans were not suitable for the soils of the new world. This was the driving force behind the 1930s Dust Bowl in mid-western USA. Similar consequences of agricultural intensification and soil degradation occurred in other temperate regions in the Canadian prairies (Marchildon, 2009), Patagonia, Argentina (Peri et al., 2018), Chile (Dube et al., 2018), Australia (Reid and Moore, 2008), New Zealand (Kemp et al., 2018), China (Chang et al., 2018), Himalayan region of India (Kumar et al., 2018), and Europe (Dupraz et al., 2018). As a result, modern agriculture played a significant role in increasing the global total of marginal and degraded land that is now substandard for long-term food production (Amundson, 2022).

Due to increasing environmental degradation associated with single-crop systems, a resurgence of agroforestry practices occurred in the early 1970s. The renewed interest in agroforestry led to knowledge of which agroforestry practices are most effective for specific environmental conditions (King, 1987). However, King (1987) also noted that at this time research was scant and uncoordinated with a lack of reliable field data. Once the International Centre for Research on Agroforestry (World Agroforestry Centre) was established in 1977, organized and coordinated research began (King, 1987). Since then, our understanding of the soil–microbe–crop–atmospheric continuum advanced, and so did our understanding of the dynamic interactions when trees were integrated into the agricultural landscape (Nair et al., 2009). For example, newly developed agroforestry practices like alley cropping allowed for the systematic integration of trees into the cropped landscape (King, 1987). This provided an opportunity for agroforestry to move into the modern era of highly mechanized agriculture and trees began to be viewed as providing environmental services.

Although modernized agroforestry systems addressed multiple environmental issues related to soil and water conservation (Young, 1997), restoration of marginal/degraded lands (Montagnini et al., 2011), and improved yield (Pardon et al., 2018), intensified activities of tropical deforestation raised concerns on the impact of these activities on the global carbon cycle and climate change (Sánchez, 2000). To address this growing global issue, Dixon (1995) suggested that agroforestry systems could be used to rehabilitate marginal/degraded lands to sequester carbon and help mitigate climate change. Dixon (1995) estimated that for each hectare of agroforestry production, up to 20 ha of deforestation may be prevented. In fact, Dixon et al. (1993) suggested that the potential of agroforestry systems to store carbon was between 15 Mg C ha^{-1} and 198 Mg C ha^{-1}, whereas Schroeder (1994) estimated 63 Mg C ha^{-1}, in temperate regions. The wide range of the potential of carbon sequestration in agroforestry systems suggested by Dixon et al. (1993) was due to regional differences, species composition, tree age, local environmental factors, and management practices of the agroforestry system. Consequently, agroforestry was integrated into the Kyoto Protocol due to its capacity to sequester carbon. Indeed, the 2000 IPCC report on land use, land-use change, and forestry recognized agroforestry systems as having the greatest potential for carbon sequestration of all land-use types (Watson et al., 2000). This is because trees function as an active carbon sink for many years by assimilating carbon in stems, branches, and roots until they die (Nair et al., 2009).

Agroforestry systems typically use fast-growing tree species, some of which have nitrogen-fixing capabilities (Nair et al., 2009). This enhances nutrient cycling and organic matter input from leaves/needles in addition to crop residues, causing a greater accumulation of SOC compared to single-crop agroecosystems that rely on carbon input from crop residues only (Oelbermann et al., 2004). For example, a 13-year-old hybrid poplar (*Populus* × *canadensis*) alley crop in southern Canada had an annual input of 537 g C m^{-2} compared to a 480 g C m^{-2} in a row crop system (Oelbermann et al., 2005). In the alley crop, 117 g C m^{-2} was derived from autumnal litterfall from the 13-year-old trees. Furthermore, Oelbermann et al. (2005) suggested that this quantity of carbon input from autumnal litterfall increases with increasing tree age. Thus, agroforestry systems create a synergistic interaction between the soil, crops, trees, and the atmosphere to help sustain agricultural productivity with the co-benefit of continuously sequestering carbon for the purpose of climate mitigation (Oelbermann et al., 2004). However, until 30 years ago, the majority of information based on the carbon sequestration potential of agroforestry systems (c.f., Dixon et al., 1993, Schroeder, 1993) was based on models and the use of allometric equations (Kürsten and Burschel, 1993). But the capacity of agroforestry systems to sequester carbon, based on actual field data, from temperate regions was missing. Research output based on field data began to appear ~20 years ago (Oelbermann et al., 2004). Since then, numerous publications on carbon sequestration in various temperate agroforestry systems have appeared (Peichl et al., 2006; Bambrick et al., 2010; Udawatta and Jose, 2012; Wotherspoon et al., 2014; Mayer et al., 2022). However, Udawatta et al. (2022) found that our current understanding of carbon sequestration of various temperate agroforestry practices still remains rudimentary. Therefore, the objective of this chapter is to provide an overview of the current state of research and challenges associated with carbon sequestration studies in above and belowground tree components, in soil, and the long-term stabilization of carbon in soil in agroforestry systems commonly adopted in temperate environments. We will conclude by providing insights into future research directions.

2 System-level carbon dynamics in temperate agroforestry

At the system-level, agroforestry systems store carbon in above- and belowground components (Oelbermann et al., 2006). Aboveground components include standing biomass in woody stems and branches and biomass returned to the soil from crop residues and woody litter, whereas belowground components include woody and crop roots and soil humus (Fig. 1) (Oelbermann et al., 2006; Borden et al., 2014). Integrating woody and non-woody crops on the same land unit in agroforestry systems allows for more efficient capturing of growth resources for increased biomass growth and subsequent increase in biomass and soil carbon sequestration (Nair and Nair, 2003; Lorenz and Lal, 2018). Moreover, both woody and non-woody components in agroforestry systems contribute to carbon accumulation in above- and belowground systems (Nair et al., 2009). In this respect, agroforestry systems accumulate more carbon than sole-cropped systems, woody plantations, or pasture systems without woody components (Nair et al., 2009). Sharrow and Ismail (2004) observed 8.17 Mg ha^{-1} year^{-1} and 5.77 Mg ha^{-1} year^{-1} greater system-level carbon in an 11-year-old Douglas fir (*Pseudotsuga menziesii*) silvopasture system than in pure stands

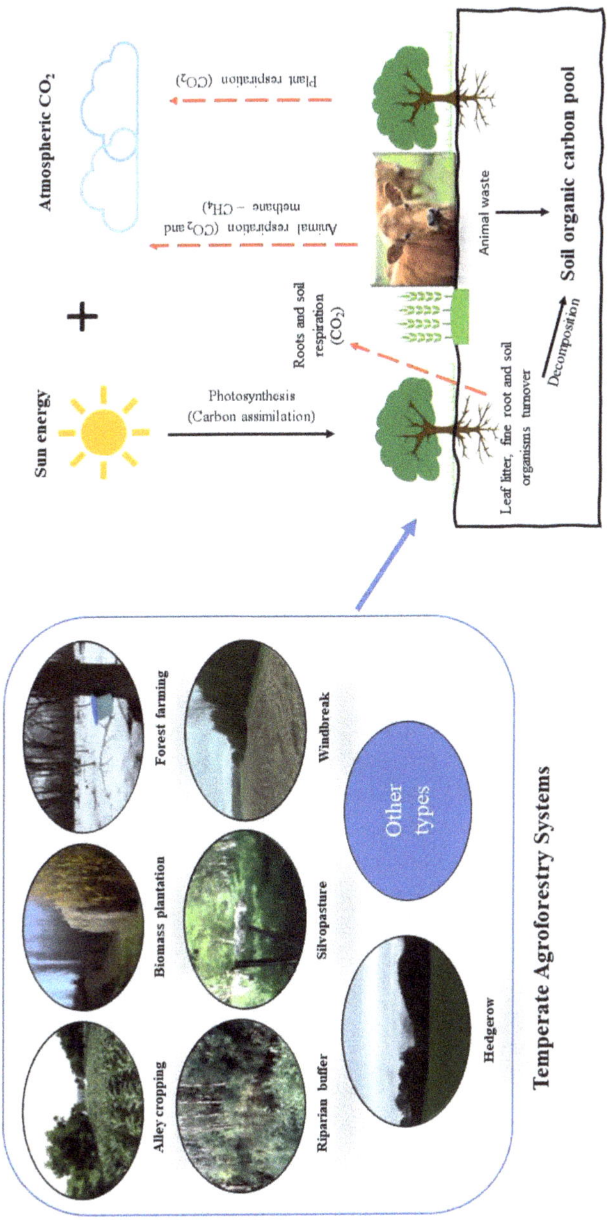

Figure 1 A schematic representation of carbon cycle in agroforestry systems.

of Douglas-fir and pasture. Gordon and Thevathasan (2004) discovered that in southern Canada, silvopastoral systems incorporating fast-growing tree species, such as hybrid poplar, can sequester 2.7 to 3 times more carbon at the system-level than monoculture pasture systems. In a 13-year-old alley cropping system in southern Canada, Peichl et al. (2006) identified a system-level total carbon pool of 96.5 Mg C ha^{-1} and 75.3 Mg C ha^{-1} to a 20 cm soil depth, using hybrid poplar and Norway spruce, respectively, in the agroforestry system, in comparison to 68.5 Mg C ha^{-1} in a barley sole crop. Similarly, Wotherspoon et al. (2014) determined system-level total carbon pools of 113.4, 99.4, 99.2, 91.5, and 91.3 Mg C ha^{-1} in a 25-year-old alley cropping system in southern Canada, utilizing hybrid poplar, white cedar, red oak, black walnut, and Norway spruce, respectively, compared to 71.1 Mg C ha^{-1} in the sole crop.

Unlike in tropical agroforestry systems, research on carbon dynamics in temperate agroforestry systems is relatively recent and still evolving (Chang et al., 2018). Carbon sequestration which refers to the capturing, removing, and storing of atmospheric carbon dioxide (CO_2) in long-term reservoirs such as terrestrial systems (e.g. vegetation and soil) (Fig. 1) is crucial to climate change mitigation (Nair et al., 2009). This natural process of capturing and storing carbon in terrestrial ecosystems is more cost-effective and sustainable compared to technological approaches (e.g. using engineering techniques to capture and inject industrial CO_2 into the ocean and deep geological strata) (Lal, 2008).

Studies evaluating system-level carbon stocks in temperate agroforestry systems are few. This may be because trees are generally thought to contain lower carbon than the soil (Peichl et al., 2006). Also, carbon assimilated in tree biomass is relatively labile compared to the soil carbon pool and may depend on the fate of the products derived from tree biomass (Upson and Burgess, 2013; Lorenz and Lal, 2018). For these reasons and other challenges associated with quantifying biomass carbon pools, the majority of carbon studies in temperate agroforestry systems largely focus on soil carbon with comparatively few studies quantifying biomass carbon.

In forest systems, the soil is estimated to contain about 60% of the system's carbon with 30% in aboveground parts (Nair et al., 2010). However, woody biomass in agroforestry systems can contribute significantly to the overall system's carbon sequestration potential. For example, a meta-analysis by Drexler et al. (2021) revealed that hedgerows in temperate regions stored 104 Mg ha^{-1} more carbon than croplands. Out of this amount, biomass contributed 87 Mg ha^{-1}, representing 84% of the system's total carbon stock. Ma et al. (2020) also found that global agroforestry systems, on average, had 46.1 Mg ha^{-1} more carbon in tree biomass than crop and pastureland without trees. This suggests that at the system-level, biomass carbon accumulation plays a vital contribution to the carbon sequestration potential of agroforestry systems.

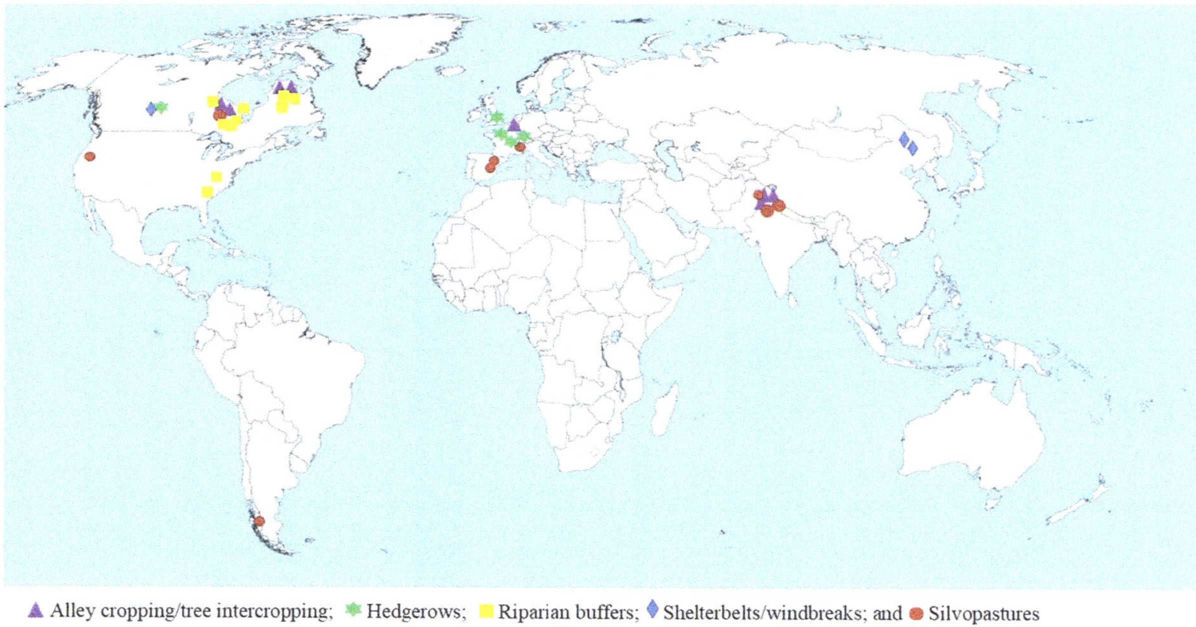

Figure 2 The global distribution of the different temperate agroforestry systems in the study sites used for our analysis of carbon sequestration rates in this study.

To understand the carbon sequestration benefits of temperate agroforestry systems, we review studies (Fig. 2; Table 1) that quantify carbon stocks in aboveground standing biomass, belowground biomass (roots), and soil and analyze carbon sequestration at the system-level for agroforestry systems in temperate regions at the global scale (Table 2). We focus our analyses on alley cropping/intercropping, riparian buffers, shelterbelts/windbreaks, silvopastures, and hedgerows (Table 1). Although, we acknowledge there are other diverse kinds of agroforestry systems in temperate regions in Asia, particularly, in the Indian Himalayan region (Kumar et al., 2018). These regionally specific agroforestry practices include agrihorticulture, agrihortisilviculture, agrisilviculture, agrisilvihorticulture, hortiagriculture, silvipasture, among others, which embrace the principles of tree-based intercrops with crops and/or pasture (Kumar et al., 2018). The nomenclature of these agroforestry systems is based on the structure and functions of the various components (e.g. food, fruits, fodder, and timber production) (Goswani et al., 2014). For this reason and for the purpose of this chapter, we categorize these different agroforestry practices from the Indian Himalayan region under intercropping or silvopasture systems, depending on the components involved (Table 1).

2.1 *Carbon sequestration in aboveground standing live biomass*

Agroforestry systems with trees have higher carbon sequestration potential in aboveground than treeless systems or agroforestry systems with shrubs and grasses as the perennial components (Ofosu et al., 2022). This is because trees in general store about 50–60% of their carbon in aboveground biomass components whereas perennial pasture grasses store only 10% in aboveground, with the rest allocated belowground (Houghton and Hackler 2000; Sharrow and Ismail 2004). For instance, between trees and riparian grass buffer systems, Ofosu et al. (2022) observed higher carbon (2.31 Mg C ha^{-1} $year^{-1}$) in aboveground standing biomass in treed riparian buffers compared to grassed riparian buffers (0.05 Mg C ha^{-1} $year^{-1}$) in southern Canada. Tufekcioglu et al. (2003) also found 4–8 times higher aboveground biomass carbon in poplar (~20 Mg C ha^{-1}) than in switchgrass (*Panicum virgatum* L.) (5 Mg C ha^{-1}) and cool-season grass (2.5 Mg C ha^{-1}) riparian buffers in Iowa, USA. Assimilated carbon from atmospheric CO_2 through photosynthesis becomes locked in aboveground biomass components of stems, branches, and foliage (Nair et al., 2009). Whereas carbon assimilated in tree stems can offer long-term carbon sequestration as wood products (e.g. construction timber, furniture, and wood crafts) (Nair et al., 2009; Udawatta et al., 2022), carbon from tree foliage can be stored over the long-term in the soil through litter input as soil organic matter (SOM) (Fig. 1) (Oelbermann et al., 2004).

Table 1 System-level carbon stocks in different temperate agroforestry systems and adjacent land uses from different locations (only studies published in the past 20 years that included aboveground, belowground (root) and soil carbon pools were considered)

Agroforestry system	Location	Age	Dominant tree/shrub species	Carbon stocks in agroforestry system (Mg ha^{-1})*			Annual detrital carbon inputs	Adjacent land use	Carbon stocks in adjacent land use (Mg ha^{-1})		Source
		Years		Aboveground	Belowground	Soil	Mg ha^{-1} year^{-1}		Biomass	Soil	
Agri-horticulture (*Intercropping*)	Himachal Pradesh, India	NA†	Apple (*Malus domestica*), Japanese Persimmon (*Diospyros kaki*), pear (*Pyrus communis*)	33.38	7.46	56.34 (0–30)‡	0.39	NA	NA	NA	Bhardwaj et al. (2023)
	Northwestern Himalaya, India	NA	Mango (*Mangifera indica*), citrus (*Citrus* spp.), plum (*Prunus domestica*), peach (*Prunus persica*), pear	13.94	3.52	22.73 (0–30)	NA	NA	NA	NA	Sharma et al. (2023)
	Northwestern Himalaya, India	21	Apple	11.08	3.67	151.77 (0–100)		Agriculture	3.34	132.75 (0–100)	Chisanga et al. (2018)
	Himachal Pradesh, India	24	Not provided	32.37	16.68	42.51 (0–40)	1.26	Agriculture	12.22	34.72 (0–40)	Rajput et al. (2017)
	Himachal Pradesh, India	~30	Gooseberry (*Emblica officinalis*), citrus, peach	9.36	2.79	90.07 (0–40)		Agriculture	8.24	65.43 (0–40)	Goswami et al. (2014)
Agri-horti-silviculture (*Intercropping*)	Himachal Pradesh, India	NA	Armenian plum (*Prunus armeniaca*), nettle tree (*Celtis australis*), oak (*Quercus ilex*), white willow (*Salix alba*), Himalayan poplar (*Populus ciliate*), Bhutan pine (*Pinus wallichiana*), Chilghoza pine (*Pinus gerardiana*)	45.08	11.13	58.75 (0–30)	0.42	NA	NA	NA	Bhardwaj et al. (2023)
Agri-silviculture (*Intercropping*)	Himachal Pradesh, India	NA	Armenian plum, oak, Himalayan poplar, Chilghoza pine, apple	48.66	14.44	52.06 (0–30)	0.45	NA	NA	NA	Bhardwaj et al. (2023)
	Northwestern Himalaya, India	NA	Australian red cedar/Toon tree (*Toona ciliata*), bhimal tree (*Grewia optiva*), orchid tree (*Bauhinia variegate*), red silk cotton tree (*Bombax ceiba*), *Albizia chinensis*, *Ficus palmata*), mulberry (*Morus alba*), Nettle tree, Himalayan poplar	19.56	4.95	25.37 (0–30)	NA	NA	NA	NA	Sharma et al. (2023)
	Himachal Pradesh, India	~30	Bhimal tree, fig tree (*Ficus* spp.), Nettle tree	8.19	2.45	92.65 (0–40)		Agriculture	8.24	65.43 (0–40)	Goswami et al. (2014)
Agri-silvihorti-culture (*Intercropping*)	Himachal Pradesh, India	~30	Nettle tree, bhimal, fig tree, Australian red cedar/toon tree, pomegranate (*Punica granatum*), Armenian plum	11.38	3.40	95.46 (0–40)		Agriculture	8.24	65.43 (0–40)	Goswami et al. (2014)

(*Continued*)

Table 1 (*Continued*)

Agroforestry system	Location	Age	Dominant tree/shrub species	Carbon stocks in agroforestry system (Mg ha^{-1})*			Annual detrital carbon inputs	Adjacent land use	Carbon stocks in adjacent land use (Mg ha^{-1})		Source
		Years		Aboveground	Belowground	Soil	Mg ha^{-1} year^{-1}		Biomass	Soil	
Agro-horti-silviculture (*Intercropping*)	Northwestern Himalaya, India	NA	Himalayan poplar, nettle tree, deodar cedar (*Cedrus deodara*), Indian horse chestnut (*Aesculus indica*), Indian willow (*Salix tetrasperma*), apple, plum, pear, English walnut (*Juglans regia*)	23.74	6.10	25.92 (0–30)	NA	NA	NA	NA	Sharma et al. (2023)
Horti-agriculture (*Intercropping*)	Northwestern Himalaya, India	NA	Apple, pear, hazel (*Corylus jacquemontii*), English walnut	18.04	4.69	22.25 (0–30)	NA	NA	NA	NA	Sharma et al. (2023)
Agro-silvohorti-culture (*Intercropping*)	Northwestern Himalaya, India		Australian red cedar tree, bhimal tree, Albizia tree (*Albizia chinensis*), Chinaberry tree (*Melia composita*) *Leucaena leucocephala*, mango, citrus, pear, plum, peach	17.87	4.48	31.66 (0–30)	NA	NA	NA	NA	Sharma et al. (2023)
Alley cropping/ Intercropping Systems	Northern Himalaya, India	11	Apple	14.31	3.82	45.15 (0–20)	NA	Agriculture	4.58	34.94 (0–20)	Zahoor et al. (2021)
	France	6	Hybrid walnut (*Juglans* × *intermedia* Carr.)	0.02	0.01	96.10 (0–30)	NA	Agriculture	NA	90.90 (0–30)	Cardinael et al. (2017)
		6	Hybrid walnut	0.07	0.03	82.70 (0–30)	NA	Agriculture	NA	78.70 (0–30)	
		41	Black walnut (*Juglans nigra*)	19.85	5.55	96.40 (0–20)	NA	Agriculture	NA	65.10 (0–20)	
		18	Hybrid walnut	26.64	6.61	289.40 (0–60)	NA	Agriculture	NA	263.80 (0–60)	
		18	Hybrid walnut	10.88	2.99	317.90 (0–100)	NA	Agriculture	NA	294.30 (0–100)	
	Southern Quebec, Canada	10	Hybrid poplar (*Populus* × *canadensis*), red oak (*Quercus rubra* L.), and black cherry (*Prunus serotina* Ehrh.)	31.09	7.25	148.50 (0–30)	0.04	Agriculture	2.28	131.90 (0–30)	Winans et al. (2016)
		10	Hybrid poplar, red oak, and white ash (*Fraxinus americana* L.)	70.75	15.95	159.90 (0–30)	0.06	Non-tree-based forage	3.19	167.90 (0–30)	

	Himachal Pradesh, India	20	Apple and plum	36.42	11.16	41.87 (0–40)	2.38	Agriculture	15.08	35.77 (0–40)	Rajput et al. (2015)
	Southern Ontario, Canada	25	Poplar (*Populus* spp.)	21.06	5.94	86.86 (0–40)	2.45	Agriculture	1.40	71.08 (0–40)	Wotherspoon et al. (2014)
		25	Red oak	11.84	4.16	83.77 (0–40)	1.60				
		25	Black walnut	12.30	2.70	76.84 (0–40)	2.25				
		25	Norway spruce (*Picea abies* L.)	9.49	3.51	78.33 (0–40)	1.94				
		25	White cedar (*Thuja occidentalis*)	14.88	1.12	83.23 (0–40)	0.88				
	Southern Ontario, Canada	13	Hybrid poplar	14.80	3.20	78.50 (0–20)	1.20	Agriculture	3.50	65.00 (0–20)	Peichl et al. (2006)
		13	Norway spruce	7.10	2.20	66.00 (0–20)	0.12				
	Northwestern Himalaya, India	50	Black locust (*Robinia* spp.), Apple	30.25	8.14	125.37 (0–100)		Agriculture	3.34	132.75 (0–100)	Chisanga et al. (2018)
	Himachal Pradesh, India	~30	Nettle tree, Pomegranate, Pear, Bhimal tree	11.13	3.32	84.14 (0–40)		Agriculture	8.24	65.43 (0–40)	Goswami et al. (2014)
Hedgerows	Western France	20–120	Chestnut (*Castanea sativa*), Elderberry, English oak (*Quercus robur*), Hornbeam (*Carpinus betulus*), Hazel (*Corylus avellana*), European ash (*Fraxinus excelcior*), Beech (*Fagus sylvatica*), Black locust, Rowan (*Sorbus aucuparia*), Elm (*Ulmus minor*), Maple (*Acer campestre*), Hawthorn (*Crataegus monogyna*)	1.20–21.60 Mg C 100 m^{-1}	0.30–6.10 Mg C 100 m^{-1}	109.05–155.4 (0–90)	NA	NA	NA	NA	Viaud and Kunnemann (2021)
	Central Alberta, Canada	27	Wood rose (*Rosa woodsii* Lindl.), Raspberry (*Rubus idaeus* L.), Aspen (*Populus tremuloides* Michx.), Balsam poplar (*Populus balsamifera* L.)	126.49	35.51	315.79 (0–100)	1.64	Agriculture	3.85	168.30 (0–100)	Gross et al. (2022)
	Southeast England	100	Blackthorn (*Prunus spinosa*)	600.00	110.00	210.00 (0–50)	35.04	NA	NA	NA	Crossland (2015)
		100	Hawthorn	179.00	50.00	990.00 (0–50)	32.30	NA	NA	NA	
		100	Hazel	125.00	49.00	100.00 (0–50)	20.85	NA	NA	NA	

(*Continued*)

Table 1 *(Continued)*

Agroforestry system	Location	Age		Carbon stocks in agroforestry system (Mg ha⁻¹)*			Annual detrital carbon inputs	Adjacent land use	Carbon stocks in adjacent land use (Mg ha⁻¹)		Source
		Years	Dominant tree/shrub species	Aboveground	Belowground	Soil	Mg ha⁻¹ year⁻¹		Biomass	Soil	
Riparian buffers	Southern Ontario, Canada	50	Norway spruce	184.60	55.30	271.80 (0–60)	2.30	Agriculture	NA	113.60 (0–60)	Ofosu et al. (2022)
		156	White cedar, sugar maple (*Acer saccharinum*), birch (*Betula alleghaniensis*), black walnut	207.70	62.10	658.30 (0–60)	2.70	Agriculture	NA	101.30 (0–60)	
		156	Hybrid poplar White cedar, BirchAsh	276.30	82.70	578.30 (0–60)	3.60	Agriculture	NA	157.10 (0–60)	
		32	Sugar maple, black walnut, hybrid poplar	243.10	72.70	356.90 (0–60)	4.50	Agriculture	NA	157.10 (0–60)	
		35	Bentgrass (*Agrostis* spp.) and purple-stemmed aster (*Symphyotrichum puniceum*).	1.70	3.20	286.00 (0–60)	0.90	Agriculture	NA	164.90 (0–60)	
	Southern Ontario, Canada	37–60	Sugar maple and Ash	171.78	73.62	229.61 (0–30)	NA	Agriculture	NA	97.40 (0–30)	Vijayakumar et al. (2020)
		34–60	White cedar and white pine (*Pinus strobus*)	137.76	59.04	155.75 (0–30)	NA	Agriculture	NA	78.95 (0–30)	
		6–8	Sugar maple and Black walnut	12.18	5.22	124.07 (0–30)	NA	Agriculture	NA	118.45 (0–30)	
		6–8	White spruce and white pine	4.90	2.10	131.41 (0–30)	NA	Agriculture	NA	100.00 (0–30)	
	Southern Quebec, Canada	9–200	Eastern hemlock (*Tsuga canadensis*), white cedar, grey birch (*Betula populifolia*), sugar maple, hybrid poplar	25.00–115.00	4.80–36.80	83.00–176.20 (0–60)	2.30–12.30	Pasture	3.7	122.39 (0–60)	Fortier et al. (2013; 2015)
	North Carolina, USA	>50	Black gum (*Nyssa biflora*), red oak, sweetgum (*Liquidambar styraciflua*), red maple (*Acer rubrum*)	115.66	40.64	39.15 (0–10)	0.93	Agriculture	0.85	16.25 (0–10)	Rheinhardt et al. (2012)
		25–50	Red maple, sweetgum, loblolly pine (*Pinus taeda*)	44.51	15.64	33.40 (0–10)	0.70–1.39				
		5–25	loblolly pine, sweetgum, red maple, black willow (*Salix nigra*)	45.40	15.95	28.55 (0–10)	0.50–2.48				
		0–5		1.63	0.57	41.45 (0-10)	7.76				

	Northeastern Ontario, Canada	95	Black spruce (*Picea mariana*), white spruce (*Picea glauca*), balsam fir (*Abies balsamea*), whitebirch (*Betula Papyrifera*), mountain maple (*Acer spicatum*), speckled alder (*Alnus rugosa*), and beakedhazel (*Corylus cornuta*)	63.56	20.29	52.65 (0–75)	0.38	Forest	81.41	68.02 (0–75)	Hazlett et al. (2005)
	South Carolina, USA	5	Blackberry (*Rubus* spp.), smooth alder (*Alnus serrulata*), and buttonbush (*Cephalanths occidentalis* L.)	5.35	1.85	20.26 (0–10)	0.07	NA	NA	NA	Giese et al.(2003)
		11	Willow, Waxmyrtle (*Myrica cerifera* L.), smooth alder, and buttonbush	17.14	2.77	17.47 (0–10)	1.64	NA	NA	NA	
		15	Willow, red maple, smooth alder, Waxmyrtle, river birch (*Betula nigra* L.), sweetgumand persimmon (*Diospyros virginiana* L.)	17.23	3.04	15.84 (0– 10)	1.19	NA	NA	NA	
		63	Matured bottomlandhardwood riparian forest with several mixed species	101.77	4.36	12.04 (0–10)	1.35	NA	NA	NA	
Shelterbelts	Central Alberta, Canada	29	Balsam poplar, and willow	130.03	23.57	210.11 (0–100)	0.46	Agriculture	4.53	175.96 (0–100)	Gross et al. (2022)
	The Three-North, China	25	Mixed woodlands and shrublands	11.19	12.93	70.32	0.17	NA	NA	NA	Chu et al. (2019)
Pastoral-silviculture (*Silvopasture*)	Northwestern Himalaya, India	NA	Himalayan poplar, nettle tree, deodar cedar (*Cedrus deodara*), pear	13.89	3.47	28.15 (0–30)	NA	NA	NA	NA	Sharma et al. (2023)
Horti-pastoral (*Silvopasture*)	Northwestern Himalaya, India	NA	Apple, pear, English walnut	15.92	4.02	23.96 (0–30)	NA	NA	NA	NA	Sharma et al. (2023)
Silvopasture systems	Northwestern Himalaya, India	NA	Australian red cedar, *Albizia chinensis*, orchid tree, nettle tree, pear, *Ficus palmata*, Mulberry, Chir pine (*Pinusroxburghii*), *Zizyphus nummularia, Dodonea viscosa*	19.09	4.90	35.77 (0–30)	NA	NA	NA	NA	Sharma et al. (2023)
	Northwestern Himalaya, India	70	Chilgoza pine (*Pinus gerardiana*)	42.33	9.75	108.97 (0–100)		Agriculture	3.34	132.75 (0–100)	Chisanga et al. (2018)
	France	26	Wild cherry (*Prunus avium*)	36.69	9.13	541.70 (0–50)	NA	Agriculture	NA	540.00 (0–50)	Cardinael et al. (2017)
	Northern Extremadura, Spain	13	Hybrid walnut	14.72	10.57	91.20 (0–100)	NA	Hybrid walnut plantation with ploughing	26.13	81.49 (0–100)	López-Díaz et al. (2017)

(*Continued*)

Table 1 (*Continued*)

Agroforestry system	Location	Age	Dominant tree/shrub species	Carbon stocks in agroforestry system (Mg ha^{-1})*			Annual detrital carbon inputs	Adjacent land use	Carbon stocks in adjacent land use (Mg ha^{-1})		Source
		Years		Aboveground	Belowground	Soil	Mg ha^{-1} year^{-1}		Biomass	Soil	
	Himachal Pradesh, India	52	NA (silvipasture)	40.70	20.96	38.54 (0–40)	0.80	Agriculture	12.22	34.72 (0–40)	Rajput et al. (2017)
	Himachal Pradesh, India	~30	Acacia (*Acacia catechu*), nettle tree, Chir pine, pear, Himmalayan oak	2.59	0.77	115.45 (0–40)		Agriculture	8.24	65.43 (0–40)	Goswami et al. (2014)
								Grassland	1.23	80.15 (0–40)	
	Patagonia, Chile	18	Ponderosa pine (*Pinus ponderosa*)	22.73	11.81	193.76 (0–40)	NA	*Pinus ponderosa* plantation	47.74	149.25 (0–40)	Dube et al. (2012)
								Natural pasture	4.65	177.10 (0–40)	
	Spain	11	Monterey pine (*Pinus radiata*)	73.60	19.33	138.25 (0–25)	2.58	NA	NA	NA	Fernández-Núnez et al. (2010)
		11	Silver birch	29.99	4.17	157.21 (0–25)	NA	NA	NA	NA	
	Southern Ontario, Canada	15	Hybrid poplar	18.21	4.94	15.21 (0–5)	1.92	Ryegrass monoculture	7.96	14.24 (0–5)	Gordon and Thevathasan (2004)
	Oregon, USA	11	Douglas-fir (*Pseudotsuga menziesii*)	10.71	1.53	95.89 (0–45)	0.11	Pasture	1.00	102.52 (0–45)	Sharrow and Ismail (2004)

†NA, not determined in study, hence data not available.
*Where C concentration was not provided, biomass C was assumed as 50% of total biomass.
‡Soil depth (cm) is provided in parentheses.

Table 2 Estimated carbon sequestration rate in aboveground, belowground, soil, and at the system-level for different agroforestry systems in temperate regions

	Tree/shrub density	Agroforestry	Carbon stocks					
Agroforestry system	Average No. of Tree/shrub ha^{-1}	component	Minimum	Maximum	Average	Average age	Carbon sequestration rate	Total system-level carbon
			Mg C ha^{-1}			(Years†)	Mg C ha^{-1} year^{-1}	
Alley cropping/ Intercropping systems	240	Aboveground	0.02	70.75	20.50	24	0.85	
		Belowground (roots)	0.01	16.68	5.60		0.23	
		Soil	22.25	317.90	90.60		0.64 (0–50) 0.87 (0–100)	**1.72 (0–50)** **1.95 (0–100)**
Riparian buffers	1424	Aboveground	1.63	276.30	89.61	49	1.83	
		Belowground (roots)	0.57	82.70	28.13		0.57	
		Soil	12.04	658.30	165.61		3.08 (0–50) 2.75 (0–100)	**5.48 (0–50)** **5.15 (0–100)**
Shelterbelts	1200	Aboveground	11.19	130.03	70.61	27	2.62	
		Belowground (roots)	12.93	23.57	18.25		0.68	
		Soil	70.32	210.11	140.22		1.18 (0–100)	**4.48 (0–100)**
Silvopasture systems	200	Aboveground	2.59	73.60	26.24	27	0.97	
		Belowground (roots)	0.77	20.96	8.10		0.30	
		Soil	15.21	541.70	121.85		0.45 (0–50) 0.20 (0–100)	**1.72 (0–50)** **1.47 (0–100)**
Hedgerows	7533	Aboveground	125.00	600.00	257.62	82	3.14	
		Belowground (roots)	35.51	110.00	61.13		0.75	
		Soil	100.00	990.00	403.95		5.46 (0–100)	**9.35 (0–100)**

†Average age for each system was calculated by averaging the years for that system as reported in Table 1. Where years are provided in range from Table 1, average years were calculated from the primary source of the study. Age for systems with NA were assumed as 30 years based on average age of similar agroforestry practices in the Indian Himalayan Region.

‡Annual SOC sequestration rates for each agroforestry system were calculated as:

$$\text{Annual SOC Sequestration Rate} = \left(\frac{SOC\ stocks_{Agroforestry} - SOC\ stocks_{Adjacent\ land\ use}}{Age\ of\ Agroforestry\ system} \right)$$

. Soil depth (cm) is provided in parenthesis.

*NA, Annual SOC sequestration rates could not be determined due to the absence of adjacent land use in the study

Temperate agroforestry systems are estimated to sequester 1.9 Pg C $year^{-1}$ in aboveground components (Oelbermann et al., 2004) but to date, only a few studies have quantified biomass carbon sequestration potentials of temperate agroforestry systems. The paucity of data on aboveground biomass carbon in temperate agroforestry systems may be attributed to challenges associated with tree biomass estimations (Sinacore et al., 2017). Estimating aboveground biomass is challenging because the most accurate approach for generating reliable biomass estimates involves destructive sampling and determining carbon concentrations in the various aboveground biomass components (Ketterings et al., 2001; Kaonga and Bayliss-Smith, 2009). This approach is, however, complex, time-consuming, destructive, and labor intensive which restricts its use to small areas and small tree sample sizes (Ketterings et al., 2001; Kaonga and Bayliss-Smith, 2009).

Due to the large number of tree species that are sometimes available in a particular agroforestry system and the challenges associated with destructively sampling large-size trees to accurately determine aboveground biomass, biomass regression models and sometimes remote sensing techniques such as LiDAR are used in estimating aboveground biomass (Drake et al., 2003; Sinacore et al., 2017). Also, in most of these studies, carbon is estimated based on the assumption that biomass contains 45–50% carbon (Goswani et al., 2014; Udawatta et al., 2022). Moreover, in some of these studies, biomass carbon of associated crops is often neglected. Given trees in agroforestry systems influence understory or companion crops in many diverse ways (Lorenz and Lal, 2018), neglecting carbon stocks in these secondary biomass components may result in a significant underestimation of the carbon sequestration potential of agroforestry systems.

Most of the studies that have quantified aboveground biomass carbon stocks in temperate agroforestry systems are concentrated in temperate North America and in the Indian Himalayan region (Table 1). Results from these studies are widely variable where the carbon sequestration potential of aboveground standing biomass ranges from 0.85 Mg C ha^{-1} $year^{-1}$ in alley cropping/intercropping systems to 3.14 Mg C ha^{-1} $year^{-1}$ in hedgerows (Table 2). Cardinael et al. (2018) determined that aboveground carbon sequestration rates, based on a global literature review for temperate agroforestry systems, ranges from 0.52 to 3.11 Mg C ha^{-1} $year^{-1}$. Nair and Nair (2003) also determined that the aboveground biomass carbon sequestration rates for different agroforestry systems ranged from 0.29 to 15.21 Mg C ha^{-1} $year^{-1}$. The wide range determined by Nair and Nair (2003) was attributed to differences in management, species, age, and differences in ecological and edaphic factors.

Udawatta et al. (2022) reported aboveground biomass carbon sequestration rates in alley crops (2.6 Mg ha^{-1} $year^{-1}$), riparian buffers (3.0 Mg ha^{-1} $year^{-1}$), and silvopasture (4.9 Mg ha^{-1} $year^{-1}$) in temperate North America.

These aboveground biomass carbon rates are higher than the 0.9 Mg C ha^{-1} $year^{-1}$, 1.8 Mg C ha^{-1} $year^{-1}$, and 1.0 Mg C ha^{-1} $year^{-1}$ that we calculate for alley crops, riparian buffers, and silvopastures, respectively, for temperate regions at the global scale (Table 2). Whereas the analysis by Udawatta et al. (2022) considered relatively matured agroforestry systems in only USA and Canada, our synthesis covers all global temperate regions including the dry and cold temperate region of the Indian Himalayas (e.g. cold deserts) that incorporate diverse tree species of various age classes and types of agroforestry systems.

Furthermore, local environmental and edaphic conditions, tree species, and age of the agroforestry system impact biomass productivity, and therefore carbon sequestration rates (Ma et al., 2020). Schroeder (1994) estimated that aboveground biomass carbon sequestration rate for temperate riparian buffers with 30-year rotation cycle was 2.1 Mg C ha^{-1} $year^{-1}$. However, Udawatta and Jose (2012) estimated a carbon sequestration rate of 2.5 Mg C ha^{-1} $year^{-1}$ for mature riparian buffers in temperate North America with 50-year rotation cycle. Vijayakumar et al. (2020) also found that mature (34–60 years) riparian buffers had a higher (1.1–4.1 Mg ha^{-1} $year^{-1}$) carbon sequestration rate in aboveground biomass than younger (6–8 years) riparian buffer systems (0.3–1.1 Mg ha^{-1} $year^{-1}$) in southern Canada.

In general, as an agroforestry system matures, the biomass of the woody components and understory vegetation increases, resulting in an overall increase in biomass carbon stocks (Udawatta and Jose, 2012). Since tree growth follows a sigmoidal pattern (Weiner and Thomas, 2001), the accumulation of biomass carbon is expected to occur with age until the tree reaches its maximum growth. However, the time taken for a tree to reach its full growth, and its biomass carbon sequestration rate, depends on the species. Balian and Naiman (2005) observed that ~90% of aboveground biomass accumulation occurred during the first 20–40 years in riparian buffer with willow (*Salix* spp.) and red alder (*Alnus rubra*) as the dominant species. Yet, when sitka spruce (*Picea sitchensis*) dominated the riparian buffer landscape, biomass carbon accrual by the system did not reach equilibrium until after 300 years. In a tree-based intercropping study in southern Canada, Wotherspoon et al. (2014) observed that fast-growing hybrid poplar reached its full carbon sequestration potential 25 years after establishment compared to the Norway spruce (*Picea abies*) which could continue to sequester atmospheric CO_2 until after 60 years. Nair et al. (2009) suggested the use of slow-growing rather than fast-growing tree species in agroforestry systems if carbon sequestration is the intended purpose of the agroforestry system. Apart from accumulating carbon over the long-term, slow-growing species have high specific-gravity wood and are deemed valuable for use as wood products (e.g. construction timber, furniture, and wood crafts) which constitute longer-term sink for fixed carbon (Nair et al., 2009).

Agroforestry systems such as riparian buffers and hedgerows tend to sequester more carbon in their biomass than systems which integrate annual row crops on the same land space such as alley cropping or intercropping (Thevathasan et al., 2018). We determine that hedgerows and riparian buffers have a 269% and 115% higher aboveground biomass carbon sequestration rate than alley cropping/intercropping system, respectively (Table 2). Thevathasan et al. (2018) suggested that riparian buffers have greater potential for aboveground biomass carbon sequestration than alley cropping systems because the absence of associated annual crops in riparian systems allows for a greater tree density. Also, since riparian plantings are located in the terrestrial-aquatic ecotone, competition for moisture is normally not a limiting factor compared to other agroforestry systems (Thevathasan et al., 2018). Accordingly, riparian buffer systems have a greater potential for biomass carbon sequestration with the selection of right mixture of species. Udawatta and Jose (2012) explained that diverse species mixture and the presence of different functional groups such as trees and grasses in riparian buffers and hedgerows enhance the effective capturing of resources which promote carbon sequestration in these systems. Ma et al. (2020) found agroforestry systems with multiple tree species contained greater biomass carbon stocks and accumulated biomass carbon faster than systems with single tree species. This implies that using diverse and complementary species in the implementation of agroforestry systems can yield greater carbon sequestration benefits than using single species.

2.2 Belowground (root) biomass carbon sequestration

Roots in agroforestry systems play significant role in long-term carbon sequestration by storing carbon in deeper soil layers (Nair, 2011). For example, roots can constitute up to 50% of total tree and shrub biomass (Axe et al., 2017). Similar to aboveground standing biomass carbon sequestration studies, quantification of belowground biomass carbon in roots of temperate agroforestry systems are scarce. In a meta-analysis to assess the contribution of hedgerows in carbon sequestration, Drexler et al. (2021) observed that only one of the nine studies measured carbon stocks in roots. They suggested the need for more field-based measurements of belowground biomass carbon stocks of hedgerows to derive more thorough estimates. This is because roots contribute substantially to the overall carbon stock of hedgerows.

The observation by Drexler et al. (2021) is not just unique in hedgerows but applies to the other temperate agroforestry systems. Until recently, most root biomass studies in temperate agroforestry focused on competitive interactions between trees and crops for soil resources (e.g. Clinch et al., 2009; Borden et al., 2017). The absence of studies incorporating the potential of root carbon sequestration is due to the major challenges associated with studying root systems. Dube et al. (2012) emphasized that root biomass carbon studies are dominated by logistical challenges including a lack of suitable and appropriate field equipment for excavation and methodological difficulties. For example, during root excavation process, roots break easily, or the soil may not be sufficiently excavated to expose the entire root system (Sinacore et al., 2017). Hence, it requires an enormous effort and time to obtain accurate estimates of root biomass carbon in agroforestry systems (Sinacore et al., 2017). These challenges and efforts required in root biomass studies can serve as disincentives to include root in biomass carbon studies. Alternatively, below- to aboveground conversion factors are usually used to estimate belowground biomass carbon in roots of agroforestry systems (Nair, 2011).

The use of generalized root (belowground) to shoot (aboveground) ratios for belowground carbon estimations is, however, problematic because root to shoot ratios vary widely and can range from 0.20 to 4.50 in temperate biomes, depending on factors such as environmental and edaphic factors, management system, species, stand age, and tree/shrub density (Mokany et al., 2006). Root to shoot ratios of up to 0.50 have been used to estimate belowground biomass carbon in temperate agroforestry systems. For instance, Kort and Turnock (1998) used root to shoot ratios of 0.30, 0.40, and 0.50 to estimate belowground biomass carbon for coniferous, deciduous, and shrub species of shelterbelts in the Canadian prairies. Agroforestry systems with regular aboveground disturbance such as pruning can increase root biomass and root-to-shoot ratio (Drexler et al., 2021). Axe et al. (2017) measured root-to-shoot ratio of 0.94 in hedgerow in the UK. The greater root-to-shoot ratio in hedgerows was attributed to periodic trimming and coppicing of aboveground components (Axe et al., 2017; Drexler et al., 2021). For this reason, using generalized root-to-shoot ratios rather than field quantification of root biomass can result in underestimation of belowground biomass carbon sequestration in temperate agroforestry system, especially where management systems, species and stand densities differ.

Annual root biomass carbon sequestration rates for the different agroforestry systems across temperate regions we have evaluated in our review range from 0.23 Mg C ha^{-1} $year^{-1}$ in alley cropping/intercropping systems to 0.75 Mg C ha^{-1} $year^{-1}$ in hedgerows (Table 2). These root biomass carbon sequestration rates are within the 0.11 to 1.0 Mg C ha^{-1} $year^{-1}$ determined by Cardinael et al. (2018) for temperate agroforestry systems. Our 0.23 Mg C ha^{-1} $year^{-1}$ root carbon rate for alley cropping/intercropping is within the higher range of the 0.11 to 0.23 Mg C ha^{-1} $year^{-1}$ root carbon accumulation rate determined by Wotherspoon et al. (2014) in 25-year-olds tree species in tree-based intercropping systems in southern Ontario. In the riparian buffer system, we calculate 0.57 Mg C ha^{-1} $year^{-1}$ root biomass carbon sequestration rate which is higher than the 0.18 Mg C ha^{-1} $year^{-1}$ observed by Udawatta et al. (2022) for North American riparian buffer systems. Udawatta and Jose (2012) also estimated belowground biomass carbon sequestration rate of 0.09 Mg C ha^{-1} $year^{-1}$ for riparian buffer systems with 50-year harvest cycle.

Root biomass carbon accumulation rates, like in aboveground biomass, depend on the system's age, species, tree density, environmental and edaphic factors, and management impact such as aboveground disturbance (Vijayakumar et al., 2020; Drexler et al., 2021; Viaud and Kunnemann, 2021). Whereas root biomass carbon accumulation can increase with the system's age, the rate of carbon accumulation follows the aforementioned sigmoidal growth pattern of trees (Weiner and Thomas, 2001). Viaud and Kunnemann (2021) observed, in a study in Gourin, northwestern France, that a hedgerow accumulated 4.35 Mg C ha^{-1} in its root biomass after 60 years compared to 3.3 Mg C ha^{-1} and 0.7 Mg C ha^{-1} in a 45- and 20-year-old hedgerow, respectively. Also, the 60-year-old hedgerow, which was pruned 30 years prior to the study, with *Castanea sativa* species was 135% lower in root biomass carbon than the unpruned hedgerow with *Castanea sativa* and *Corylus avellana* in the same location (Viaud and Kunnemann, 2021).

Agroforestry systems such as silvopasture and riparian buffers also show a greater carbon sequestration rate in root biomass than forest and tree plantations. Hazlett et al. (2005) observed 1.84 Mg C ha^{-1} more carbon in belowground biomass of riparian buffers than adjacent boreal forest in Canada. Sharrow and Ismail (2004) also found 1.53 Mg C ha^{-1} in the belowground biomass of 11-year-old Douglas-fir silvopastural system compared to 0.85 Mg C ha^{-1} in the Douglas-fir plantation. Whereas some studies provided evidence of higher carbon sequestration in root biomass of agroforestry systems, others either observed no difference or lower carbon sequestration rate in root biomass of agroforestry systems. For example, López-Díaz et al. (2017) did not observe any significant difference in root biomass carbon in a grazed hybrid walnut (*Juglans major × regia* mj 209xra) system in Spain compared to mowed and ploughed systems. Dube et al. (2012), on the other hand, found 82% more carbon in root biomass in a ponderosa pine (*Pinus ponderosa*) plantation compared to a pine-based silvopasture system in Patagonia, Chile. However, tree density was two times greater in the pine plantation compared to the silvopasture system (Dube et al., 2012). This suggests that difference in tree density between the two systems may have played an important role with the lower carbon sequestration rate in the silvopasture system than the pine plantation. Among the factors that impacted carbon sequestration potential of agroforestry systems, Nair and Nair (2003) concluded that tree density played a significant role. It is, therefore, important that the design and implementation of any agroforestry system takes into consideration the optimum tree density required to yield maximum carbon sequestration benefits.

2.3 Soil organic carbon sequestration

Total carbon in the terrestrial ecosystem is estimated at 2860 Pg (Lal, 2008). Of this amount ~80% (2300 Pg) is contained in the soil to 1 m depth, with SOC constituting 1550 Pg C (Lal, 2008). Batjes (2014) estimates SOC to 0.3 m depth at 700 Pg C. The 2300 Pg total carbon contained in the soil to a 1 m depth is three times greater than that in the atmosphere (760 Pg C) and 4.1 times greater than that stored in vegetation (560 Pg C) (Lal, 2008). Therefore, the soil represents an important component for carbon storage in terrestrial ecosystems.

Stabilizing atmospheric CO_2 will require land management systems that reduce soil carbon loss and enhance SOC storage (Lorenz and Lal, 2014). Among these land management strategies are agroforestry systems which have been recognized under afforestation and reforestation programs for soil carbon sequestration (Nair et al., 2009). Compared to treeless systems, Ma et al. (2020) determined that agroforestry systems increased SOC stocks by an average of 26%. Higher soil carbon sequestration rates in the soil of agroforestry systems was attributed to an increase in organic matter input from above- and belowground components to the soil through litterfall, root turnover and rhizodeposition (Oelbermann et al., 2004). For example, Cardinael et al. (2012) determined that SOC increased by 48% in a willow alley cropping system in southern Canada compared to a 27% increase in the willow monocrop.

Soil carbon storage in agroforestry systems can range from 30 to 300 Mg C ha^{-1} to 1 m depth and depends on the type of agroforestry system, soil properties, and the quantity and quality of organic input into the soil from within the system (Nair et al., 2010). Ofosu et al. (2022) even reported 466.3 Mg C ha^{-1} in tree buffer systems in southern Canada to a 60 cm soil depth compared to 132.2 Mg C ha^{-1} in an adjacent agricultural field. A meta-analysis by Mayer et al. (2022) in temperate agroforestry systems revealed that SOC stocks were 47 Mg C ha^{-1} (0–20 cm) and 32 Mg C ha^{-1} (20–40 cm) compared to 40 Mg C ha^{-1} (0–20 cm) and 29 Mg C ha^{-1} (20–40) in adjacent treeless systems. Soil organic carbon sequestration rates of 1–10 Mg C ha^{-1} year^{-1} (0–50 cm) were also reported by Crossland (2015) for unmanaged hedgerows in England.

Generally, soil carbon sequestration rates in agroforestry systems are higher in upper soil horizons than in deeper soil horizons (Cardinael et al., 2015; Mayer et al., 2022). For example, Mayer et al. (2022) found a SOC sequestration rate of 0.21 Mg C ha^{-1} year^{-1} at a depth of 0–20 cm compared to 0.15 Mg C ha^{-1} year^{-1} at a depth of 20–40 cm in temperate agroforestry systems. Across eight different riparian buffer types and ages in southern Canada, a study by Vijayakumar et al. (2020) revealed SOC stocks in the 0–30 cm depth averaged 135.64 Mg C ha^{-1} compared to 114.68 Mg C ha^{-1} in the 30–60 cm depth. However, the amount of carbon a soil stores depends on the previous land use management practices and the type of agroforestry system that is established (Mayer et al., 2022). Analyzing soil carbon sequestration rates in three different temperate agroforestry systems, Mayer et al. (2022) observed that hedgerows had the highest SOC sequestration rates in 0–20 cm (0.32 Mg C ha^{-1} $year^{-1}$) and 20–40 cm depths (0.28 Mg C ha^{-1} $year^{-1}$), followed by alley cropping systems with 0.26 Mg C ha^{-1} $year^{-1}$ (0–20 cm) and 0.23 Mg C ha^{-1} $year^{-1}$ (20–40 cm) whereas a SOC loss occurred in silvopasture systems at 0–20 cm (–0.17 Mg C ha^{-1} $year^{-1}$) and 20–40 cm (–0.03 Mg C ha^{-1} $year^{-1}$). Ma et al. (2020) observed that converting agricultural land to agroforestry systems yielded a greater soil carbon benefit than converting grassland to agroforestry. As such, it was possible that the SOC loss observed in the silvopasture system by Mayer et al. (2022) was because the adjacent land use for the silvopasture was grassland. For example, Sharrow and Ismail (2004) found a decrease in SOC stocks to a 45 cm depth when pasture (103 Mg C ha^{-1}) was converted to silvopasture (96 Mg C ha^{-1}) in Oregon, USA. Cardinael et al. (2018) also observed SOC losses ranging from –0.60 to –0.34 Mg C ha^{-1} $year^{-1}$ when grasslands were converted to silvopasture systems in temperate regions across the world. Hence, it is important that comprehensive information about previous land use and land use history is provided in SOC studies in agroforestry systems. This will provide greater and better understanding on how converting different land use to agroforestry systems can impact long-term soil carbon benefits.

Nair (2011) also explained that soil carbon storage in agroforestry systems changes with distance from the tree. For example, Cardinael et al. (2015) observed higher carbon addition to 0–50 cm depth within the tree row (near the trees) compared to 0–30 cm depth in the tree inter-row (away from trees). Higher SOC within tree rows was linked to the lack of soil disturbance in these zones, which created conditions similar to undisturbed systems (Cardinael et al., 2015). Thevathasan et al. (2018), however, explained that the spatial distribution of SOC in alley crops is age related. They suggested that as trees grow, more litter is distributed further from the tree (Thevathasan et al., 2018). In a poplar tree-based intercropping system in southern Canada, Thevathasan and Gordon (1997) determined that 80% of litter fell within 2.5 m of the tree rows with 20% falling in the rest of the field when poplar trees were 7–9 years old. Conversely, only 47% of the litter fell within 2.5 m of the trees when the poplar trees matured to 15 years (Thevathasan et al., 2018). Oelbermann et al. (2006), on the other hand, found higher carbon inputs from litterfall at 1 m (9.46 Mg C ha^{-1}) distance from tree row of a 13-year-old hybrid poplar alley cropping system in Ontario, Canada, than at 11.5 m (2.19 Mg C ha^{-1}) from tree row. These conflicting results suggest that the processes involved in the spatial distribution of SOC in alley cropping systems are still not well understood. Therefore, long-term field studies are required to fully understand what influences the spatial distribution of SOC in alley cropping systems to inform appropriate soil sampling protocols in agroforestry systems.

Although, soil carbon benefits in agroforestry systems have been widely evaluated, most of these studies took place at a 0–50 cm soil depth (Sharrow and Ismail, 2004; Bambrick et al., 2010; Biffi et al., 2022). This is of concern because roots from woody species extend beyond 50 cm depth and thus, the effect of agroforestry systems on soil carbon stocks may not have been accurately captured (Nair, 2011). Cardinael et al. (2015) observed that out of the 209 Mg C ha^{-1} SOC stocks measured in the 0–200 cm soil depth in an alley cropping system in Montpellier, France, only 40 Mg C ha^{-1}, representing less than 20% of the SOC stocks, was contained in the 0–30 cm layer. Haile et al. (2008) also found higher SOC contribution from trees in a slash pine (*Pinus elliottii*) and bahiagrass (*Paspalum notatum*) silvopasture system throughout a 125 cm deep soil profile compared to a non-treed pasture in Florida, USA. The SOC stocks in the silvopasture system were 491 Mg C ha^{-1} (0–50 cm) and 101 Mg C ha^{-1} (50–125 cm) compared to 450 Mg C ha^{-1} (0–50 cm) and 62 Mg C ha^{-1} (50–125 cm) for the non-treed pasture (Haile et al., 2008). However, the impact of tree roots on soil carbon storage in deeper soil layers may not always be positive. Although, Upson and Burgess (2013) determined higher SOC stocks (0–40 cm) in a poplar alley cropping system than in an arable system without trees in England, the reverse was observed for deeper soil layers up to 150 cm. This implies that inadequate sampling depths in agroforestry systems can result in an overestimation of the soil carbon sequestration potential. For instance, from our calculated SOC sequestration rates (Table 2), the few studies that report SOC stocks to depths greater than 50 cm show marginal sequestration rates with even lower SOC rates in lower depths. In riparian buffers, the SOC rate within the upper 50 cm is 3.1 Mg C ha^{-1} $year^{-1}$ compared to 2.8 Mg C ha^{-1} $year^{-1}$ for the upper 100 cm (Table 2). Similarly, silvopasture systems show a higher SOC rate in the upper 50 cm depth (0.5 Mg C ha^{-1} $year^{-1}$) than the upper 100 cm (0.2 Mg C ha^{-1} $year^{-1}$) (Table 2).

Upson and Burgess (2013) explained that at lower depths, a positive priming effect from root exudates could cause a depletion of old SOC. Also, conditions may be more favorable at greater soil depths for enhanced organic matter oxidation (Upson and Burgess, 2013). Whether SOC accumulates or decreases in lower soil depths will depend on several factors including clay content of the soil. For example, Mayer et al. (2022) found that clay content had a positive impact on SOC at lower depths in temperate agroforestry systems. More field studies are required to fully understand the SOC sequestration rate of the various temperate agroforestry systems at greater soil depths. Furthermore, soil has a finite capacity (saturation) to assimilate carbon (Chung et al., 2010), and therefore SOC sequestration in agroforestry systems will not continue indefinitely. Ma et al. (2020) observed that on average, the soil carbon sequestration rate was higher in tropical agroforestry systems than in temperate agroforestry systems. It will generally take 45 years for soils in temperate agroforestry systems to reach equilibrium compared to 14 years in tropical agroforestry systems (Oelbermann, 2002). This implies that temperate agroforestry systems can continue to store more carbon in the soil over the long-term.

Managing agroforestry systems to increase SOC sequestration should consider increasing species diversity and tree density (Nair et al., 2009; Ma et al., 2020). For instance, integrating nitrogen-fixing trees can improve biomass productivity and improve SOC sequestration and stabilization (Kumar et al., 1998; Prescott, 2010). Ma and Chen (2016) determined that enhancing species diversity and richness in forests, grasslands, and croplands increased fine root biomass by 28% and annual production by 45% compared to their monoculture stands. Although, increasing species richness and tree density can increase SOC sequestration (Nair et al., 2009), interspecific competition for growth resources from increased tree density and species can affect biomass production, impacting SOC sequestration (Lorenz and Lal, 2018). The choice of complimentary species and the optimum tree density in the design of temperate agroforestry systems is critical for SOC sequestration and this needs further empirical and field studies.

2.4 Soil organic carbon stabilization

Not only do tree-based systems contribute to higher carbon accumulation than treeless systems, but trees also promote the stability of the contributed carbon to the soil by enhancing long-term stabilization of SOM (Haile et al., 2008; Baah-Acheamfour et al., 2014; Vijayakumar et al., 2023). Lorenz and Lal (2014) proposed that to successfully use agroforestry systems as a climate change mitigation strategy, a reduction in SOC loss, and enhanced SOC stabilization is imperative. With SOM being the largest reservoir of SOC in the soil, any change in the size and the turnover rate of SOM impacts SOC pools (Lehmann and Kleber, 2015).

The mechanisms involved in SOM stabilization are not well understood. Different processes and pathways have been proposed to control SOM preservation and stabilization (e.g. Lehmann et al., 2020; Cotrufo and Lavallee, 2022). SOM stabilization is facilitated by biochemical recalcitrance, physical protection, and chemical stabilization through organo-mineral interaction with the soil matrix and environmental factors (Cotrufo and Lavallee, 2022). These stabilization processes limit microbial access to SOM and inhibit the physiological and metabolic capabilities of the soil microbial community involved in plant litter decomposition (Lehmann et al., 2020; Cotrufo and Lavallee, 2022). The biochemical recalcitrance in SOM stabilization is defined by the composition of complex mixtures of organic compounds, which vary in their degradability (Cotrufo and Lavallee, 2022). These complex mixtures of organic compounds may result from the inherent property of the plant litter or are formed during organic matter decomposition (Lorenz and Lal, 2005). Although, the importance of biochemical recalcitrance in SOM stabilization has been questioned (Lehmann and Kleber, 2015), von Lutzow et al. (2006) explained that biochemical recalcitrance is only important during the early stages of decomposition and in active surface soils. Lorenz and Lal (2014) explained that physical protection of SOM and organo-mineral and metal interactions are the two most important groups of processes for long-term SOC stability. This is because physical protection may preserve organic inputs for decades to centuries, whereas organo-mineral complexes or organo-metal interactions may be responsible for stabilization of most of the non-charred SOC for centuries to millennia (Lorenz and Lal, 2014). However, because biochemical recalcitrance plays a crucial role in preserving organic inputs at the early stages of decomposition and in active surface soils (von Lutzow et al., 2006), this delay in organic matter decomposition gives time for physical and organo-mineral interactions to form (Six et al., 2000). As such, mechanisms in SOC stabilization in agroforestry systems involves the combination of both short- and long-term processes (Nair et al., 2010).

Studies evaluating SOM chemistry in temperate agroforestry systems are lacking. Dhillon et al. (2017) reported varying effects of tree species on SOM chemistry in shelterbelts in Saskatchewan, Canada. Whereas some species increased labile carbon forms such as carbohydrates, other species were associated with a higher abundance of recalcitrant aliphatic carbon forms (Dhillon et al., 2017). Compared to annual row cropland, An et al. (2021) and An et al. (2023) observed soil under cropland had more stable SOC than those under temperate agroforestry systems of hedgerows and shelterbelts in western Canada. They explained that SOC may be comparatively higher under agroforestry systems than in cropland (An et al., 2023). Carbon in the agroforestry systems is, however, more susceptible to losses due to higher presence of labile carbon pools (An et al., 2023). Between the hedgerow and shelterbelt, An et al. (2023) observed that the SOC under the hedgerow was more stable than shelterbelt and attributed this to differences in species composition. Tree species diversity can increase SOM degradation by influencing soil microbial community and structure (Lehmann et al., 2020). This implies that SOC stabilization of root-derived carbon through physicochemical interactions in deeper soil horizons may be more important in long-term SOC stabilization under temperate agroforestry systems than the biochemical recalcitrance of aboveground biomass-derived carbon, especially, in systems with high species diversity. Jackson et al. (2017) explained that root inputs are approximately five times more likely to be stabilized as SOC by association with aggregates or clay minerals than an equivalent mass of aboveground litter carbon. Therefore, evaluating the association of soil carbon with different soil size fractions in temperate agroforestry systems will be an important step to understanding their potential impact on long-term SOC stabilization.

Using stable carbon isotope techniques, Haile et al. (2008) observed that slash pine silvopasture systems across four sites in Florida, USA, contributed more carbon to the silt- and clay-sized soil fractions in deeper soil layers compared to an open pasture (no trees). This signified that the trees contributed more stable carbon than the open pasture. Baah-Acheamfour et al. (2014) also studied SOC distribution in three soil fractions (silt- and clay-sized <53 μm, micro-sized 53–250 μm, and macro-sized fractions 250–2000 μm) in temperate agroforestry systems with hedgerows, shelterbelts, and silvopastures across 35 sites in Alberta, Canada. The authors observed a higher SOC content in all the soil fractions in the agroforestry systems (82 g kg^{-1}) compared to treeless agricultural systems (54 g kg^{-1}). Carbon in the silt- and clay fractions, which constituted the most stable SOC, was 17% higher in the agroforestry systems than the agricultural system (Baah-Acheamfour et al., 2014). Among the agroforestry systems, SOC in the three soil fractions combined was highest in the silvopasture system (82 g kg^{-1}), followed by hedgerow (67 g kg^{-1}), and lowest in the shelterbelt (50 g kg^{-1}). Although shelterbelts had the lowest SOC in all three soil fractions, 58% of its C was located in the silt- and clay-sized fractions (Baah-Acheamfour et al., 2014). This was followed by hedgerow (52%) and the silvopasture (36%). The results from the study by Baah-Acheamfour et al. (2014) imply that soil conditions and litter quality, rather than quantity, may have influenced SOC stabilization in the shelterbelts compared to the silvopasture system.

Also, Oelbermann et al. (2006) observed a greater average annual SOC accumulation rate in a tropical alley cropping system in Costa Rica, due to greater biomass input, compared to a temperate alley cropping system in southern Canada. The temperate system, however, was more efficient in stabilizing the carbon added from crop residues and leaf litter than the tropical system (Oelbermann et al., 2006). This was because SOM mineralization was slower in the cold temperate system compared to the warm and humid tropical system (Oelbermann et al., 2006). Cotrufo and Lavallee (2022) explained that cold conditions inhibit the physiological functions of the soil microbial community, enhancing SOC stabilization. Therefore, climatic conditions in temperate regions could make a substantial contribution to enhancing SOC stability compared to tropical agroforestry systems. More field studies are required to fully understand the mechanisms involved in SOC stabilization under temperate agroforestry systems and their implications for long-term SOC sequestration.

3 Projections of potential carbon sequestration in temperate agroforestry systems

Currently, there is a high degree of uncertainty regarding the land area available for agroforestry or those currently under agroforestry practices. This uncertainty poses a challenge in estimating precise carbon sequestration potentials for various agroforestry systems on a global scale. Adding to this layer of challenge is the lack of coordinated research and, in some cases, inconsistencies in reporting results from carbon studies in agroforestry systems. Despite these challenges and shortcomings, we have endeavoured to provide projections on potential carbon sequestration in temperate agroforestry types globally, drawing from estimates calculated in different studies conducted in temperate regions (Table 1). It is crucial to emphasize that these are projections and should be regarded as such.

As in many regions, data on land under agroforestry management in Canada is lacking. Drever et al. (2021) attempted and estimated potential land area available for some land management options for climate change mitigation purposes in Canada, using available spatial data. Among these land options were temperate agroforestry systems of riparian buffer, silvopasture, and alley cropping. The authors projected that across the provinces of Ontario and Quebec (which together occupy approximately 25% of Canada's total land area, as per Statistics Canada (2016)), a total of 797 298 ha is available for alley cropping. This represents only 2% of the 37.9 M ha of cropped land in Canada in 2021 (Statistics Canada, 2022). If 10% of the 2021 cropped land was used for alley cropping, it could potentially mitigate 0.027 Pg CO_2 year^{-1} based on our annual carbon sequestration rate of 1.95 Mg C ha^{-1} year^{-1} to a 100 cm soil depth (Table 2).

Drever et al. (2021) also estimated the availability of 985 518 ha and 465 819 ha for silvopasture systems and riparian buffers, respectively, in Canada. These figures do not include areas already under silvopasture and riparian buffer systems. For instance, An et al. (2022) estimated that about 1.8 M ha in the Central Alberta region is already under silvopasture forests. According to the 2021 Canada land use statistics, a total of 4.8 M ha was dedicated to pasture production in Canada (Statistics Canada, 2022). Therefore, if this 4.8 M ha was converted to silvopasture systems, it could potentially help mitigate about 0.026 Pg CO_2 year^{-1} using our annual carbon sequestration rate for silvopasture systems of 1.47 Mg C ha^{-1} year^{-1} to a 100 cm soil depth (Table 2). The estimated 465 819 ha available for riparian buffers (Drever et al., 2021) could offset another 0.009 Pg CO_2 year^{-1} to a 100 cm soil depth. Moreover, a study by Thevathasan et al. (2018) revealed that approximately 212 000 km of land area in Quebec, Atlantic, and Prairie provinces of Canada were under shelterbelts. An average annual carbon accumulation rate of 3.1 Mg C km^{-1} was calculated by Amichev et al. (2016) at the system-level in prairie shelterbelts. Using this annual rate, the estimated 212 000 km shelterbelt system could potentially mitigate ~0.0024 Pg CO_2 annually. Additionally, the 24 390 ha of hedgerow forests in the Central Alberta region alone (An et al., 2022) could contribute a further 0.0008 Pg CO_2 year^{-1} in mitigation benefits. Collectively, these examples of temperate agroforestry systems in Canada, while representing only a small fraction of the actual land under agroforestry, constitute approximately 12% (0.065 Pg CO_2 year^{-1}) of the total 0.546 Pg CO_2 emissions for 2021 from all sources in Canada (Global Carbon Project, 2022).

Data from the European agricultural census for 2020 indicate that of the 157 M ha utilized agricultural area in Europe, fodder production occurred on 66 M ha, with the remaining 91 M ha under cereal, kitchen gardens, and industrial crop production (Eurostat, 2020). Also, data from an AGFORWARD project indicate that hedgerows cover about 1.78 M ha in Europe (Burges et al., 2018). If the 66.4 M ha fodder production went under silvopasture system and 10% of the 91 M ha went into silvorable agroforestry such as alley cropping or tree-based intercropping system, using our carbon sequestration rates to 100 cm depth (Table 2) will yield 0.13 Pg C year^{-1} or 0.48 Pg CO_2 year^{-1}. This yearly CO_2 sequestration rate represents ~18% of total CO_2 emissions in Europe from all sources for the year 2021.

In the UK, the area under hedgerow is estimated at 158 600 ha (Forestry Commission, 2017). Also, available land for grazing with potential for silvopasture system is 12.4 M ha and about 4.6 M ha land went into arable crop production in 2021 (Government of UK, 2022). If 10% of the 4.6 M ha cropped land went into silvorable system, we estimate 0.02 Pg C year^{-1} or 0.07 Pg CO_2 year^{-1}, representing ~20% of total CO_2 (0.35 Pg CO_2) emissions in the UK from all sources for the year 2021 (Global Carbon Project, 2022).

According to estimates by Udawatta et al. (2022), there are potentially 120 M ha of cropland and erodible non-federal land available and suitable for alley cropping in the USA. If 10% of this land (12 M ha) went into alley cropping, using our carbon sequestration rate of 1.95 Mg C ha^{-1} year^{-1} (Table 2) will contribute 0.09 Pg CO_2 year^{-1} mitigation benefit. Additionally, the potential 847 500 ha, 34 M ha, and 7.45 M ha available in USA for riparian buffers, silvopastures, and windbreaks, respectively (Udawatta et al., 2022), could potentially contribute a further 0.32 Pg CO_2 year^{-1} based on our annual carbon sequestration rates to a 100 cm soil depth (Table 2) for these agroforestry practices. Altogether, these four agroforestry systems could mitigate 0.41 Pg CO_2 annually. This value represents ~8.1% of total US CO_2 (5.007 Pg CO_2) emissions in 2021 (Global Carbon Project, 2022).

Also, 22 M ha of land in the Indian Himalayas has been indicated to be degraded (Kumar et al., 2018). This land could be made productive with the adoption of silvorable systems such as tree-based intercrop. Adoption of tree-based intercrop can yield ~0.16 Pg CO_2 $year^{-1}$. The CO_2 emissions in India in 2021 is estimated at 2.71 Pg CO_2 (Global Carbon Project, 2022). These carbon sequestration rates are indicative of the potential agroforestry practices can have in mitigating climate change through CO_2 assimilation.

Due to the slower rate of SOC turnover in colder temperate environments, compared to tropical regions, the benefits of SOC sequestration in temperate agroforestry systems may take a longer period to manifest (Oelbermann and Voroney, 2007). This implies that long-term studies are necessary to understand the carbon sequestration benefits of temperate agroforestry systems. However, conducting long-term research on a decadal or century time scale to evaluate changes in SOC in forest ecosystems and agroecosystems, including agroforestry systems, is impractical. Furthermore, there is a lack of comprehensive long-term and well-designed field studies for the various temperate agroforestry systems to conduct such research. As a result, process-based models have often been used to evaluate and project long-term changes in SOC dynamics under different land management systems. Despite this, there is a limited application of process-based models for biomass and soil carbon studies in temperate agroforestry systems. For example, Oelbermann and Voroney (2011) for the first time calibrated and used the century model to accurately predict long-term SOC changes in a temperate alley cropping system in southern Canada. They found that SOC stocks increased steadily over 100 years once alley cropping systems were initiated, and that most of this carbon accumulated in the slow fraction (Oelbermann and Voroney, 2011). Francaviglia et al. (2012) used the RothC model to evaluate changes in SOC stocks in an agrosilvopastoral system in Italy. They observed that agroforestry systems were more efficient in SOC storage than agricultural systems over a 90-year period (Francaviglia et al., 2012). Additionally, they found that the RothC model accurately predicted SOC changes compared to measured values (Francaviglia et al., 2012). Palma et al. (2018) also demonstrated that modifying the century and RothC models could accurately quantify and predict long-term changes in soil and biomass carbon stocks in hedgerows in the UK. Similarly, Crossland et al. (2015) integrated the RothC into the Yield-SAFE model and accurately quantified soil and biomass carbon stocks in temperate agroforestry systems in the UK and Europe. In both studies by Palma et al. (2018) and Crossland et al. (2015), they found that the biomass and soil carbon values predicted by the models were consistent with measured data. These successful applications of process-based models suggest that they can be used as tools to accurately predict the long-term changes in system-level carbon stocks in temperate agroforestry systems. Thus, it is essential to integrate field studies on system-level carbon stocks with process-based models to determine long-term carbon changes and to further understand potential impacts of agroforestry system management practices on carbon dynamics.

While these examples of temperate agroforestry systems have shown evidence of greater potential for carbon sequestration, significant obstacles persist in their adoption and maintenance on agricultural and other landscapes. A recent study by Amichev et al. (2020) observed that between 2008 and 2016, a total of 2491.2 km of shelterbelts in Saskatchewan, Canada, containing about 190.7 Gt C, had been removed. Farmers have cited the difficulty of operating large farming equipment in crop fields, reduced land area for crop production, and the time, labor, and cost involved in maintaining shelterbelts as some of the major reasons for removing existing shelterbelts in the Canadian prairies (Rempel et al., 2017; Amichev et al., 2020; Drever et al., 2021). A 2014 shelterbelt survey in Manitoba and Saskatchewan, Canada, found that crop producers were more likely than livestock producers to remove shelterbelts to make room for irrigation pivots and large machinery or to expand fields (Rural Development Institute, 2014). Thevathassan et al. (2018) also highlighted the lack of policy incentives and land ownership as additional barriers to adopting and maintaining trees on agricultural lands in Canada. The study by Thevathasan et al. (2018) explained that corporate farms may be less interested in tree planting or other environmental practices that are not perceived as profit-generating activities. Additionally, farming on rented land may pose obstacles to long-term conservation practices involving trees (Thevathasan et al., 2018). The implications of these findings underscore the need for concerted efforts to boost interest in adopting and maintaining agroforestry systems in temperate regions if the full potential benefits of carbon sequestration in these land uses are to be realized.

A proposed solution to increase interest in adopting and maintaining trees on landscapes include the provision of incentives such as financial payments and tax credits to landowners or managers (Thevathasan et al., 2018; Amichev et al., 2020). For example, Mayrinck et al. (2019) argue that implementing carbon market schemes, where any carbon sequestered by agroforestry trees holds a monetary value, would integrate trees into a farm's overall budget. This integration could motivate future tree planting and enhance the maintenance of existing ones. Moreover, it is suggested that farmers may not be fully aware of the advantages of having trees on agricultural landscapes (Rempel et al., 2017). Therefore, education on the long-term environmental benefits of agroforestry practices could help farmers to better value trees on farmlands, leading to increased adoption of agroforestry as a land management option (Mayrinck et al., 2019). A study by Reimer and Prokopy (2014) also revealed that, in addition to financial and environmental benefits, the provision of expertise that reduced technical barriers was equally instrumental to farmers' participation in the USA Farm Bill conservation programs. Thevathasan et al. (2018) suggest the training of high-qualified personnel to provide technical expertise will be required to promote the adoption of agroforestry practices on farms in Canada.

4 Conclusion and future directions

Temperate agroforestry practices have the potential to provide climate change mitigation benefits through carbon sequestration, making them a promising option for agricultural landscapes. However, the carbon sequestration benefits of agroforestry systems depend on various factors such as woody species, density, age, management practices, and local environmental and edaphic factors. Analysis of global data from different temperate agroforestry systems reveals that hedgerows have the highest annual carbon sequestration rate (9.35 Mg C ha^{-1} $year^{-1}$) at the system level to 100 cm depth, followed by riparian buffers (5.15 Mg C ha^{-1} $year^{-1}$), shelterbelts/windbreaks (4.48 Mg C ha^{-1} $year^{-1}$), alley cropping/intercropping (1.95 Mg C ha^{-1} $year^{-1}$), and silvopasture systems (1.47 Mg C ha^{-1} $year^{-1}$). Biomass (above- and belowground) accounts for more than half of the system-level carbon sequestration rate in the various agroforestry systems, indicating the need for more attention to biomass carbon quantification. Based on our calculated carbon sequestration rates from various studies conducted in different agroforestry systems for temperate regions, we project that agroforestry practices could annually offset approximately 20%, 18%, 12%, and 8% of total CO_2 emissions in the UK, Europe, Canada, and the USA, respectively. We emphasize that the number of reviewed literatures does not provide sufficient information to accurately estimate reliable values for carbon sequestration rates. Additionally, there is a huge degree of uncertainty in the available land area for agroforestry practices. Hence, these projections are just projections and should be regarded as such. Also, agroforestry can contribute to stable carbon pools in the soil, more field studies are required to understand how different agroforestry systems can affect long-term SOC stabilization and the underlying mechanisms. Furthermore, integration of data from field studies into process-based models will help understand how various types of agroforestry systems impact long-term system-level carbon dynamics.

Even though there is increasing research on the carbon sequestration benefits of temperate agroforestry systems, more work is required. The following are some areas that need further consideration to improve our understanding of the carbon sequestration benefits of temperate agroforestry systems.

1 In most studies, crucial information such as tree age, species, tree density, and tree spacing are not provided or clearly defined. Moreover, land management history before the conversion to agroforestry is sometimes not provided. In the absence of baseline data, the inclusion of adjacent land use systems can be useful in estimating and reporting carbon sequestration rates rather than stocks of a particular agroforestry system. This can provide better understanding of a system's potential in mitigating climate change.

2 The importance of temperate agroforestry systems in carbon sequestration has gained greater prominence in the Himalayas in India and North America than other temperate regions. Therefore, more carbon sequestration studies in temperate regions, especially in Europe should be given equal attention.
3 Among the various recognized agroforestry systems in temperate regions, data on forest farming is limited. This may arise from the difficulty in delineating what constitute forest farming from other agroforestry practices. For instance, harvesting sap from maple trees in riparian buffers for maple syrup production is a common practice in southern Ontario, Canada. Managing maple trees for syrup can be classified as forest farming whereas their occupation in riparian zones make them riparian buffers. In short, forest farming has not received the studies other systems have received and more efforts are needed in understanding carbon sequestration benefits of these systems.
4 There is currently no reliable data on how much land area is under agroforestry systems or available for agroforestry practices in temperate regions. This makes it challenging in providing precise estimates on the carbon sequestration potential of these systems. Reliable land area estimates are therefore required to make it possible in estimating the potential carbon sequestration benefits of temperate agroforestry systems.
5 Most agroforestry systems do not account for carbon stocks in companion crops, litterfall, and standing biomass. These components contribute to carbon accumulation at the system-level. Udawatta et al. (2022) equally mention the lack of greenhouse gas emission and source-sink data from agroforestry systems. These data are useful in determining the carbon dynamics and net gains for the adoption and implementation of an agroforestry systems. Hence, more field studies are needed to quantify greenhouse gas emissions in temperate agroforestry systems.
6 The only consistency in soil carbon quantification in agroforestry systems is the inconsistency in soil sampling depths and sometimes reporting soil carbon results. Whereas some studies report in carbon stocks (e.g. Mg ha^{-1}), others report carbon content (g kg^{-1} or % C) without providing important parameters such as soil bulk density or soil weight. This makes it difficult to compare results. Even if reporting in carbon concentrations, soil depth and bulk density values or soil weight must be specified to allow for easy comparison with other studies.

5 Where to look for further information

The following books and articles provide a good overview of temperate agroforestry systems and carbon benefits:

- Nair, P. R., Kumar, B. M. and Nair, V. D. (2021), 'An introduction to agroforestry: four decades of scientific developments'. 2nd Edition, Springer Nature, pp. 1–666.
- Garrett, H. E. G., Jose, S. and Gold, M. A. (2022), 'North American Agroforestry'. Vol. 194. John Wiley and Sons, pp. 1–566.
- Gordon, A.M., Newman, S.M. and Coleman, B.R.W. (2018), 'Temperate Agroforestry Systems'. 2nd Edition, CABI International, Wallingford, UK, pp. 1-313.
- Lorenz, K. and Lal, R. (2014), 'Soil organic carbon sequestration in agroforestry systems. A review', *Agronomy for Sustainable Development*, 34, 443–454.
- Nair, P. K. R and Garrity, D. (2012), 'Agroforestry - The future of global land use', *Advances in Agroforestry 9*. Springer Netherlands, pp. 1–541.
- Nair, P. K. R. and Nair, V. D. (2003), 'Carbon storage in North American agroforestry systems'. In J. Kimble, L. S. Heath, R. A. Birdsey, and R. Lal (Eds), *The Potential of U.S. Forest Soils to Sequester Carbon and Mitigate the Greenhouse Effect*, CRC Press, Boca Raton, USA, pp. 333–346.
- Udawatta, R.P. and Jose, S. (2012), 'Agroforestry strategies to sequester carbon in temperate North America', *Agroforestry Systems*, 86, 225–42.

Key research and up-to-date research on soil carbon benefits of temperate agroforestry systems can be found at the following organizations:

- Association for Temperate Agroforestry (https://www.aftaweb.org/)
- European Agroforestry Federation (http://www.europeanagroforestry.eu/)

6 References

Amichev, B. Y., Bentham, M. J., Kurz, W. A., Laroque, C. P., Kulshreshtha, S., Piwowar, J. M. and van Rees, K. C. (2016), 'Carbon sequestration by white spruce shelterbelts in Saskatchewan, Canada: 3PG and CBM-CFS3 model simulations', *Ecological Modelling*, 325, 35–46.

Amichev, B. Y., Laroque, C. P. and van Rees, K. C. (2020), 'Shelterbelt removals in Saskatchewan, Canada: implications for long-term carbon sequestration', *Agroforestry Systems*, 94, 1665–1680.

Amundson, R. (2022), 'Soil biogeochemistry and the global agricultural footprint', *Soil Security*, 6, 100022. https://doi.org/10.1016/j.soisec.2021.100022

An, Z., Bernard, G. M., Ma, Z., et al. (2021), 'Forest land-use increases soil organic carbon quality but not its structural or thermal stability in a hedgerow system', *Agriculture, Ecosystems & Environment*, 321, 107617.

An, Z., Bork, E. W., Duan, X., Gross, C. D., Carlyle, C. N. and Chang, S. X. (2022), 'Quantifying past, current, and future forest carbon stocks within agroforestry systems in central Alberta, Canada', *GCB Bioenergy*, 14(6), 669–680.

An, Z., Pokharel, P., Plante, A. F., Bork, E. W., Carlyle, C. N., Williams, E. K. and Chang, S. X. (2023), 'Soil organic matter stability in forest and cropland components of two agroforestry systems in western Canada', *Geoderma*, 433, 116463.

Axe, M. S., Grange, I. D. and Conway, J. S. (2017), 'Carbon storage in hedge biomass – A case study of actively managed hedges in England', *Agriculture, Ecosystems & Environment*, 250, 81–88.

Baah-Acheamfour, M., Carlyle, C. N., Bork, E. W., Chang, S. X. (2014), 'Trees increase soil carbon and its stability in three agroforestry systems in central Alberta, Canada', *Ecological Management*, 328, 131–139.

Bainbridge, D. A., Virginia, R. A. and Jarrell, W. M. (1990), 'Honey mesquite: a multi-purpose tree for arid lands', *Nitrogen Fixing Tree Association*, 90(7), 2–4.

Balian, E. V. and Naiman, R. J. (2005), 'Abundance and production of riparian trees in the lowland floodplain of the Queets River, Washington', *Ecosystems*, 8, 841–861.

Bambrick, A. D., Whalen, J. K., Bradley, R. L., Cogliastro, A., Gordon, A. M., Olivier, A. and Thevathasan, N. V. (2010), 'Spatial heterogeneity of soil organic carbon in tree-based intercropping systems in Quebec and Ontario, Canada', *Agroforestry Systems*, 79, 343–353.

Batjes, N. H. (2014), 'Total carbon and nitrogen in the soils of the world', *European Journal of Soil Science*, 65(1), 10–21.

Bhardwaj, D. R., Salve, A., Kumar, J., Kumar, A., Sharma, P. and Kumar, D. (2023), 'Biomass production and carbon storage potential of agroforestry land use systems in high hills of north-western Himalaya: an approach towards natural based climatic solution', *Biomass Conversion and Biorefinery*, 1–14.

Biffi, S., Chapman, P. J., Grayson, R. P. and Ziv, G. (2022), 'Soil carbon sequestration potential of planting hedgerows in agricultural landscapes', *Journal of Environmental Management*, 307, 114484.

Borden, K. A., Isaac, M. E., Thevathasan, N. V., Gordon, A. M. and Thomas, S. C. (2014), 'Estimating coarse root biomass with ground penetrating radar in a tree-based intercropping system', *Agroforestry Systems*, 88, 657–669.

Borden, K. A., Thomas, S. C. and Isaac, M. E. (2017), 'Interspecific variation of tree root architecture in a temperate agroforestry system characterized using ground-penetrating radar', *Plant and Soil*, 410, 323–334.

Burgess, P. J., den Herder, M., Dupraz, C., Garnett, K., Giannitsopoulos, M., Graves, A. R., Hermansen, J. E., Kanzler, M., Liagre, F., Mirck, J., Moreno, G., Mosquera-Losada, M. R., Palma, J. H. N., Pantera, A. and Plieninger, T. (2018), 'AGFORWARD PROJECT Final Report. Cranfield University: AGFORWARD', https://www.agforward.eu/documents/AGFORWARD%20Final%20Report%2028%20Feb%202018.pdf (Accessed April 17, 2023).

Cardinael, R., Chevallier, T., Barthès, B. G., et al. (2015), 'Impact of alley cropping agroforestry on stocks, forms and spatial distribution of soil organic carbon–A case study in a Mediterranean context', *Geoderma*, 259, 288–299.

Cardinael, R., Chevallier, T., Cambou, A., et al. (2017), 'Increased soil organic carbon stocks under agroforestry: a survey of six different sites in France', *Agriculture, Ecosystems & Environment*, 236, 243–255.

Cardinael, R., Thevathasan, N., Gordon, A., Clinch, R., Mohammed, I. and Sidders, D. (2012), 'Growing woody biomass for bioenergy in a tree-based intercropping system in southern Ontario, Canada', *Agroforestry Systems*, 86, 279–286.

Cardinael, R., Umulisa, V., Toudert, A., Olivier, A., Bockel, L. and Bernoux, M. (2018), 'Revisiting IPCC Tier 1 coefficients for soil organic and biomass carbon storage in agroforestry systems', *Environmental Research Letters*, 13(12), 124020.

Castleden, H., Garvin, T. and Huu-ay-aht First Nation (2009), 'Hishuk Tsawak" (Everything is one/connected): a Huu-ay-aht worldview for seeing forestry in British Columbia, Canada', *Society & Natural Resources*, 22(9), 789–804. https://doi.org/10.1080/08941920802098198

Chang, S. X., Wang, W., Zhu, Z., Wu, Y. and Peng, X. (2018), 'Temperate agroforestry in China'. In A. M. Gordon, S. M. Newman ad B. R. W. Coleman (Eds), *Temperate Agroforestry Systems*. 2nd Edition, CAB International, Wallingford, UK, pp. 173–194.

Chisanga, K., Bhardwaj, D. R., Pala, N. A. and Thakur, C. L. (2018), 'Biomass production and carbon stock inventory of high-altitude dry temperate land use systems in North Western Himalaya', *Ecological Processes*, 7(1), 1–13.

Chu, X., Zhan, J., Li, Z., Zhang, F. and Qi, W. (2019), 'Assessment on forest carbon sequestration in the Three-North Shelterbelt Program region, China', *Journal of Cleaner Production*, 215, 382–389.

Chung, H., Ngo, K. J., Plante, A. and Six, J. (2010), 'Evidence for carbon saturation in a highly structured and organic-matter-rich soil', *Soil Science Society of America Journal*, 74(1), 130–138.

Crossland, M. (2015), 'The carbon sequestration potential of hedges managed for woodfuel', *The Organic Research Centre, Elm Farm. https://www.organicresearchcentre.com/manage/authincludes/article_uploads/project_outputs/TWECOM%20ORC%20Carbon%20report%20v1.0.pdf* (Accessed April 17, 2023).

Clinch, R. L., Thevathasan, N. V., Gordon, A. M., Volk, T. A. and Sidders, D. (2009), 'Biophysical interactions in a short rotation willow intercropping system in southern Ontario, Canada', *Agriculture, Ecosystems & Environment*, 131(1–2), 61–69.

Cotrufo, M. F. and Lavallee, J. M. (2022), 'Soil organic matter formation, persistence, and functioning: a synthesis of current understanding to inform its conservation and regeneration', *Advances in Agronomy*, 172, 1–66.

Dhillon, G. S., Gillespie, A., Peak, D. and van Rees, K. C. (2017), 'Spectroscopic investigation of soil organic matter composition for shelterbelt agroforestry systems', *Geoderma*, 298, 1–13.

Dixon, R. K. (1995), Agroforestry systems: sources and sinks of greenhouse gases? Agroforestry Systems, 31, 99–116.

Dixon, R. K., Winjum, J. K. and Schroeder, P. E. (1993), Conservation and sequestration of carbon: the potential of forest and agroforestry management practices. *Global Environmental Change*, 3, 159–173.

Drake, J. B., Knox, R. G., Dubayah, R. O., Clark, D. B., Condit, R., Blair, J. B. and Hofton, M. (2003), 'Above-ground biomass estimation in closed canopy neotropical forests using lidar remote sensing: factors affecting the generality of relationships', *Global Ecology and Biogeography*, 12(2), 147–159.

Drever, C. R., Cook-Patton, S. C., Akhter, F., et al. (2021), 'Natural climate solutions for Canada', *Science Advances*, 7(23), eabd6034.

Drexler, S., Gensior, A. and Don, A. (2021), 'Carbon sequestration in hedgerow biomass and soil in the temperate climate zone', *Regional Environmental Change*, 21(3), 74.

Dube, F., Espinosa, M., Stolpe, N. B., Zagal, E., Thevathasan, N. V. and Gordon, A. M. (2012), 'Productivity and carbon storage in silvopastoral systems with Pinus ponderosa and Trifolium spp., plantations and pasture on an Andisol in Patagonia, Chile', *Agroforestry Systems*, 86, 113–128.

Dube, F., Stolpe, N. B., Zagal, E., Figueroa, C. R., Concha, C., Neira, P., Carrasco, C., Schwenke, J. M., Schwenke, V. and Müller, B. (2018), 'Novel agroforestry systems in temperate Chile'. In A. M. Gordon, S. M. Newman and B. R. W. Coleman (Eds), *Temperate Agroforestry Systems*. 2nd Edition, CABI International, Wallingford, UK, pp. 237–251.

Dupraz, C., Lawson, G.J., Lamersdorf, N., Papanastasis, V.P., Rosati, A. and Ruiz-Mirazo, J. (2018), 'Temperate agroforestry: the European way'. In A. M. Gordon, S. M. Newman, ad B. R. W. Coleman (Eds), *Temperate Agroforestry Systems*. 2nd Edition, CABI International, Wallingford, UK, pp. 98–152.

Eurostat. (2020), 'Agri-environmental indicators'. https://ec.europa.eu/eurostat/statistics-explained/index.php?title=Agri-environmental_indicator_-_cropping_patterns (Accessed April 17, 2023)

Fernández-Núñez, E., Rigueiro-Rodríguez, A. and Mosquera-Losada, M. R. (2010), 'Carbon allocation dynamics one decade after afforestation with *Pinus radiata* D. Don and *Betula alba* L. under two stand densities in NW Spain', *Ecological Engineering*, 36(7), 876–890.

Forestry Commission. (2017), 'Tree cover outside woodland in Great Britain National Forest Inventory', April 2017. https://cdn.forestresearch.gov.uk/2022/02/fr_tree_cover_outside_woodland_in_gb_statistical_report_2017.pdf (Accessed April 17, 2023).

Fortier, J., Truax, B., Gagnon, D. and Lambert, F. (2013), 'Root biomass and soil carbon distribution in hybrid poplar riparian buffers, herbaceous riparian buffers and natural riparian woodlots on farmland', *Springer Plus*, 2(1), 1–19.

Fortier, J., Truax, B., Gagnon, D. and Lambert, F. (2015), 'Biomass carbon, nitrogen and phosphorus stocks in hybrid poplar buffers, herbaceous buffers and natural woodlots in the riparian zone on agricultural land', *Journal of Environmental Management*, 154, 333–345.

Francaviglia, R., Coleman, K., Whitmore, A. P., Doro, L., Urracci, G., Rubino, M. and Ledda, L. (2012), 'Changes in soil organic carbon and climate change–application of the RothC model in agro-silvo-pastoral Mediterranean systems', *Agricultural Systems*, 112, 48–54.

Giese, L. A., Aust, W. M., Kolka, R. K. and Trettin, C. C. (2003), 'Biomass and carbon pools of disturbed riparian forests', *Forest Ecology and Management*, 180(1–3), 493–508.

Global Carbon Project. (2022), 'Carbon dioxide emissions worldwide in 2010 and 2021, by select country (in million metric tons)', Statista. Statista Inc. https://www.statista.com/statistics/270499/co2-emissions-in-selected-countries/ (Accessed April 17, 2023).

Gordon, A. M. and Thevathasan, N. V. (2004), 'How much carbon can be stored in Canadian agroecosystems using a silvopastoral approach?'. In M. R. Mosquera-Losada, J. McAdam, and A. Rigueiro-Rodriguez (Eds), *Silvopastoralism and sustainable land management*. CAB International, Wallingford, UK. pp. 210–218.

Goswami, S., Verma, K. S. and Kaushal, R. (2014), 'Biomass and carbon sequestration in different agroforestry systems of a Western Himalayan watershed', *Biological Agriculture & Horticulture*, 30(2), 88–96.

Government of UK (2022), 'Agriculture in the United Kingdom 2021', Department for Environment, Food and Rural Affairs. UK. https://assets.publishing.service.gov.uk/government/uploads/system/uploads/attachment_data/file/1094493/Agriculture-in-the-UK-27jul22.pdf (Accessed April 17, 2023).

Gross, C. D., Bork, E. W., Carlyle, C. N. and Chang, S. X. (2022), 'Agroforestry perennials reduce nitrous oxide emissions and their live and dead trees increase ecosystem carbon storage', *Global Change Biology*, 28(20), 5956–5972.

Haile, S. G., Nair, P. R. and Nair, V. D. (2008), 'Carbon storage of different soil-size fractions in Florida silvopastoral systems', *Journal of Environmental Quality*, 37(5), 1789–1797.

Hazlett, P. W., Gordon, A. M., Sibley, P. K. and Buttle, J. M. (2005), 'Stand carbon stocks and soil carbon and nitrogen storage for riparian and upland forests of boreal lakes in northeastern Ontario', *Forest Ecology and Management*, 219(1), 56–68.

Hoffman, K.M., Christianson, A.C., Dickson-Hooyle, S., Copes-Gerbitz, K., Nikolakis, W., Diabo, D.A., McLeod, R., Mitchell, H.J., Al Mamun, A., Zahara, A., Mauro, N., Gilchrist, J., Myers Ross, R. and Daniels, L.D. (2022), The right to burn: barriers and opportunities for Indigenous-led fire stewardship in Canada. *FACETS*, 7, 464–481. https://doi.org/10.1139/facets-2021-0062

Houghton, R. A. and Hackler, J. L. (2000), 'Changes in terrestrial carbon storage in the United States. 1: The roles of agriculture and forestry', *Global Ecology and Biogeography*, 9(2), 125–144.

Jackson, R. B., Lajtha, K., Crow, S. E., Hugelius, G., Kramer, M. G. and Piñeiro, G. (2017), 'The ecology of soil carbon: pools, vulnerabilities, and biotic and abiotic controls', *Annual Review of Ecology, Evolution, and Systematics*, 48, 419–445.

Kaonga, M. L. and Bayliss-Smith, T. P. (2009), 'Carbon pools in tree biomass and the soil in improved fallows in eastern Zambia', *Agroforestry Systems*, 76(1), 37–51.

Kemp, P. D., Hawke, M. F. and Knowles, R. L. (2018), 'Temperate agroforestry systems in New Zealand'. In A. M. Gordon, S. M. Newman, ad B. R. W. Coleman (Eds), *Temperate Agroforestry Systems*. 2nd Edition, CAB International, Wallingford, UK, pp. 224–236.

Ketterings, Q. M., Coe, R., van Noordwijk, M. and Palm, C. A. (2001), 'Reducing uncertainty in the use of allometric biomass equations for predicting above-ground tree biomass in mixed secondary forests', *Forest Ecology and Management*, 146(1–3), 199–209.

King, K.F.S. (1987), 'The history of agroforestry'. In H. A. Steppler and P. K. R. Nair (Eds), *Agroforestry: A Decade of Development*. International Centre for Research in Agroforestry (ICRAF), Nairobi, Kenya, pp. 3–13.

Kort, J. and Turnock, R. (1998), 'Carbon reservoir and biomass in Canadian prairie shelterbelts', *Agroforestry Systems*, 44(2–3), 175–186.

Kumar, B. M., Handa, A. K., Dhyani, S. K. and Arunachalam, A. (2018), 'Agroforestry in the Indian Himalayan region: an overview'. In A. M. Gordon, S. M. Newman, ad B. R. W. Coleman (Eds), *Temperate Agroforestry Systems*. 2nd Edition, CAB International, Wallingford, UK, pp. 153–172.

Kumar, B. M., Kumar, S. S. and Fisher, R. F. (1998), 'Intercropping teak with Leucaena increases tree growth and modifies soil characteristics', *Agroforestry Systems*, 42, 81–89.

Kürsten, E. and Burschel, P. (1993), 'CO_2-mitigation by agroforestry', *Water, Air, & Soil Pollution,* 70, 533–544.

Lal, R. (2008), 'Carbon sequestration', *Philosophical Transactions of the Royal Society B*, 363, 815–830.

Lehmann, J., Hansel, C. M., Kaiser, C., et al. (2020), 'Persistence of soil organic carbon caused by functional complexity', *Nature Geoscience*, 13(8), 529–534.

Lehmann, J. and Kleber, M. (2015), 'The contentious nature of soil organic matter', *Nature*, 528(7580), 60–68.

Lorenz, K. and Lal, R. (2005), 'The depth distribution of soil organic carbon in relation to land use and management and the potential of carbon sequestration in subsoil horizons', *Advances in Agronomy*, 88, 35–66.

Lorenz, K. and Lal, R. (2014), 'Soil organic carbon sequestration in agroforestry systems: a review', *Agronomy for Sustainable Development*, 34, 443–454.

Lorenz, K. and Lal, R. (2018), *Carbon Sequestration in Agricultural Ecosystems*, Springer International Publishing, Cham.

López-Díaz, M. L., Benítez, R. and Moreno, G. (2017), 'How do management techniques affect carbon stock in intensive hardwood plantations?', *Forest Ecology and Management*, 389, 228–239.

Luick, R. (2009), 'Wood pastures in Germany'. In A. Rigueiro-Rodriguez, J. McAdam, and M. Mosquera-Losado (Eds), *Agroforestry in Europe: Current Status and Future Prospects*. Springer Science, Dordrecht, NL, pp. 359–376.

Ma, Z. and Chen, H. Y. (2016), 'Effects of species diversity on fine root productivity in diverse ecosystems: a global meta-analysis', *Global Ecology and Biogeography*, 25(11), 1387–1396.

Ma, Z., Chen, H. Y., Bork, E. W., Carlyle, C. N. and Chang, S. X. (2020), 'Carbon accumulation in agroforestry systems is affected by tree species diversity, age and regional climate: a global meta-analysis', *Global Ecology and Biogeography*, 29(10), 1817–1828.

Marchildon, G. P. (2009), 'The prairie farm rehabilitation administration: climate crisis and federal–provincial relations during the Great Depression', *Canadian Historical Review*, 90(2), 275–301. https://doi.org/10.3138/chr.90.2.275

Mayer, S., Wiesmeier, M., Sakamoto, E., Hübner, R., Cardinael, R., Kühnel, A. and Kögel-Knabner, I. (2022), Soil organic carbon sequestration in temperate agroforestry systems – A meta-analysis. *Agriculture, Ecosystems & Environment*, 323, 107689. https://doi.org/10.1016/j.agee.2021.107689

Mayrinck, R. C., Laroque, C. P., Amichev, B. Y. and van Rees, K. (2019), 'Above-and below-ground carbon sequestration in shelterbelt trees in Canada: a review', *Forests*, 10(10), 922.

Mokany, K., Raison, R. J. and Prokushkin, A. S. (2006), 'Critical analysis of root: shoot ratios in terrestrial biomes', *Global Change Biology*, 12(1), 84–96.

Montagnini, F., Francesconi, W. and Rossi, E. (2011), *Agroforestry as a Tool for Landscape Restoration*. Nova Science Publishers, UK.

Nair, P. R. (2011), 'Agroforestry systems and environmental quality: introduction', *Journal of Environmental Quality*, 40(3), 784–790.

Nair, P. K. R. and Nair, V. D. (2003), 'Carbon storage in North American agroforestry systems'. In J. Kimble, L. S. Heath, R. A. Birdsey, and R. Lal (Eds), *The Potential of U.S. Forest Soils to Sequester Carbon and Mitigate the Greenhouse Effect*, CRC Press, Boca Raton, USA, pp. 333–346.

Nair, R. P. K., Mohan Kumar, B. and Nair, V. D. (2009), 'Agroforestry as a strategy for carbon sequestration', *Journal of Plant Nutrition and Soil Science*, 172, 10–23. https://doi.org/10.1002/jpln.200800030

Nair, P. K. R., Nair, V. D., Kumar, B. M. and Showalter, J. M. (2010), 'Carbon sequestration in agroforestry systems', *Advances in Agronomy*, 108, 237–307.

Nerlich, K., Graeff-Hönninger, S. and Claupein, W. (2013), 'Agroforestry in Europe: a review of the disappearance of traditional systems and development of modern agroforestry practices, with emphasis on experiences in Germany', *Agroforestry Systems*, 87, 475–492. https://doi.org/10.1007/s10457-012-9560-2

Oelbermann, M. (2002), *'Linking carbon inputs to sustainable agriculture in Canadian and Costa Rican agroforestry systems'*, Ph.D. Thesis. Department of Land Resource Science, University of Guelph, p. 208.

Oelbermann, M. and Voroney, R. P. (2007), 'Carbon and nitrogen in a temperate agroforestry system: using stable isotopes as a tool to understand soil dynamics', *Ecological Engineering*, 29(4), 342–349.

Oelbermann, M. and Voroney, R. P. (2011), 'An evaluation of the century model to predict soil organic carbon: examples from Costa Rica and Canada', *Agroforestry Systems*, 82, 37–50.

Oelbermann, M., Voroney, R. P. and Gordon, A. M. (2004), 'Carbon sequestration in tropical and temperate agroforestry systems: a review with examples from Costa Rica and Southern Canada', *Agriculture, Ecosystems & Environment*, 104, 359–377.

Oelbermann, M., Voroney, R. P., Kass, D. C. L. and Schlönvoigt, A. M. (2005), 'Above- and below-ground carbon inputs in 19-, 10-, and 4-year old Costa Rican alley cropping systems', *Agriculture, Ecosystems & Environment,* 105, 163–172.

Oelbermann, M., Voroney, R. P., Thevathasan, N. V., Gordon, A. M., Kass, D. C. and Schlönvoigt, A. M. (2006), 'Soil carbon dynamics and residue stabilization in a Costa Rican and southern Canadian alley cropping system', *Agroforestry Systems*, 68(1), 27–36.

Ofosu, E., Bazrgar, A., Coleman, B., Deen, B., Gordon, A., Voroney, P. and Thevathasan, N. (2022), 'Diverse temperate riparian buffer types promote system-level carbon sequestration in Southern Ontario, Canada', *The Forestry Chronicle*, 98(1), 103–118.

Palma, J. H., Crous-Durán, J., Graves, A. R., et al. (2018), 'Integrating belowground carbon dynamics into Yield-SAFE, a parameter sparse agroforestry model', *Agroforestry Systems*, 92, 1047–1057.

Pardon, P., Reubens, B., Mertens, J., Verheyen, K., De Frenne, P., De Smet, G., Van Waes, C. and Reheul, D. (2018), Effects of temperate agroforestry on yield and quality of different arable intercrops, *Agricultural Systems*, 166, 135-151 https://doi.org/10.1016/j.agsy.2018.08.008.

Peichl, M., Thevathasan, N. V., Gordon, A. M., Huss, J. and Abohassan, R. A. (2006), 'Carbon sequestration potentials in temperate tree-based intercropping systems, Southern Ontario, Canada', *Agroforestry Systems*, 66, 243–257.

Peri, P. L., Caballé, G., Hansen, N. E., Bahamonde, H. A., Lencinas, M. V., von Müller, A. R., Ormaechea, S., Gargaglione, V., Soler, R., Sarasola, M., Rusch, V., Borrelli, L., Fernández, M. E., Gyenge, J., Tejera, L. E., Lloyd, C. E. and Martínez Pastur, G. (2018), 'Silvopastoral systems in Patagonia, Argentina'. In A. M. Gordon, S. M. Newman, and B. R. W. Coleman (Eds), *Temperate Agroforestry Systems*. 2nd Edition, CAB International, Wallingford, UK, pp. 252–273.

Prescott, C. E. (2010), 'Litter decomposition: what controls it and how can we alter it to sequester more carbon in forest soils?', *Biogeochemistry*, 101, 133–149.

Rajput, B. S., Bhardwaj, D. R. and Pala, N. A. (2015), 'Carbon dioxide mitigation potential and carbon density of different land use systems along an altitudinal gradient in north-western Himalayas', *Agroforestry Systems*, 89, 525–536.

Rajput, B. S., Bhardwaj, D. R. and Pala, N. A. (2017), 'Factors influencing biomass and carbon storage potential of different land use systems along an elevational gradient in temperate northwestern Himalaya', *Agroforestry Systems*, 91, 479–486.

Reeg, T., Möndel, A., Brix, M. and Konold, W. (2008), Conservation in agricultural landscape– new options through modern agroforestry systems? *Nature Landscape,* 83, 261–266.

Reid, R. and Moore, R. (2018), 'Agroforestry systems in temperate Australia'. In A. M. Gordon, S. M. Newman and B. R. W. Coleman (Eds), *Temperate Agroforestry Systems*. 2nd Edition, CAB International, Wallingford, UK, pp. 195–223.

Reimer, A. P. and Prokopy, L. S. (2014), 'Farmer participation in US Farm Bill conservation programs', *Environmental Management*, 53, 318–332.

Rempel, J. C., Kulshreshtha, S. N., Amichev, B. Y. and Van Rees, K. C. (2017), 'Costs and benefits of shelterbelts: a review of producers' perceptions and mind map analyses for Saskatchewan, Canada'. *Canadian Journal of Soil Science*, 97(3), 341–352.

Rheinhardt, R., Brinson, M., Meyer, G. and Miller, K. (2012), 'Integrating forest biomass and distance from channel to develop an indicator of riparian condition', *Ecological Indicators*, 23, 46–55.

Rural Development Institute. (2014). '2014 shelterbelt survey prairie producers' use of and attitudes towards shelterbelts', Brandon University. https://www.brandonu.ca/rdi/files/2011/02/AGGP-Survey-Report-2014.pdf (Accessed 25 November 2023).

Sánchez, P.A. (2000), Linking climate change research with food security and poverty reduction in the tropics. *Agriculture, Ecosystems & Environment*, 82, 371–383.

Schroeder, P. (1994), 'Carbon storage benefits of agroforestry systems', *Agroforestry Systems*, 27, 89–98.

Sharma, H., Pant, K. S., Bishist, R., Gautam, K. L., Dogra, R., Kumar, M. and Kumar, A. (2023), 'Estimation of biomass and carbon storage potential in agroforestry systems of north western Himalayas, India', *CATENA*, 225, 107009.

Sharrow, S. H. and Ismail, S. (2004), 'Carbon and nitrogen storage in agroforests, tree plantations, and pastures in western Oregon, USA', *Agroforestry Systems*, 60, 123–130.

Sinacore, K., Hall, J. S., Potvin, C., Royo, A. A., Ducey, M. J. and Ashton, M. S. (2017), 'Unearthing the hidden world of roots: Root biomass and architecture differ among species within the same guild', *PLoS One*, 12(10), e0185934.

Six, J. A. E. T., Elliott, E. T. and Paustian, K. (2000), 'Soil macroaggregate turnover and microaggregate formation: a mechanism for C sequestration under no-tillage agriculture', *Soil Biology and Biochemistry*, 32(14), 2099–2103.

Statistics Canada. (2016), '2016 Census of Agriculture', https://www.statcan.gc.ca/en/ca2016 (Accessed April 17, 2023).

Thevathasan, N. V., Coleman, B., Zabek, L., Ward, T. and Gordon, A. M. (2018), 'Agroforestry in Canada and its role in farming systems'. In A. M. Gordon, S. M. Newman, and B. R. W. Coleman (Eds), *Temperate Agroforestry Systems*. 2nd Edition, CAB International, Wallingford, UK, pp. 7–49.

Thevathasan, N. V., Coleman, B., Zabek, L., Ward, T. and Gordon, A. M. (2018), 'Agroforestry in Canada and its role in farming systems'. In A. M. Gordon, S. M. Newman, and B. R. W. Coleman (Eds), *Temperate Agroforestry Systems*. 2nd Edition, CAB International, Wallingford, UK, pp. 7–49.

Thevathasan, N. V. and Gordon, A. M. (1997), 'Poplar leaf biomass distribution and nitrogen dynamics in a poplar-barley intercropped system in Southern Ontario, Canada', *Agroforestry Systems*, 37, 79–90.

Tufekcioglu, A., Raich, J. W., Isenhart, T. M. and Schultz, R. C. (2003), 'Biomass, carbon and nitrogen dynamics of multi-species riparian buffers within an agricultural watershed in Iowa, USA', *Agroforestry Systems*, 57, 187–198.

Udawatta, R. P., Walter, D. and Jose, S. (2022), 'Carbon sequestration by forests and agroforests: a reality check for the United States', *Carbon Footprints*, 1, 8. https://dx.doi.org/10.20517/cf.2022.06

Udawatta, R.P. and Jose, S. (2012), 'Agroforestry strategies to sequester carbon in temperate North America', *Agroforestry Systems*, 86, 225–42.

Upson, M. A. and Burgess, P. J. (2013), 'Soil organic carbon and root distribution in a temperate arable agroforestry system', *Plant and Soil*, 373, 43–58.

Viaud, V. and Kunnemann, T. (2021), 'Additional soil organic carbon stocks in hedgerows in crop-livestock areas of western France', *Agriculture, Ecosystems & Environment*, 305, 107174.

Vijayakumar, S., Bazrgar, A. B., Coleman, B., Gordon, A., Voroney, P. and Thevathasan, N. (2020), 'Carbon stocks in riparian buffer systems at sites differing in soil texture, vegetation type and age compared to adjacent agricultural fields in southern Ontario, Canada', *Agriculture, Ecosystems & Environment*, 304, 107149.

Vijayakumar, S., Bazrgar, A. B., Gordon, A., Voroney, P. and Thevathasan, N. (2023), 'Soil organic carbon stabilization based on physical fractionation method in tree based riparian and adjacent agricultural systems in southern Ontario, Canada'. *Agroforestry Systems*, 1–15. https://doi.org/10.1007/s10457-023-00913-4

von Lützow, M., Kögel-Knabner, I., Ekschmitt, K., Matzner, E., Guggenberger, G., Marschner, B. and Flessa, H. (2006), 'Stabilization of organic matter in temperate soils: mechanisms and their relevance under different soil conditions–a review', *European Journal of Soil Science*, 57(4), 426–445.

Watson, R.T., Noble, I.R., Bolin, B., Ravindranath, N.H., Verardo, D.J. and Dokken, D.J. (2000), *Land use, Land-use Change and Forestry*. Cambridge University Press, Cambridge, England.

Weiner, J. and Thomas, S. C. (2001), 'The nature of tree growth and the "age-related decline in forest productivity"', *Oikos*, 94(2), 374–376.

Williams, P. A., Gordon, A. M., Garret, H. E. and Buck, L. (1997), 'Agroforestry in North America and its role in farming systems'. In A. M. Gordon and S. M. Newman (Eds), *Temperate Agroforestry Systems*. 1st Edition, CAB International, Wallingford, UK, pp. 9–85.

Winans, K. S., Whalen, J. K., Rivest, D., Cogliastro, A. and Bradley, R. L. (2016), 'Carbon sequestration and carbon markets for tree-based intercropping systems in Southern Quebec, Canada', *Atmosphere*, 7(2), 17.

Worms, P. (2021), A history on the forgotten but enduring practice of agroforestry and the role it can play in shaping the future of Europe's sustainable agriculture. *Revolve*, 41. https://revolve.media/agroforestry-the-age-old-future-of-europes-agriculture/

Wotherspoon, A., Thevathasan, N. V., Gordon, A. M. and Voroney, R. P. (2014), 'Carbon sequestration potential of five tree species in a 25-year-old temperate tree-based intercropping system in Southern Ontario, Canada', *Agroforestry systems*, 88, 631–643.

You, X. L. (1991), 'Mixed cropping with trees in ancient China'. In Z. H. Zhu, M. T. Cai, S. J. Wang, and Y. X. Jiang (Eds), *Agroforestry Systems in China*. IDRC Canada and CAF, China, pp. 8–9.

Young, A. (1997), *Agroforestry for Soil Management*. CAB International, Wallingford, UK.

Zahoor, S., Dutt, V., Mughal, A. H., Pala, N. A., Qaisar, K. N. and Khan, P. A. (2021), 'Apple-based agroforestry systems for biomass production and carbon sequestration: implication for food security and climate change contemplates in temperate region of Northern Himalaya, India', *Agroforestry Systems*, 95, 367–382.

Chapter 3

Ecosystem accounting to value ecosystem assets and services provided by agroforestry

Anthony P. O'Grady, Stephen B. Stewart, Aysha Fleming, Chris Ware and Daniel S. Mendham, CSIRO Environment, Australia; Sorada Tapsuwan, CSIRO Environment and Swinburne University of Technology, Australia; Tim Capon, Suzanne M. Prober and Karel Mokany, CSIRO Environment, Australia; Sue Ogilvy, Integrated Futures Pty Ltd, Australia; Martin Moroni, Forestry Tasmania, Australia; Jacqueline R. England, CSIRO Environment, Australia; Shaun T. Brooks and Thomas P. Baker, University of Tasmania, Australia

1 Introduction

Much of the global economy is dependent on natural capital. Natural capital is defined as the stock of renewable and non-renewable resources that combine to provide flows of services and benefits for humanity (Natural Capital Coalition, 2016). These benefits are generally derived from the ecosystem services that natural capital generates. In many situations, the benefits that humanity receives from these ecosystem services can be significant. Recent analysis suggests that over half the global gross domestic product (GDP) is highly or moderately dependent on the goods and services that nature provides (Dasgupta, 2021). Increasingly, climate change and ongoing loss of nature are being seen as significant risks not just to ecosystems but to the economic

http://dx.doi.org/10.19103/AS.2024.0141.07

systems that these ecosystem goods and services underpin. Nature-related risks represent six of the top ten risks to the stability of the global economy over the next decade (World Economic Forum, 2023), and there is an urgent need to adopt production systems that are more resilient to this changing risk profile to ensure the ongoing sustainability of production. Agricultural systems are highly dependent on natural capital, and agricultural producers are important stewards of large reserves of natural capital. Agricultural producers have a significant opportunity to manage their natural capital in a manner that increases the sustainability and transparency of agricultural produce within global supply chains and to take advantage of the emerging demand for market-driven natural capital opportunities.

Agricultural landscapes are under increasing pressure to increase food, fibre and energy production, while simultaneously increasing nature and biodiversity outcomes (Hajkowicz and Moody, 2010). Agroforestry has been recognised as playing an important role in protecting or increasing natural capital (Dudley and Alexander, 2017; Benton et al., 2021) and enhancing ecosystem services within agricultural systems (Nair, 1993; Nair, 2012). There are many types of agroforestry systems in practice around the globe (van Noordwijk et al., 2020), but it can be broadly defined as the deliberate retention and/or incorporation of long-lived woody perennials into the production system with the intent of lifting farm enterprise and landscape productivity by enhancing the flows of ecosystem services to the farm enterprise (private benefits) and/or to the broader community (public benefits) (Nair, 1993). Agroforestry has the potential to prevent or remediate environmental degradation, enhance agricultural productivity, increase carbon sequestration and storage, promote healthy soils and provide habitat for the maintenance of biodiversity and diversify agricultural incomes (Jose, 2009).

Quantitative information on the natural capital benefits associated with agroforestry remains relatively sparse, despite decades of research and much qualitative work. Many benefits are self-evident and quantifiable, e.g. carbon sequestration and habitat provision for biodiversity (Paul et al., 2013; Torralba et al., 2016), but the key recipients of such benefits are typically external to the farm enterprise. There is a strong belief that there are benefits to the landowner that agroforestry provides, including increases in crop and livestock productivity (Baker et al., 2018) or the livestock and human welfare and amenity values (Polyakov et al., 2014); however, these beliefs are often based on information that is largely observational and/or anecdotal (Gregory, 1995; Baker et al., 2018; England et al., 2020). The lack of ability to quantify and account for the full range of public and private benefits often hampers the adoption of agroforestry practices and constrains analysis of costs and benefits associated with agroforestry (Fleming et al., 2019).

Regardless of the challenges in quantifying natural capital values, recognition of the benefits that agroforestry provides for agricultural systems is growing. Further, agroforestry systems are increasingly being promoted as nature-based or natural climate solutions (Brown et al., 2018; Terasaki Hart et al., 2023; Telwala, 2023) and can contribute to emerging demand and markets for nature repair (Convention on Biological Diversity, 2022; TNFD, 2023). The UN Environment Program (UNEP, 2022) noted agroforestry as the nature-based solution requiring the greatest investment (US$3600 B by 2050) to limit global warming to 1.5°C, in line with the Paris Agreement. Nature-based solutions are actions that protect, conserve, restore and sustainably use and manage natural resources or modified terrestrial, freshwater coastal and marine ecosystems while simultaneously addressing significant social economic and environmental challenges (IUCN, 2020). The potential of agroforestry as a nature-based solution is largely unquantified and most likely unrealised. For example, Terasaki Hart et al. (2023) found a large gap between the actual and potential climate mitigation benefits of agroforestry on all continents except Africa.

Recognition of agroforestry as a nature-based solution comes with the additional responsibility that the outcomes need to be verified and tracked through time (IUCN, 2020). There is a pressing need to develop approaches to more formally recognise the services and benefits associated with agroforestry. Natural capital accounting has recently emerged as an approach to address this problem. Derived from environmental and economic accounting, e.g. the System of Environmental Economic Accounting (United Nations, 2014; United Nations, 2021), natural capital accounting promotes the recognition of natural and modified ecosystems as assets. The adoption of an accounting approach is deliberate and designed to enable formal recognition of the environmental contributions to the economy and society more broadly or when framed appropriately provide the capacity to quantify the natural contributions to the environmental and financial performance of an enterprise (Ogilvy, 2015; Hein et al., 2016; Barker, 2019; Ogilvy et al., 2022).

In this chapter, we demonstrate how individual agroforestry projects could be recognised as assets (albeit natural capital assets) on the balance sheet of an enterprise, using a case study in Tasmania, Australia. Quantifying the natural capital benefits associated with various agroforestry systems will increase the adoption of agroforestry by land managers (Fleming et al., 2022). Ultimately, our aim is to see formal integration of the financial and natural capital performance of farm businesses and to provide farmers and their advisors with the frameworks required to better analyse the costs and benefits associated with agroforestry.

Section 2 of this chapter highlights some of the ecosystem services commonly associated with selected agroforestry systems (focusing on belts that provide shelter to crops and livestock) and provides a high-level overview of approaches to quantifying these for accounting purposes. Section 3 demonstrates how the values associated with these services might be represented on the enterprise balance sheet. Section 4 summarises stakeholder perspectives and insights gained from this form of presentation. Finally, we provide a discussion and some suggestions for future research directions.

2 Key ecosystem services associated with agroforestry

Agroforestry systems provide bundles of ecosystem services with both private and public benefits. Different systems and configurations of trees will have differing levels of provision of these services. For example, shelter belts composed predominantly of exotic species may provide more forest and wood products but lower habitat conditions for biodiversity compared to a block of mixed native species. In order to recognise benefits associated with these varying configurations in a formal accounting process, it is important that ecosystem services be quantified in a manner that faithfully represents the ecosystem services generated by the ecosystem asset (AASB, 2022). In this study, we focussed on a subset of the possible spectrum of ecosystem services associated with agroforestry systems (Table 1). While these selected ecosystem services are not exhaustive, they are representative of the range of ecosystem services commonly associated with agroforestry. Each of the ecosystem services

Table 1 Example ecosystem services associated with agroforestry in temperate agricultural systems

Category	Ecosystem service	Beneficiary	Description
Provisioning (biotic)	Biomass provisioning	Private	Cultivated plants for nutrition, materials or energy, predominantly wood products
Regulating and maintenance (biotic)	Life cycle maintenance, habitat and gene pool protection	Public/private	Habitat provision for biodiversity
	Global climate regulation services	Public/private	Carbon sequestration to help stabilise climate
	Local (micro- and meso-climate regulation services)	Private	Shade and shelter for crop and livestock production
	Pollination	Private/public	Pollination of crops
Cultural (biotic)	Visual amenity services	Private/public	Characteristics of nature that enhance aesthetic experiences
	Other cultural services	Private/public	Characteristics of nature that enhance cultural experiences

Source: Haines-Young and Potschin-Young (2018). Note that this is not an exhaustive list, but a list of key services that we have chosen to account for here.

chosen requires its own metrics, some are relatively easy to quantify (such as from the sale of forest products), and some require the development of robust metrics that are scalable and can be implemented for individual agroforestry assets, allowing integration into the farm's balance sheet.

2.1 Biomass provisioning services

Biomass provisioning services from agroforestry systems include food, forage, wood products (solid wood, wood chips, firewood, etc.) and oils and fuels (Báder et al., 2023). The demand for wood products, in particular from agroforestry, is likely to grow significantly. This growth in demand for wood products is being driven by rising populations, increasing substitution of carbon-intensive materials coupled with a declining supply of native forest timbers and limited opportunities for the establishment of new large-scale plantation assets (O'Grady and Mitchell, 2018; Monckton and Mendham, 2022).

Standing wood volume in agroforestry assets can be estimated using site-specific measures of stem diameter and height through allometric equations, or it can be modelled using a range of empirical or process-based models. To estimate standing wood volumes in our case study, a modified version of the dynamic forest growth model, CABALA (Battaglia et al., 2004), was employed, with modifications to reflect the challenges of agroforestry systems, particularly shelterbelts (England et al., 2018b). Indicative volumes of harvested log products were then estimated as 85% of standing volumes (Stewart and O'Grady, 2020). An alternative approach could rely on the application of functions that relate log size to product recovery, but there is little published information currently available to inform this approach for agroforestry systems.

2.2 Habitat provision for biodiversity

Biodiversity is an ecosystem attribute critical to a wide range of ecosystem services that are recognised as being globally important (Dasgupta 2021). Biodiversity is an emergent property of the biophysical environment (including climate, soils, condition and configuration) ecological interactions and co-evolutionary history. As such, this presents a complex measurement challenge. Habitat-based assessments of biodiversity have emerged as a valuable tool in accounting for biodiversity (Horner et al., 2022; Mokany et al., 2022). Habitat-based biodiversity assessments rely on observations of changes in the extent, condition and configuration of habitat (King et al., 2021). Many of these characteristics can be derived from remote sensing and can provide scalable, consistent and comparable reporting for areas through time (Mokany et al., 2022). Agroforestry systems provide habitat for biodiversity as they influence the extent, composition and connectivity of habitats within the landscape. The contribution these systems make to biodiversity can be estimated as a function of their similarity to extant ecosystems in conditions where human influence on biodiversity is minimal, i.e. difference from the reference state. Different species compositions, configurations and ages of agroforestry systems will convey different levels of similarity to the original native vegetation. These can be expressed as an index of 0–1, where ideally, 1 represents the full suite of native species in the reference, or other easier to assess but less optimal measures such as species richness or habitat condition scores can be used as proxies (Prober et al., 2025). Using this approach, a measure of the effective habitat area (E_{Ha}) of an agroforestry asset (ha) can be produced by multiplying the extent of the asset (Ext_A, ha) by the average habitat condition (Cd_A), as defined above, for each area on a farm, such that

$$E_{Ha} = Ext_A \times Cd_A \tag{1}$$

This approach can also be used to forecast changes in habitat conditions using state and transition modelling approaches to estimate the upper limit of change in habitat conditions, and how this change is likely to occur over time since planting (Ogilvy et al., 2022). This approach has a number of advantages over inventory approaches to biodiversity accounting, e.g. Jones (1996) or Horner and Davidson (2020), as the metric is simple, repeatable and potentially more robust than measures built on more complex biodiversity indicators.

2.3 Global climate regulation services

Commercial and environmental plantings have the potential to increase on-farm carbon storage in woody biomass through afforestation, integration of belt plantings or by increasing the length of rotations in plantings designed to be harvested (Paul et al., 2022a). The capture and storage of carbon in vegetation will play a critical role in managing changing climates in the future (Bastin et al., 2019), as emissions reductions alone will not be enough to avoid global warming exceeding the 1.5°C ambition set under the Paris Agreement (Wood et al., 2023).

Carbon stocks in agroforestry assets of varying configurations can be modelled or estimated using site-specific measures. Australia uses the stand-level carbon accounting model FullCAM as the national carbon accounting tool for reporting against the country's national greenhouse gas emissions. The model is also applied at project scales under some methods of the Australian Carbon Credit Units (ACCUs) Scheme. Predictions of carbon stocks in FullCAM are predominantly influenced by an input layer of the maximum potential biomass for a location (Roxburgh et al., 2019) and calibrated yield curve parameters, including for 12 categories of plantations and 10 categories of environmental and mallee plantings (Paul and Roxburgh, 2020; Paul et al., 2022b). Where calibrations for agroforestry assets are insufficient, such as for shelter belts specifically designed for the production of commercial timber, further refinement of carbon sequestration estimates can be based on stand inventory (Marais et al., 2022; England et al., 2018a) and scaling of biomass/carbon through the application of allometric equations, expansion factors, and conversion to carbon or CO_2 equivalents (Paul et al., 2016).

2.4 Microclimate regulation services

Shade and shelter benefits of trees for the agricultural part of the enterprise are one of the key ecosystem services that can convey a substantial private benefit, depending on the configuration of the trees and the responsiveness of the agricultural system (Mendham, 2018). These benefits include improved crop and pasture productivity (Baker et al., 2018), reduced physical damage to crops (Cleugh, 1998), mitigation of extreme weather (i.e. heat waves, frost) (Cleugh and Hughes, 2002; Baker et al., 2021) and provision of shade for livestock (Giro et al., 2019; Gregory, 1995). The scale of microclimate regulation, and the subsequent agricultural benefit that occurs, varies depending on weather conditions (Kort, 1988), biophysical characteristics of trees (e.g. age, height, porosity and species) (Cleugh, 1998; Marais et al., 2022) and landscape configuration. This means that there are few generalisable approaches to quantifying these benefits. A recent global quantitative analysis developed a typical productivity response to the shelter provided by windbreaks and alleys, with a competition zone typically occurring within 0–2 tree heights, followed by a zone of benefit (increased pasture/crop production compared to unsheltered control) occurring from 2 to about 20 tree heights (Baker et al., 2018).

2.5 Pollination services

Pollination is a critical ecosystem function that is thought to be enhanced by agroforestry (Bentrup et al., 2019; Centeno-Alvarado et al., 2023). Over 85% of the world's flowering species are dependent on animal-mediated pollination services (Ollerton et al., 2011), and it has been estimated that approximately 35% of global food production is also dependent on pollination by animals (Porto et al., 2020). Thus, agriculture is highly dependent on the pollination services provided by nature. Declines in pollinator diversity and abundance are fuelling rising concern (Potts et al., 2010; IPBES, 2016), with climate change, habitat loss and fragmentation, disease and the use of pesticides all contributing to global pollinator decline (Dicks et al., 2021).

Ecosystem service models typically focus on pollinator abundance rather than the actual process of pollination. The Lonsdorf model (Lonsdorf et al., 2009) is commonly used to model pollinator abundance and has been adapted for various research and ecosystem services modelling applications (Joseph et al., 2020; Sharp et al., 2020; Villa et al., 2014). The model uses information on forage distance, habitat suitability, and floral resource availability to estimate where pollinators are most likely to be active across the landscape. Alternate approaches based on species distribution and richness modelling have also been developed but so far have not been widely applied (Perennes et al., 2021).

2.6 Visual amenity services

Trees and forests have been shown to provide health and well-being benefits to people through multiple pathways including the provision of visual and aesthetic improvements (Lin et al., 2018). The Helliwell System (Helliwell, 2008) provides a measure for valuing the visual amenity of trees using factors including tree size, location, life expectancy, suitability of the tree in the landscape and the number of other trees in the surrounding area to evaluate a tree value. The Helliwell System, combined with sophisticated spatial mapping methods, has been applied to examine the visual amenity values of trees and forests in rural areas in Scotland (van der Horst, 2006). However, assigning dollar values to the amenity benefits of trees and forests is controversial. The Helliwell System, despite being used to provide evidence in court, is still based on expert judgement on the value of the tree rather than the market value of the tree. Other methods, such as the hedonic property price approach (see e.g. Tapsuwan et al., 2009), are unable to disentangle the visual amenity value of trees from other amenities that they provide. Other methods that might be able to separate out the amenity value are mostly based on stated willingness to pay of people to preserve the visual amenity value of trees (see e.g. Grala et al., 2012) rather than how much people have actually paid. Another economic approach that has been used to assign monetary values to visual amenity is the replacement cost approach. The premise behind using the replacement cost approach is that the enhancement of visual amenities of trees and forests can have a significant impact on people's mental and physical health (Barton and Pretty, 2010; Bowler et al., 2010), which in turn reduces the burden on the cost of medical care to society (Ulrich et al., 1991; Shaw et al., 2006). The difficulty in objectively measuring the visual amenity of forests lies in two main areas: the standard of measurement of aesthetic quality and the method for eliciting the monetary value of the aesthetic quality.

2.7 Other cultural services

Cultural services from trees in the landscape include the 'nonmaterial benefits people obtain from ecosystems through spiritual enrichment, cognitive development, reflection, recreation, and aesthetic experiences' (Millennium Ecosystem Assessment, 2005). In addition to visual amenity services, other cultural services include values for education, heritage and sense of place. This definition recognises the contributions of ecosystem to formal and informal education, their contributions to historically or culturally significant ecosystems and landscapes, and the 'sense of place' that people associate with features of their environment and aspects of the ecosystem (Millennium Ecosystem Assessment, 2005; de Groot and Ramakrishnan, 2005).

Systematic reviews of papers that attempt to evaluate, define, measure or assess cultural ecosystem services find that, although cultural services are experienced and appreciated more directly by people than some other kinds of services, they are the most difficult type of ecosystem service to quantify in monetary, or even quantitative terms (Horcea-Milcu et al., 2013; Hølleland et al., 2017). For this reason, cultural services are sometimes sacrificed in favour of other types of services that are easier to value in economic or ecological terms (Horcea-Milcu et al., 2013). The flow of cultural services is influenced by the interaction of stakeholders and the environment, including the managers who influence the supply of cultural services and the beneficiaries who receive the services (Xiao et al., 2017). People's values, beliefs and norms mediate their interpretation of sense data and other information they receive about the environment and are therefore likely to influence how they value cultural services (Stern et al., 1999). This means that some provisioning services can also be perceived as having a traditional or heritage value (e.g. Asah et al., 2014).

As a way to integrate the visual and cultural amenity values of trees, Polyakov (2014) explored the effect of the proportion of native forest on land value across a 3 million ha region of north-Central Victoria, Australia. They found that increasing the area of trees on a property was associated with an initial increase in land value, followed by a decrease, and peaked at between 20% tree cover (for a 1000 ha property) and around 37% tree cover for a 10 ha property. They found that trees increased property values by a maximum of 5% and 16% at the 1000 ha and 10 ha properties, respectively. This work provides a guide for separating out the amenity value provided by the trees compared to the amenity value of other factors (landscape features, residence quality, proximity to services, etc.).

3 Integration into ecosystem accounts for agroforestry

3.1 Accounting approach

Ecosystem accounting provides a standardised approach for the measurement and monitoring of ecosystems and their contribution to the economy. The capacity to value goods and services that ecosystems provide is designed to enhance economic decision-making. Historically, decisions related to the allocation of resources or management of the underlying ecosystem have occurred without consideration of the capacity of the ecosystem to continue providing goods and services, leading to a distortion of markets and ongoing depletion and degradation of natural capital. The primary methodology for ecosystem accounting developed by the United Nations Statistical Division is the System of Environmental Economic Accounting-Ecosystem Accounting (SEE-EA) (United Nations, 2021). Designed primarily to meet the needs of countries to develop national environmental accounts and inform the System of National Accounts there has been increased focus on the application of these accounting principles at enterprise scales (Ogilvy, 2015; Ogilvy, 2020; Stewart and O'Grady, 2020). The conceptual model for ecosystem accounting is shown in Fig. 1. Ecosystem accounting relies on measuring changes in the extent and condition of ecosystem assets through time. Ecosystem assets represent the basic accounting unit in SEEA-EA and are defined as contiguous spaces of a specific ecosystem type characterised by a distinct set of biotic and abiotic components and their interactions (United Nations, 2021). The definition is not constrained to ecosystems in a relatively natural state and can include highly modified ecosystems (e.g. a wheat paddock or an individual shelter belt). Changes in the extent and condition of ecosystem assets directly influence the flows of ecosystem services that these assets generate, and thus the value of the underlying ecosystem asset base and the benefits that are derived from the assets.

An ecosystem accounting system is comprised of five core accounts (United Nations, 2021):

- Ecosystem extent accounts record changes in the area or stock of ecosystem assets within an ecosystem accounting area (e.g. a farm enterprise) between accounting periods (e.g. 1, 2, 5 years).
- Ecosystem condition accounts record the condition of ecosystem assets that are under control or management. Changes in the extent and condition of ecosystem provide valuable insights into the health of the ecosystem and the capacity of these assets to continue the production of the ecosystem goods and services the enterprise is reliant on.
- Ecosystem flow accounts record the supply of ecosystem services by ecosystem assets. These are presented in two formats, accounts that describe the physical flows and accounts that describe ecosystem service flows in monetary values. Physical ecosystem services accounts provide

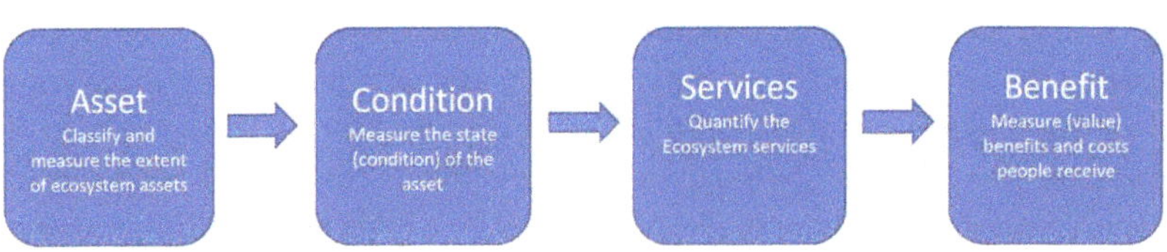

Figure 1 Ecosystem accounting conceptual model with the quantifiable components of each step.

Published by Burleigh Dodds Science Publishing Limited, 2025.

the foundation for the valuation of the ecosystem assets (monetary ecosystem service account).
- The monetary ecosystem asset account records changes (additions/reductions) in the stock of ecosystem accounts in monetary terms, including contributions to this valuation associated with ecosystem enhancement or degradation. The value of the ecosystem asset can be approximated as the net present value of the future flows of ecosystem services, enabling translation of a diverse set of ecosystem services with differing physical units into a common unit, in our case study, the currency of interest, e.g. AUD.

An important feature of ecosystem accounting is that it enables the contributions of ecosystems to be recognised as part of a broader set of environmental economic accounts in physical and/or monetary terms enabling the valuation of nature's contributions to economic position or performance. Increasingly, the concepts of ecosystem accounting are being adapted to enterprise scales, commonly referred to as natural capital accounting, with a growing emphasis on the recognition of the ecosystem assets that produce goods and services for the enterprise. Doing so may provide valuable insights into the contributions that ecosystems make to an organisation's financial position (Ogilvy, 2015; Houdet et al., 2020). This is likely to be particularly important in agricultural contexts as agricultural businesses are typically highly dependent on natural capital (Ogilvy, 2015; Smith et al., 2021). As such, changes in the extent or condition of ecosystem assets under stewardship may have significant impacts on the generation of the ecosystem services that the entity is reliant on, leading to increased operational or capital costs or lost revenue. More broadly, natural capital accounts provide a framing of natural capital information in a manner that facilitates comparisons between organisations environmental stewardship in a consistent and comparable format (Ogilvy et al., 2022). This is becoming increasingly important as a growing number of organisations grapple with climate- and nature-related financial disclosures that span supply-chain boundaries (TNFD, 2023).

To some extent, existing accounting standards already incorporate some aspects of natural capital. For example, the IAS 41 standard defines agricultural activities as the 'management by an entity of the biological transformation and harvest of biological assets for sale or conversion into agricultural produce or into additional biological assets' (AASB, 2015). Standards also exist that cover the human-produced capital of an agricultural entity, such as property, plant and equipment (IAS 16) or inventory (IAS 2). However, these provide an incomplete picture of an entity's assets (Fig. 2). Natural capital accounting provides an opportunity to present an extended view of the entity's assets and the goods and services associated with that asset base (Fig. 3).

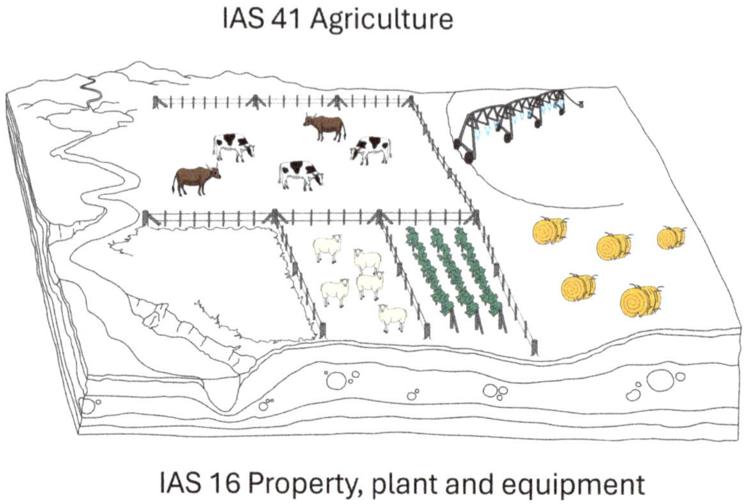

Figure 2 An illustration of the present coverage of financial accounting standards for agriculture. Present financial accounting standards already cover crops and livestock and property, plant and equipment. However, the property valuation does not formally or explicitly include detail of the ecosystem assets on the land. Source: Reproduced with permission from *Farming for the Future*.

In financial accounts, assets are defined as a present economic resource controlled by an entity because of past events. This control confers a right to potentially produce economic benefits (AASB, 2022). Decisions on recognition of assets on an entity's statement of financial position (i.e. balance sheet) require that such recognition conveys useful information about the asset and any resultant income, expenses or changes in equity of the entity, that the recognition is relevant to the user of the information and that the asset can be measured in a manner that provides a faithful representation of the asset and any resultant changes in the financial position of the entity (AASB, 2022). These characteristics of useful information apply equally to decisions to recognise natural capital assets, and it is equally important that this rigor is applied in the recognition of natural capital assets particularly if natural capital accounting is to inform and or be integrated into the entity's financial statements (Fig. 4).

The conceptualisation of the equivalent of a statement of financial position for natural capital, i.e. the ecological balance sheet, has been articulated in detail by Ogilvy (2015). Examples of natural capital accounts and environmental profit and loss statements prepared for individual businesses are emerging (Forico, 2020; Ogilvy et al., 2022), suggesting that broader adoption of financial accounting concepts and standards for the presentation of useful information on natural capital could enhance decision-making both within the organisations and for stakeholders with an interest in the financial and environmental position of the entity (e.g. lenders and buyers of agricultural produce).

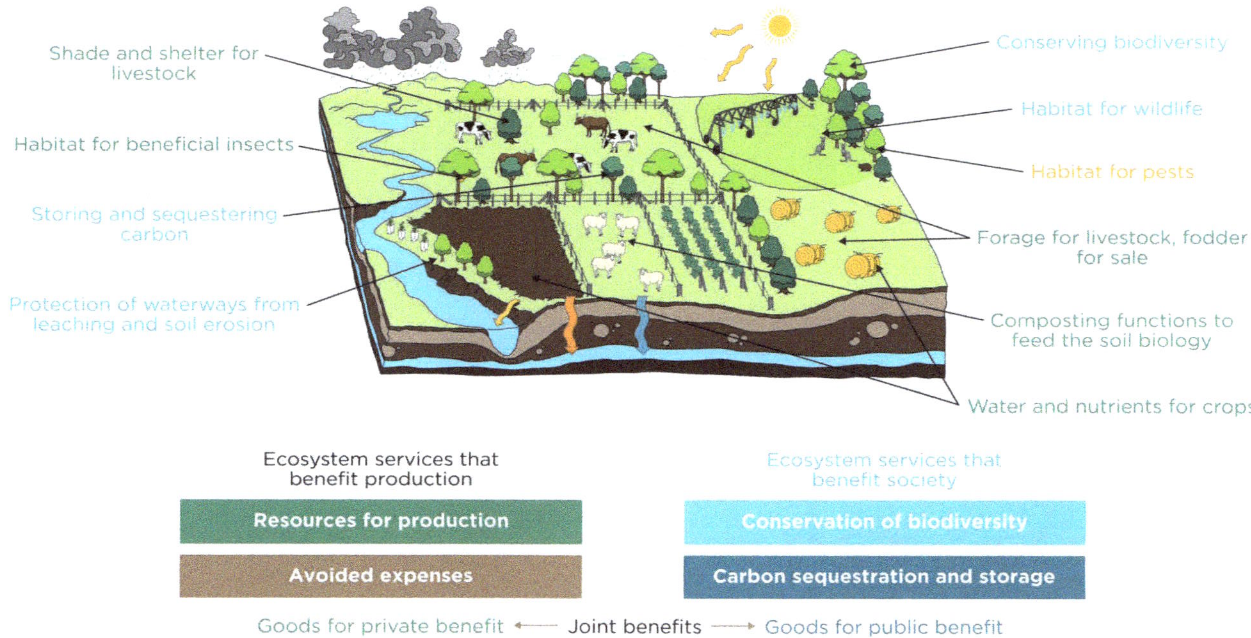

Figure 3 A natural capital accounting perspective of an agricultural enterprise that illustrates the ecosystem services – public and private economic goods – being delivered by different ecosystem assets of the property. Joint benefits indicate the maintenance of ecosystem assets for private benefit that also produces public benefit. Source: Reproduced with permission from *Farming for the Future*.

Figure 4 Temperate agroforestry can build natural capital on farm, generating a range of ecosystem services that increase both the productivity and amenity of temperate farming systems. Photo credits (a, b, c). A. Gray, (d) A. O'Grady, (e). S. Brooks.

Here, we present an example of how the value being generated by the trees on a farm – the private benefits received by the farming business could be presented on a farm business balance sheet using approaches to valuation already commonly used by farm accountants (Table 2). Where:

- Natural capital assets are represented as long-term productive ecological resources (in this case the trees) attached to the land controlled by the entity, and where changes in the composition or condition of these resources can enhance or diminish economic services to the entity (Ogilvy, 2015).
- Natural capital liabilities represent unmet nature-based obligations, e.g. to restore degraded or overused assets to a predetermined condition. A liability is recognised in the financial accounts when satisfaction of the obligation is expected to result in an outflow of economic resources for the entity (Ogilvy et al., 2018; Horner et al., 2022).
- Natural capital equity is the difference between the value of the entity of its natural capital assets and natural capital liabilities (Smith et al., 2023).

Table 2 Example presentation of an ecologically extended financial statement that 'itemises' the ecosystem assets of a farm

Ecologically extended financial balance sheet entries	Notes	2015	2020
Assets			
Agricultural assets livestock (fair value)	1	$1 517 091	$1 548 366
Agricultural assets – trees – timber (fair value)	2	$15 600	$76 825
Agricultural assets – land (market value)	3	*$4 898 217*	$5 284 884
Residual – amenity values	4	$1 130 358	$1 219 589
Agricultural assets – ecosystems (value-in-use)	5		
Ecosystems – paddocks for crop and livestock production	6	$3 438 528	$3 509 412
Ecosystem – trees – reference woodland biodiversity stewardship	7		$200 000
Ecosystem – trees – storm refuge for livestock	8	$252 907	$252 907
Ecosystem – trees – shelter for crops/pasture	9	$34 100	$67 320
Ecosystem – trees – carbon sequestration	10	$26 677	$20 007
Ecosystem – trees – pollination	11	$15 648	$15 648
Ecosystem – trees – amenity		$244 910	$264 244
Sub-total ecosystems	12	*$3 767 590*	*$4 065 295*
Total assets (less market value of land)	**13**	**$6 415 590**	**$6 833 249**

Per normal accounting practice. The increase in value is based on an increase in the number of livestock. Source: Reproduced with permission from EcoFarmCo NC Reports and Accounts prepared by Integrated Futures Pty Ltd. (unpublished).

To demonstrate how these values may appear in a financial balance sheet for a farm business, we have prepared a set of values. The values presented are for a hypothetical farm business and, following the convention for financial accounting, the methods used are explained in notes to the accounts. To provide a context for the relative economic contribution that the trees might be providing to the business, some values of the types of assets that are normally seen on a farm balance sheet according to current accounting standards are provided. However, methods for valuing ecosystem assets of a farm business are still emerging and experimental. The illustrations provided here to do not imply that these have been finalised in accounting standards. The hypothetical farm business is EcoFarmCo which is managed by Jill and Joe Einstein. They started accounting for their natural capital in 2015 at which time they realised changes to the ecosystem assets of the farm would assist them in decreasing their greenhouse gas emissions and increasing carbon storage and biodiversity of the farm. Table 2 sets out these changes in an experimental ecologically extended financial balance sheet.

Notes to Table 2:

1. Under the requirements of the present accounting standards for agriculture, the value of livestock as assets of a farm business is already normally included on the farm balance sheet. The values are estimated at the Fair Value of the livestock.
2. Under the requirements of the present accounting standards for timber, the value of standing timber as assets of a farm business can be included on the farm balance sheet. The values are estimated at the Fair Value of the standing timber (sale price less costs to sell). The increase in value between 2015 and 2020 reflects the increase in the size and condition of the timber. These values for the hypothetical farm reflect the value of 52 ha of radiata pine plantings that could be sold as peeler logs in 2015 and as small sawmill rounds in 2020 plus the potential value in 2020 of an additional 40-ha radiata pine block planted in 2016 that could supply a Christmas tree market in 2020.
3. Under the requirements of present accounting standards, the value of the farm property is already included on the farm balance sheet. The value is usually the present market price of the property and is commonly estimated by a property valuer who uses a comparison to recently sold similar properties to provide the estimate.
4. This value is an experimental value and is calculated as the residual of the property value (in note 3) less the sum of the experimental agricultural ecosystem asset values described in notes 6–14 inclusive. It might be interpreted as the value of the amenity of the property,

its proximity to town, the features of the house and garden or other intangible (hedonic) values. Note that this approach to an experimental ecologically extended balance sheet may reveal, via a negative residual value, situations in which the value in use of the individual ecosystem assets may be higher than the present market value of a property.

5 Agricultural assets – ecosystems (value-in-use) is the asset concept developed in this project that may provide a way for the value of agricultural ecosystems to be included on the balance sheet of a farm business. The goal is to 'itemise' and make visible the contribution that ecosystems found on a farm are probably generating for the farm business. This can be thought of as being similar to explaining how many bedrooms and bathrooms are in a house that is being offered for sale. The heading provides visibility that the subordinate line items are a different asset concept.

6 Experimental valuation: net present value (NPV), to reflect the interannual variation of prices and seasonal condition of net cash flows from livestock and crop production. The valuation method for the 2200 ha of grazed area, pastures and croplands of the farm is consistent with the methods used in financial accounting when assets are valued for what they produce for the business, their 'value in use' rather than what they would be worth if they were sold. In this example, the value is estimated from the net income or amount of money that would be earned if the farm business was paying for access to the land for livestock grazing (known as agistment in agriculture) (Ogilvy, 2018) . The farm business could sustainably carry 12 040 sheep in 2015 and was able to increase this to 12 289 sheep in 2020 because of improvements it made to the condition of its grazing areas. The estimates use a daily agistment rate of $0.30 and assume that the net income to the farm business is 0.5 of this. The NPV uses a 10-year period and a risk-adjusted rate of 0.14 (Ogilvy, 2018).

7 Experimental valuation: the hypothetical farm business was able to secure a contract for environmental stewardship payments of $20 000 per year for 10 years. This was possible due to the excellent condition of the grassy woodlands of the property. The valuation of this in 2020 is at the total value of the contract at its inception. The asset value would be decremented by $20 000 per year until the value reached zero and this asset would be removed from the balance sheet.

8 Experimental valuation: NPV of avoided loss of livestock in severe weather. It is well known that significant stands of trees can provide shelter from severe weather and avoid significant losses of livestock. The estimates here are based on the availability of trees that are suitably arranged to provide a storm refuge. The estimate here is based on the

avoided loss of 10% of lambs and adult sheep worth an average value of $63.00 net income per head. The NPV uses a 10-year period and a risk-adjusted rate of 0.14.

9 Experimental valuation: In 2015, the value in this case study is based on the cost to acquire or cost to establish 31 ha of shelter belts. While this might normally be treated as an expense in farm accounting, it is not inconsistent with the methods of valuing many other assets (such as cars and tractors) which are commonly initially recorded at historical cost and then depreciated. The 2020 value is the NPV of the additional pasture and crop production attributable to the shelter belts. Described earlier in this chapter, there is a zone near trees where crops and pastures tend to grow more vigorously (be more productive) due to the shelter from wind. The estimates provided here for 2020 multiply the area sheltered by the trees as they grow with the average amount of the additional productivity estimated by scientific models. We used 0.7%. This amount is multiplied by the value of the additional pasture, in this case study we use $0.30 per day consistent with the valuation logic used in 7. The NPV uses a 10-year period and a risk-adjusted rate of 0.14.

10 Experimental valuation based on the consumption of expected total economic value from establishing a block of trees for a contract for ACCU production. Standard accounting methods have been applied assuming that the asset in 2015 has its greatest potential value due to its ability to store carbon. The quantity of inflows of economic value to the business from this asset diminishes as it approaches the point at which it cannot store further carbon.

11 Experimental valuation: NPV of providing 'agistment' for bees. Beekeepers who supply pollination services to orchards and crops in spring and summer commonly seek places to put bees over winter so they can recover in numbers and health. Trees can provide very valuable floristic resources that are suitable for the overwintering of bees and landholders commonly agree to allow beekeepers to place hives on their farms over winter. Most commonly, this is done for free (or in exchange for honey). If the landholder was being paid an amount of money for this 'bee agistment' service, they might wish to recognise the value of these assets on their balance sheet. This experimental value uses the same valuation concepts as described in notes 7 and 9 using a net annual cash flow of $3000.00, a 10-year period and a risk-adjusted rate of 0.14.

12 Sub-total ecosystems is the sum of the ecosystem asset valuations.

13 The total asset valuation is the sum of the assets excluding the land (market value).

Newly emerging, standards and conventions being developed for natural capital accounting can help a farm accountant explicitly include the contributions that trees make to the productivity, profitability and resilience of a farm business. People who are making economic decisions about changes to the land including farmers, financiers and markets may use this information to work together to improve farmers' ability to access financial capital to enable them to invest in trees on farms.

4 How are accounts viewed by stakeholders and what is the pathway to adoption

Natural capital accounting could help to shift the globe towards more sustainable and resilient management of the range of natural and production landscapes, especially if it is adopted broadly across all sectors of the economy. To achieve this transformation, farm accountants and advisors are especially influential stakeholders to drive the early stages of development and implementation of natural capital accounts. About 55% of Australia's natural capital is managed by the agricultural sector, comprised of thousands of small businesses (ABARES, 2023). These small businesses frequently work with agronomists, advisors and accountants to prepare farm management plans, apply for funding and loans and report to stakeholders. As well as this, stakeholders in the broader fields of finance and accounting, agriculture, forestry, government, community, conservation, science and research are also stakeholders in the adoption pathway of natural capital accounts (Fleming et al., 2022). How natural capital accounts are perceived by these stakeholders is an important factor in whether they will be regarded as relevant, useful, cost effective and practical.

The concept of natural capital can be variably understood by stakeholders and some may already consider natural capital in a range of ways in their decisions (Fleming et al., 2019). Consistent ways to measure, include and represent natural capital in accounts can be difficult to produce, because of the different objectives, scales and access to data that different decision-makers require. However, as digital technology and scientific understanding improve, ways to understand, measure and calculate natural capital are improving and natural capital accounts are now more feasible to prepare (Ogilvy et al., 2022). At the same time, there is increasing demand for natural capital information at both local and aggregated scales to help monitor progress on natural restoration initiatives, resilience and sustainability assessments and national commitments to global agreements such as the Paris Agreement, the Convention on Biological Diversity and Sustainable Development Goals, as discussed above (Dasgupta, 2021).

Previous work that interviewed over 60 stakeholders has identified that they perceived natural capital accounting to be a helpful method to consider agricultural ecosystem assets as part of a farm business (Fleming et al., 2022). Accounts can be useful to transparently and consistently monitor and report on farm condition to identify opportunities for investment and improvement and track progress over time as well as to identify and manage risks. Natural capital accounts can be used to report to banks, insurance companies, investment companies, conservation groups, government and the public and new buyers.

To be practical for end users, natural capital accounts need to be simple to understand. In addition, they need to use reliable, consistent methods, that resist manipulation, be cost effective and be supported by financial services and insurance providers (Fleming et al., 2022). Natural capital accounting is relatively new, so stakeholder engagement is still emerging, and there is a need (and opportunity) for the implementation pathways of natural capital accounts to be co-designed and co-developed with end users across a range of sectors.

Digital technology continues to develop quickly, and this will impact on how natural capital accounts are produced and maintained. Developments in tools and methods to collect data, as well as software systems that can record and present data, are both opportunities to further streamline, reduce cost and aid the interpretation of natural capital accounts (Stewart et al., 2022). While there are opportunities for efficiency and standardisation improvements from digital technology, there are also considerations of equity (access to data), legal considerations (data ownership and use), skills and training (data collection and interpretation) and methods for maintaining accuracy and rigour that need to be managed as part of the shift towards increasing the use of digital technology (Eastwood et al., 2019; Jakku et al., 2019).

There is a real opportunity for natural capital accounts to be used at a range of scales, from individual entities through to regional and national accounts, to achieve different economic, environmental and social objectives. Involving a wide range of end users in the development of natural capital accounts is an important way that diverse use cases can be achieved from the same set of accounts (Kujala and Kauppinen, 2004). Natural capital accounts as a way to bring people together across sectors, disciplines and world views is an area of work that is yet to be explored, but there is potential for the accounts to work as a convergence point between many different points of view, such as conservation and industry groups, indigenous and western approaches, and national and local scale planning (Franco-Torres et al., 2020).

The range of potential uses of natural capital accounts means that coordination is required to direct and align the different approaches to development, use and implementation and to establish support and information networks. While it can be a difficult job to be aware of all of the different projects, initiatives, tools and approaches being used to develop natural capital accounts in Australia, let alone internationally, knowledge brokering, coalitions and national roadmaps are important to avoid unnecessary confusion and competition amongst stakeholders (Green, 2022).

It remains to be seen whether natural capital accounts are adopted alongside business-as-usual approaches to farm management, regional monitoring and national reporting or whether the use of natural capital accounting underpins a more transformative shift towards new business models for responsible investment in natural landscape management and restoration (Fleming et al., 2022). However, awareness of natural capital accounting and stakeholder perceptions of the relevance and useability of natural capital accounting is only one side of the adoption pathway, the other side is the momentum that arises from the outcomes that the use of natural capital accounts produces. How the accounts result in behaviour change and tangible outcomes to economic productivity, sustainability and resilience of farms and natural landscape management will likely determine how far natural capital accounts are integrated into the society. To be broadly adopted, natural capital accounts cannot become an administrative burden on farmers or be seen as a process that farmers are forced to comply with, but rather as a means to achieving other directly beneficial outcomes.

Natural capital accounting can support and improve accountability and responsibility over landscape management decisions, across sectors and scales (Fleming et al., 2023). It offers a way to achieve transparent record keeping over what natural ecosystem assets exist, their extent and condition and support decision-making, planning, risk analysis, investment and monitoring over time, enabling a range of people, not just individual landowners, to improve natural capital extent and condition. This requires a shift in a range of decision-maker roles, responsibilities and processes, towards evidence-based decisions, accountability for outcomes and new forms of green investment, as well as better coordination among and between sectors that have a role in environmental governance, which are the aspirations, if not yet the achievements, of global agreements like the Sustainable Development Goals (Biermann et al., 2022).

5 Discussion: the role of natural capital accounting

Natural capital will increasingly feature in the way the business is conducted in the coming decades. Post the United Nations' Convention on Climate Change and with the recent ratification of the Kunming-Montreal Global Biodiversity Framework at COP15, all parts of the economy are being urged to be proactive in the adoption of actions to reduce the key drivers of nature loss: land use change, resource use, pollution, invasive species and climate change (TNFD, 2023). Due to its fundamental role as the foundation of many global supply chains, and stewardship of extensive and diverse natural capital, agriculture will play an important role in this global transition with agroforestry featuring as a key strategy to drive this transition (The Nature Conservancy, 2024). The concept of nature-positive agriculture is emerging quickly (OECD, 2016) However, the evidence base for many of the key benefits associated with agroforestry is limited, a situation made even more complex by the highly site and context-specific nature of many ecosystem services. This has tended to constrain the adoption of agroforestry as the value proposition is not always obvious (Fleming et al., 2019).

As with financial accounting, natural capital accounting should serve two broad objectives. Primarily natural capital accounting should provide information that is useful to the management of the entity. As such it provides a mechanism for tracking the entity's stewardship of natural capital in a manner that facilitates proactive management and provides evidence that investments in natural capital are generating desired returns to the entity. In this respect, natural capital accounting shares many similarities with financial accounting and can provide the required information for the entity to assess its position in relation to its goals. One of the barriers to the uptake of agroforestry is a lack of empirical evidence in relation to its benefits (Fleming et al., 2023). Recognition of both the private and public benefits associated with agroforestry in a formal accounting sense can enable the entity's management to monitor and track their agroforestry assets and if adopted at scale can be used to develop and refine practices that enable farmers to optimise the configuration and the amount of agroforestry on the farm to better meet the entity's financial and environmental objectives.

Secondly, natural capital accounting should assist producers in communicating their natural capital stewardship and/or the benefits associated with their investments in natural capital. There is a rapidly growing demand for broader disclosure of environmental credentials, driven by initiatives such that the Taskforce on Nature-related Financial Disclosures (TNFD, 2023). While primarily targeted at large corporate organisations, the adoption of the whole of supply chain perspectives may increase pressure on producers to also adopt reporting systems that meet the needs of actors within their supply chains. Furthermore, there is growing evidence that financial institutions will increasingly price natural capital risk into their lending practices. Producers that can demonstrate good natural capital stewardship may be eligible for more favourable terms (EY, 2023). Natural capital assets, e.g. agriculture and forestry, are becoming more attractive to institutional investors as they look to de-risk their portfolios. Natural capital accounting at enterprise scales will be fundamental to building the trust in these emerging markets and mechanisms designed to address the global scale challenges of climate change and nature loss (Nicolle, 2023).

Despite this, challenges remain for the broader adoption of natural capital accounting. As an information system, natural capital accounting is dense and is still technically complex. Agroforestry in particular can require high resolution data at fine spatial scales and the capacity to cost-effectively quantify ecosystem services at high spatial and temporal resolution is constrained. To some extent, demand for information on natural capital is outstripping the capacity to deliver, particularly in a format that satisfies the characteristics of useful information required for more formal integration into accounts. Despite this, our work demonstrates the capacity to better value and account for the benefits of agroforestry already exists. Technological improvements in data capture and interpretation are advancing rapidly, as are approaches for compiling accounts, such as digital twins (Purcell and Neubauer, 2023; Peladarinos et al., 2023). These should bring down the costs of adoption quickly and enable greater uptake by producers. This is particularly important for ensuring equity amongst producers, especially in relation to market access as customers consciously adjust procurement policies to reflect rising market expectations to address global sustainability challenges.

In addition to tangible benefits, natural capital provides a wide range of ecosystem services benefits of the non-market nature, such as cultural services. Lack of effort to account for non-market benefits of natural capital and evaluate these benefits in monetary terms create information asymmetry which can lead to inefficient decisions and ultimately market failure. As such, to ensure the sustainable use of natural capital in the future, we must also reward the production of non-market ecosystem services, as well as, the production of marketed commodities of our natural capital (Guerry et al., 2015).

6 Conclusion and future trends

Agroforestry has been identified as one of the key nature-based solutions for tackling our dual global crises of climate change and nature loss, with the UNEP (UNEP, 2022) suggesting that agroforestry is the nature-based solution requiring the biggest investment, to the tune of US$3600B between 2022 and 2050, to keep global warming to 1.5°C. However, frameworks to establish, manage and maintain agroforestry to best meet these global imperatives are still in the development and adoption by land managers is even further away. Natural capital accounting offers a key solution to achieve best practice agroforestry and encourage adoption through explicit recognition of the public and private goods from agroforestry systems. The capacity to integrate benefits into enterprise planning is needed to accelerate the adoption of agroforestry.

As pressure on the environment grows, land managers and society are becoming increasingly aware of the need to manage environmental values, to protect and to restore them (Dasgupta, 2021), though environmental values often compete with production values. Agroforestry offers an option to minimise and manage the tradeoffs between the increasing societal demands for food, fibre, bioproducts and energy, as well as contributing to meeting the environmental and climate change outcomes that are needed from our landscapes. Establishment of agroforestry for single outcomes, such as wood production, is generally more expensive than from monocultures or extant forests due to the diversity of the system and additional management costs such as fencing and logistics. Agroforestry becomes a much more attractive system when multiple values are quantified and tracked in the enterprise balance sheet.

For the land manager, natural capital accounting can explicitly quantify and demonstrate the delivery of private and public benefits from agroforestry, including increased returns from agricultural production from sheltered paddocks, greater diversification and increased land values and resilience to extreme weather, increasing farm productivity, profitability and value. Natural capital accounting also explicitly quantifies the environmental benefits including addressing climate change and providing habitat for biodiversity. Having these values explicitly quantified, tracked and monitored further strengthens the case for, and adoption of, agroforestry, through supporting and improving accountability and responsibility over landscape management decisions, across sectors and scales (Fleming et al., 2023). It offers a way to achieve transparent record keeping over what natural ecosystem assets exist, their extent and condition and support decision-making, planning, risk analysis, investment and monitoring over time, enabling a range of people, not just individual landowners, to improve natural capital extent and condition. Robust natural capital accounting will assist in developing trust in producers' environmental credentials.

Agroforestry provides an excellent test case for the development of natural capital accounting, which is required to optimise the value of agroforestry and to recognise the many benefits needed for its business case to succeed. The natural value accounting methods developed will be relevant to other forms of land management and may assist decisions on land management as well as infrastructure developments in landscapes to optimise the provision of resources for society and the simultaneous protection of the environment.

Agriculture is operating in an increasingly complex market and where producers are likely to be required to balance production and environmental outcomes to maintain market access or to capture emerging opportunities in a carbon constrained and nature positive operating environment. Accountability is defined as 'the act of being responsible for what you do and able to give a satisfactory reason for it, or the degree to which this happens'. Natural capital accounting provides the model for underpinning this accountability.

7 Acknowledgement

This work was funded through Round 2 of the Australian Government's National Landcare Program Smart farming program. Our long-term partners in our agroforestry program include Private Forests Tasmania, Greening Australia and the University of Tasmania.

8 References

AASB 2015. *Agriculture*. Victoria, Australia: Australian Accounting Standards Board.

AASB 2022. *Conceptual Framework for Financial Reporting*. Victoria, Australia: Australian Accounting Standards Board.

ABARES 2023. *Snapshot of Australian Agriculture*. Canberra: ABARES Insights.

Asah, S. T., et al. 2014. Perception, acquisition and use of ecosystem services: Human behavior, and ecosystem management and policy implications. Ecosystem Services, 10, 180–186. doi:https://doi.org/10.1016/j.ecoser.2014.08.003

Báder, M., et al. 2023. The effect of agroforestry farming on wood quality and timber industry and its supportation by Horizon 2020. *Agroforestry Systems*, 97, 587–603.

Baker, T. P., et al. 2018. Impacts of windbreak shelter on crop and livestock production. *Crop and Pasture Science*, 69, 785–796.

Baker, T. P., et al. 2021. Temporal, environmental and spatial changes in the effect of windbreaks on pasture microclimate. *Agricultural and Forest Meteorology*, 297, 108265.

Barker, R. 2019. Corporate natural capital accounting. *Oxford Review of Economic Policy*, 35, 68–87.

Barton, J. & Pretty, J. 2010. What is the best dose of nature and green exercise for improving mental health? A multi-study analysis. *Environmental Science & Technology*, 44, 3947–3955.

Bastin, J.-F., et al. 2019. The global tree restoration potential. *Science*, 365, 76–79.

Battaglia, M., et al. 2004. Cabala: a linked carbon, water and nitrogen model of forest growth for silvicultural decision support. *Forest Ecology and Management*, 193, 251–282.

Benton, T. G., et al. 2021. *Food System Impacts on Biodiversity Loss: Three Levers for Food System Transformation in Support of Nature*. London: Chatham House.

Bentrup, G., et al. 2019. Temperate agroforestry systems and insect pollinators: a review. *Forests*, 10, 981.

Biermann, F., et al. 2022. Scientific evidence on the political impact of the sustainable development goals. *Nature Sustainability*, 5, 795–800.

Bowler, D. E., et al. 2010. A systematic review of evidence for the added benefits to health of exposure to natural environments. *BMC Public Health*, 10, 456.

Brown, S. E., et al. 2018. Evidence for the impacts of agroforestry on agricultural productivity, ecosystem services, and human well-being in high-income countries: a systematic map protocol. *Environmental Evidence*, 7, 24.

Centeno-Alvarado, D., et al. 2023. Fostering pollination through agroforestry: a global review. *Agriculture, Ecosystems & Environment*, 351, 108478.

Cleugh, H. A. 1998. Effects of windbreaks on airflow, microclimates and crop yields. *Agroforestry Systems*, 41, 55–84.

Cleugh, H. A. & Hughes, D. E. 2002. Impact of shelter on crop microclimates: a synthesis of results from wind tunnel and field experiments. *Australian Journal of Experimental Agriculture*, 42, 679–701.

Convention On Biological Diversity 2022. Decision adopted by the conference of the parties to the Convention on Biological Diversity 15/4. Kunming-Montreal Global Biodiversity Framework.

Dasgupta, P. 2021. *The economics of biodiversity: the Dasgupta review*. London: HM Treasury.

de Groot, R. & Ramakrishnan, PS. 2005. Cultural and Amenity Services. Ecosystems and Human Well-Being: Current State and Trends. Findings of the Condition and Trends Working Group. Millennium Ecosystem Assessment (Island Press, Washington, DC), pp. 455–476.

Dicks, L. V., et al. 2021. A global-scale expert assessment of drivers and risks associated with pollinator decline. *Nature Ecology & Evolution*, 5, 1453–1461.

Dudley, N. & Alexander, S. 2017. Agriculture and biodiversity: a review. *Biodiversity*, 18, 45–49.

Eastwood, C., et al. 2019. Managing socio-ethical challenges in the development of smart farming: from a fragmented to a comprehensive approach for responsible research and innovation. *Journal of Agricultural and Environmental Ethics*, 32, 741–768.

England, J., et al. 2018a. *Assessing the CABALA model for predicting shelterbelt productivity – initial stand volume comparisons*. Report prepared for Private Forests Tasmania. Melbourne: CSIRO Land and Water.

England, J., et al. 2018b. *Assessing the CABALA model for predicting shelterbelt productivity-initial stand volume comparisons*. Report prepared for Private Forests Tasmania. Australia: CSIRO Land and Water.

England, J. R., et al. 2020. Trees on farms to support natural capital: an evidence-based review for grazed dairy systems. *Science of The Total Environment*, 704, 135345.

Ey 2023. Banking for nature: aligning Australian banking with the global biodiversity framework. Australian Conservation Foundation and Ernst & Young.

Fleming, A., et al. 2022. Improving acceptance of natural capital accounting in land use decision making: barriers and opportunities. *Ecological Economics*, 200, 107510.

Fleming, A., et al. 2019. Understanding the values behind farmer perceptions of trees on farms to increase adoption of agroforestry in Australia. *Agronomy for Sustainable Development*, 39, 9.

Fleming, A., et al. 2023. Increasing trees on farms through natural capital accounting. *Rural Extension and Innovation Systems*, 19, 1–8.

Forico 2020. *Natural Capital Report 2020 of the Tasmanian Forest trust for the year ended 30 June 2020*. Launceston, Tasmania: Forico Pty Ltd.

Franco-Torres, M., et al. 2020. A framework to explain the role of boundary objects in sustainability transitions. *Environmental Innovation and Societal Transitions*, 36, 34–48.

Giro, A., et al. 2019. Behavior and body surface temperature of beef cattle in integrated crop-livestock systems with or without tree shading. *Science of The Total Environment*, 684, 587–596.

Grala, R. K., et al. 2012. Willingness to pay for aesthetics associated with field windbreaks in Iowa, United States. *Landscape and Urban Planning*, 108, 71–78.

Green, I. 2022. *Natural capital accounting as a farm management tool, barriers and opportunities*. Masters of Applied Science, University of Tasmania.

Gregory, N. G. 1995. The role of shelterbelts in protecting livestock: a review. *New Zealand Journal of Agricultural Research*, 38, 423–450.

Guerry, A. D., et al. (2015). Natural capital and ecosystem services informing decisions: From promise to practice. *Proceedings of the National Academy of Sciences*, 112(24), 7348–7355. doi:10.1073/pnas.1503751112

Haines-Young, R. & Potschin-Young, M. B. 2018. Revision of the Common International Classification for Ecosystem Services (CICES V5.1): a policy brief. *One Ecosystem*, 3. https://doi.org/10.3897/oneeco.3.e27108

Hajkowicz, S. & Moody, J. 2010. *Our future world. An analysis of global trends, shocks and scenarios*. Canberra: CSIRO.

Hein, L., et al. 2016. Defining ecosystem assets for natural capital accounting. *PLoS One*, 11, e0164460.

Helliwell, R. 2008. Amenity valuation of trees and woodlands. *Arboricultural Journal*, 31, 161–168.

Hølleland, H., et al. 2017. Cultural Heritage and Ecosystem Services: A Literature Review. Conservation and Management of Archaeological Sites, 19(3), 210-237. doi:10.1080/13505033.2017.1342069

Horcea-Milcu, I., et al. 2013. Cultural Ecosystem Services: A Literature Review and Prospects for Future Research. Ecology and Society, 18. doi:10.5751/ES-05790-180344

Horner, C. & Davidson, N. J. 2020. Accounting for biodiverse wildlife corridor plantations. *Meditari Accountancy Research*, 29(3), 502–523.

Horner, C., et al. 2022. Accounting and accountability for no net loss of biodiversity. Australian Accounting Standards Working Paper Series, Working paper No. 22-04

Houdet, J., et al. 2020. Adapting double-entry bookkeeping to renewable natural capital: an application to corporate net biodiversity impact accounting and disclosure. *Ecosystem Services*, 45, 101104.

IPBES 2016. *The assessment report of the Intergovernmental Science-Policy Platform on Biodiversity and Ecosystem Services on pollinators, pollination and food production*. S. G. Potts, et al. (ed.). Bonn, Germany: Secretariat of the Intergovernmental Science-Policy Platform on Biodiversity and Ecosystem Services.

IUCN 2020. *Guidance for using the IUCN Global Standard for Nature-based Solutions. A user-friendly framework for the verification, design and scaling up of Nature-based Solutions*. First edition ed. Gland, Switzerland: International Union for the Conservation of Nature and Natural Resources.

Jakku, E., et al. 2019. "If they don't tell us what they do with it, why would we trust them?" Trust, transparency and benefit-sharing in smart farming. *NJAS - Wageningen Journal of Life Sciences*, 90–91, 100285.

Jones, M. J. 1996. Accounting for biodiversity: a pilot study. *British Accounting Review*, 28, 281–303.

Jose, S. 2009. Agroforestry for ecosystem services and environmental benefits: an overview. *Agroforestry Systems*, 76, 1–10.

Joseph, J., et al. 2020. A spatially extended model to assess the role of landscape structure on the pollination service of Apis mellifera. *Ecological Modelling*, 431, 109201.

King, S., et al. 2021. Linking biodiversity into national economic accounting. *Environmental Science & Policy*, 116, 20–29.

Kort, J. 1988. Benefits of windbreaks to field and forage crops. *Agriculture, Ecosystems & Environment*, 22–23, 165–190.

Kujala, S. & Kauppinen, M. 2004. Identifying and selecting users for user-centered design. *Proceedings of the third Nordic conference on Human-computer interaction*. Tampere, Finland: Association for Computing Machinery.

Lin, B. B., et al. 2018. Understanding the benefits of green infrastructure in new developments: a review of green infrastructure, ecosystem services, and economic analysis frameworks ACT. Australia: CSIRO.

Lonsdorf, E., et al. 2009. Modelling pollination services across agricultural landscapes. *Annals of Botany*, 103, 1589–1600.

Marais, Z. E., et al. 2022. Shelterbelt species composition and age determine structure: consequences for ecosystem services. *Agriculture, Ecosystems & Environment*, 329, 107884.

Mendham, D. 2018. Modelling the costs and benefits of agroforestry systems–application of the Imagine bioeconomic model at four case study sites in Tasmania. CSIRO Canberra.

Millennium Ecosystem Assessment. 2005. *Ecosystems and human well-being: synthesis*. Washington, DC: Island Press.

Mokany, K., et al. 2022. Habitat-based biodiversity assessment for ecosystem accounting in the Murray–Darling Basin. *Conservation Biology*, 36, e13915.

Monckton, D. & Mendham, D. S. 2022. Maximising the benefits of trees on farms in Tasmania – a desktop review of investment opportunities to improve farm enterprise productivity, profitability and sustainability. *Australian Forestry*, 85, 6–12.

Nair, P. K. R. 1993. *An introduction to Agroforestry*. Dordrecht: Kluwer Academic Publishers.

Nair, P. K. R. 2012. Climate Change Mitigation: A Low-Hanging Fruit of Agroforestry. In: P. K. R. Nair & D. Garrity (eds.) *Agroforestry - The Future of Global Land Use*. Dordrecht: Springer Netherlands.

Natural Capitals Coalition. 2016. Natural Capital Protocol. Retrieved from https://naturalcapitalcoalition.org/natural-capital-protocol/

Nicolle, W. 2023. *Taking root: mainstreaming natural capital accounting to meet global biodiversity goals*. R. Morrow (ed.). UBS sustainability and Impact Institute.

O'grady, A. P. & Mitchell, P. J. 2018. *Agroforestry: realising the triple bottom line benefits of trees in the landscape*. Hobart, Tasmania: CSIRO.

OECD. 2016. *Farm management practices to foster green growth*. Paris: OECD.

Ogilvy, S. 2015. Developing the ecological balance sheet for agricultural sustainability. *Sustainability Accounting, Management and Policy Journal*, 6, 110–137.

Ogilvy, S. 2018. Standards-compliant accounting valuations of ecosystems. *Sustainability Accounting, Management and Policy Journal*, 9, 98–117.

Ogilvy, S. 2020. *Toward a methodology for incorporating ecological capital into the accounts of individual entities*. PhD, Australian National University.

Ogilvy, S., et al. 2018. Accounting for liabilities related to ecosystem degradation. *Ecosystem Health and Sustainability*, 4, 261–276.

Ogilvy, S., et al. 2022. A natural capital accounting framework to communicate the environmental credentials of individual wool-producing businesses. *Sustainability Accounting, Management and Policy Journal*, 13, 765–802.

Ollerton, J., et al. 2011. How many flowering plants are pollinated by animals? *Oikos*, 120, 321–326.

Paul, K. I. & Roxburgh, S. H. 2020. Predicting carbon sequestration of woody biomass following land restoration. *Forest Ecology and Management*, 460, 117838.

Paul, K. I., et al. 2022a. Carbon dynamics in tree plantings: how changes in woody biomass impact litter and soil carbon. *Forest Ecology and Management*, 521, 120406.

Paul, K. I., et al. 2013. Economic and employment implications of a carbon market for integrated farm forestry and biodiverse environmental plantings. *Land Use Policy*, 30, 496–506.

Paul, K. I., et al. 2016. Testing the generality of above-ground biomass allometry across plant functional types at the continent scale. *Global Change Biology*, 22, 2106–2124.

Paul, K. I., et al. 2022b. Sequestration of carbon in commercial plantations and farm forestry. *Trees, Forests and People*, 9, 100284.

Peladarinos, N., et al. 2023. Enhancing smart agriculture by implementing digital twins: a comprehensive review. *Sensors*, 23, 7128.

Perennes, M., et al. 2021. A hierarchical framework for mapping pollination ecosystem service potential at the local scale. *Ecological Modelling*, 444, 109484.

Polyakov, M., et al. 2014. Capitalized amenity value of native vegetation in a multifunctional rural landscape. *American Journal of Agricultural Economics*, 97, 299–314.

Porto, R. G., et al. 2020. Pollination ecosystem services: a comprehensive review of economic values, research funding and policy actions. *Food Security*, 12, 1425–1442.

Potts, S. G., et al. 2010. Global pollinator declines: trends, impacts and drivers. *Trends in Ecology & Evolution*, 25, 345–353.

Prober, S. M., et al. 2025. Accounting for the biodiversity benefits of woody plantings in agricultural landscapes: A global meta-analysis. Agriculture, Ecosystems & Environment, 381, 109453. doi:https://doi.org/10.1016/j.agee.2024.109453

Purcell, W. & Neubauer, T. 2023. Digital twins in agriculture: a state-of-the-art review. *Smart Agricultural Technology*, 3, 100094.

Roxburgh, S. H., et al. 2019. A revised above-ground maximum biomass layer for the Australian continent. *Forest Ecology and Management*, 432, 264–275.

Sharp, R., J., et al. 2020. Invest 3.10.2.post17 User's Guide. The Natural Capital Project, Stanford University, University of Minnesota, The Nature Conservancy, and World Wildlife Fund.

Shaw, K., et al. 2006. Exercise for overweight or obesity. *Cochrane Database of Systematic Reviews*, 2006, Cd003817.

Smith, G. S., et al. 2021. Materiality assessment of natural capital risks in Australian forestry. *Current Forestry Reports*, 7, 282–304.

Smith, G. S., et al. 2023. The natural capital handbook: a practical guide to corporate natural capital accounting, assessment, risk assessment and reporting. CSIRO Environment.

Stern, P. C., Dietz, T., Abel, T., & al., e. 1999. A value-belief-norm theory of support for social movements: The case of environmentalism. Human ecology revew, 81–97.

Stewart, S. & O'grady, A. P. 2020. Experimental natural capital accounts for the forestry sector in the Green Triangle. CSIRO Land and Water.

Stewart, S. B., et al. 2022. Digital tools for quantifying the natural capital benefits of agroforestry: a review. *Land*, 11, 1668.

Tapsuwan, S., et al. 2009. Capitalized amenity value of urban wetlands: a hedonic property price approach to urban wetlands in Perth, Western Australia*. *Australian Journal of Agricultural and Resource Economics*, 53, 527–545.

Telwala, Y. 2023. Unlocking the potential of agroforestry as a nature-based solution for localizing sustainable development goals: a case study from a drought-prone region in rural India. *Nature-Based Solutions*, 3, 100045.

Terasaki Hart, D. E., et al. 2023. Priority science can accelerate agroforestry as a natural climate solution. *Nature Climate Change*, 13, 1179–1190.

The Nature Conservancy. 2024. Natural Allies: nature-based solutions for climate and biodiversity. A report by the Nature Conservancy and the UBS Sustainability and Impact Institute.

TNFD 2023. Recommendations of the taskforce on nature-related financial disclosures. Taskforce on Nature-related Financial Disclosures.

Torralba, M., et al. 2016. Do European agroforestry systems enhance biodiversity and ecosystem services? A meta-analysis. *Agriculture, Ecosystems & Environment*, 230, 150–161.

Ulrich, R. S., et al. 1991. Stress recovery during exposure to natural and urban environments. *Journal of Environmental Psychology*, 11, 201–230.

UNEP. 2022. State of finance for nature. time to act: doubling investment by 2025 and eliminating nature-negative finance flows United Nations Environment Program.

United Nations. 2014. *System of environmental and economic accounting 2012-central framework*. New York: United Nations.

United Nations. 2021. System of environmental-economic accounting-ecosystem accounting (final draft). Department of Economic and Social Affairs-Statistics Division.

Van Der Horst, D. 2006. A prototype method to map the potential visual-amenity benefits of new farm woodlands. *Environment and Planning B: Planning and Design*, 33, 221–238.

Van Noordwijk, M., et al. 2020. People-centric nature-based land restoration through agroforestry: a typology. *Land*, 9, 251.

Villa, F., et al. 2014. A methodology for adaptable and robust ecosystem services assessment. *PLoS One*, 9, e91001.

Wood, S. A., et al. 2023. Mitigating near-term climate change. *Environmental Research Letters*, 18, 101002.

World Economic Forum. 2023. *Global Risks Reports 2023: Key findings*. New York: World Economic Forum.

Xiao, L., et al. 2017. A theoretical framework for researching cultural ecosystem service flows in urban agglomerations. Ecosystem Services, 28, 95–104. doi:https://doi.org/10.1016/j.ecoser.2017.09.014

Part 2

Advances in silvopastoral systems

Chapter 4

Types of silvopastoral system: an overview

Jim McAdam, Queen's University Belfast and Irish Agroforestry Forum, Ireland

1 Definition of silvopastoral systems

Agroforestry is an ancient practice which is widely incorporated globally into land use systems. It is a collective name for land use practices where trees are combined with crops and/or animals on the same unit of land and where there are significant ecological or economic interactions between the tree and the agricultural components (Lundgren and Raintree 1982).

In temperate regions, as the main agricultural land uses are livestock grazing or tillage, the most applicable agroforestry practices are *silvopastoral* systems (where trees are grown in grazed or mown pasture in a regular or varied pattern) and *silvoarable* systems (where trees or shrubs are grown in rows between an arable crop). A silvopastoral system is a land use type which is a three-dimensional complex ecological mix where seasonal and temporal evolution of the system delivers a wide range of ecosystem services. In this chapter, the definition of silvopastoral systems will be considered as those where trees are grown in grazed pasture in a regular or varied pattern. The value of the system lies in its complexity and its capacity to be resilient within an increasingly challenging climatic framework.

Silvopastoral practices, in which trees and/or shrubs are grown in pasture in cool temperate regions – namely in North America, Western Europe, Chile

http://dx.doi.org/10.19103/AS.2024.0141.09

and New Zealand – will be considered in this chapter. There are also substantial areas of Asia (particularly China) where silvopasture is widely practiced.

It is accepted, however, that in these regions, and particularly in Western Europe, many forests are grazed to a greater or lesser degree by wild or domesticated livestock. For example, in most Scandinavian countries, where tree cover is high and pressure on agricultural land is low, the most likely integration of stock and forests will be in the form of forest grazing. These forest or woodlands represent a part of the spectrum of diversity which allows trees to be integrated into farming systems and can come under the banner of silvopasture (see next section).

2 Types of silvopastoral system

In its widest context, silvopastoral systems can be considered as any situation where trees are integrated into a whole farm structure in a way that enables them to have close interaction with the farming components. Hence silvopasture represents a spectrum of activity. At one end, there could be sparse widely spaced trees, at the other, relatively dense woodlands which are opened up for grazing. In between these extremes are, for example, rows of trees widening hedges, riparian buffer strips, un-harvestable parts of fields, natural habitat with sparse amenity trees and shrubs or aesthetic planting around buildings, scattered, frequent or dense trees, etc.

However, administrators and those trying to design fundable, quantifiable systems for financial support and for land use auditing purposes, require some form of structural definition which is relatively easy to compartmentalise and cost in terms of establishment, management adjustments and benefits accrued. Unfortunately natural systems, or even evolved artificial systems, do not easily fit into a standard classification profile. The complexity dimensionality and flexibility of silvopastoralism are not suited to a synthetic classification. It would invariably exclude some tree planting and, more importantly, stifle innovation often based on intergenerational knowledge. There have been many attempts to classify silvopastoral systems and these are still relevant and viable for the administrative function they preform (Dupraz and Newman 1997; Merwin 1997; Escribano et al. 2019; Gabriel 2018; Mosquera-Losada et al. 2009; McAdam 2009).

Accepting the need for formal classification, those promoting silvopasture within the general framework of trees on farms are very keen to have the flexibility of design around the concept and hence a recognition that there is a spectrum of ways in which trees and farming systems can be integrated. Of course to avoid a multidimensional plethora of directions in ways in which this classification can occur and to enable whole-farm planning of how trees might be integrated into farms, some degree of linearity and physical delineation of boundaries needs to occur. It is proposed that the underlying definition and classification could be on the basis of tree density within any area or parcel of farmed land. This linearity in approach leads to a form of synthetic classification system of silvopastoral types encompassing a broad range of uses and benefits (Table 1).

Table 1 Proposed classification of types based on tree density along a broad spectrum of uses and benefits

Types	Description	Principal deliverables
Parklands	Sparse, widely spaced trees on grazed pasture	Biodiversity, shade, animal health and welfare
Hedgerows	Rows of trees widening hedges	Biodiversity, animal fodder, shelter, health and welfare and soil structure
Silvopasture – tree rows	Rotationally grazed paddocks or fields with rows of trees as dividers	Animal fodder, shelter, soil health and biodiversity
Sparse woodland grazing	Natural habitat areas of low-density trees	Biodiversity, animal health and welfare, landscape and timber
Riparian buffers	Usually relatively narrow areas planted with woody plants alongside watercourses	Water quality, biodiversity, fodder, soil health and stability and erosion control
Silvopasture-grazing livestock	Rows of evenly spaced trees or shrubs in pasture suitable for grazing and cutting	Diverse production, flood alleviation, extended grazing, ammonia capture, tree fodder, climate resilience–especially shading and drought– timber and fuel
Silvopasture with poultry	Poultry in relatively densely or sparsely planted trees	Ammonia capture, allows natural animal behaviour and nutrient recycling
Widened hedgerows	Sections of fields unsuitable for machinery harvesting	Diverse production, flood alleviation, extended grazing, ammonia capture and biodiversity
Mixed woodland grazing	Natural habitat areas with trees at higher density	Biodiversity, timber, animal health and welfare and soil health
Aesthetic planting	Trees around buildings	Landscape, heritage, ammonia capture and fuel
Shelterbelts	Mixed species and sizes of trees to give structural diversity and protection from wind and sun	Climate resilience, animal health and welfare, shade, soil health and pasture performance
Silvopasture suited for pigs and conservation grazing cattle	Relatively dense woodlands which are opened-up for grazing.	All livestock-related benefits, biodiversity, soil health and structure

Note: The ecosystem services and other deliverables from these types will be elaborated more fully later.

There are several ways of achieving silvopastoral systems. Two of the most common are (1) re-spacing an established woodland or planted forest by selective thinning and establishing pasture, i.e. silvopasture which has evolved from either a pure forest or forest-grazing scenario, or (2) planting trees into already established pasture. In the latter scenario, these trees will generally have to be protected from grazing initially or the pasture cut for conserved forage until the trees are above browsing height. Hence the system is essentially being covered from an agricultural viewpoint. This need not always be the case of course, the principal aim of some silvopasture may be to produce marketable products such as timber, fruit, energy or fodder. This is perfectly acceptable, as in most cases this will be accompanied by ecosystem services delivery.

3 Ecosystem services delivered by agroforestry systems

Across temperate regions generally, there are concerns over problems with grassland intensification and fragmentation of woodland habitats. From a sustainability perspective, temperate agriculture faces challenges in key areas: sustainable food production to meet increasing demand, water quality, biodiversity loss, greenhouse gas (GHG) emissions, meeting climate change mitigation obligations, tree cover, air quality, soil health, ensuring farm viability and access to markets. Any improvements and delivery of enhancement measures in these areas must be seen against a background of land use capability, financial viability, income security and rural community stability. In this context, the example of applicability of silvopasture to Ireland will be used where appropriate (McAdam 2020).

Sustainable food production to meet demand. Global population continues to rise, and increasingly "western" diets are under scrutiny for their inability to meet basic health and nutritional standards.

Water quality. Eutrophication resulting from excess nutrients (N&P) from farmland is a primary water quality challenge.

Biodiversity. Current agricultural practices place a series of pressures on the natural environment. These environmental pressures include changes to land use, loss of natural habitat – the decline in tree cover has been a major contributor – more stark habitat delineation and isolation within intensive farming systems. These, coupled with other environmental pressures listed, have had a serious impact on biodiversity in agricultural landscapes.

GHG Emissions. Food systems contribute one-third of global anthropogenic GHG emissions, roughly 16.5 GtCO2e year – from a total 54 GtCO2e year – (2020), with both pre- and post-production phases representing a high and increasing share of total emissions. The Food and Agriculture Organization of the United Nations (FAO) estimates that agricultural emissions reached 10.7 GtCO2e year – (in 2019) – while the Intergovernmental Panel on Climate Change Special Report on Climate Change and Lands estimates them at 12.0 GtCO2e year. When including only the emissions up to the farm gate (excluding land use change), the total estimated by FAO reached 7.2 GtCO2e year in 2019, with the principal source being estimated global GHG emissions due to synthetic N fertiliser manufacture, transportation, and field use. In agricultural systems, the synthetic N fertiliser supply chain was responsible for estimated emissions of 1.13 GtCO2e in 2018, representing 10.6% of agricultural emissions and 2.1% of global GHG emissions (Menegat et al. 2022).

Climate change mitigation obligations. All countries have climate change mitigation obligations to meet. Agriculture faces a range (by country) of emissions reduction targets which largely reflect the intensity of farm production in those countries.

Tree cover and forestry. While there is a variation in the metrics and definitions used to quantify hedgerow and non-forest wooded cover, there is pressure on conversion of woodland to intensive farmland in most temperate regions.

Air quality. Ammonia is a powerful air pollutant with significant implications for biodiversity and human health (Holegate et al. 2016). It is the most significant air pollutant in areas with intensive agricultural production (Wyer et al. 2022).

Soil health. Soil pH, structure and organic matter are key indicators of soil health. In fields managed for intensive grassland or tillage production, if soil pH is not within the optimum range, most productive plants will not grow to their potential, nor will they respond fully to fertility amendments. If organic matter is low or the soil structure compacted, the soil's capacity to cycle water and nutrients can be greatly diminished, resulting in either excessive leaching or surface run-off, respectively. In semi-natural upland and lowland pastures, soil conditions need to be maintained at a level which supports the native flora and fauna and maintains and prevents a decline in biodiversity.

Family farm viability. For example, relatively small, family owned and run farms are the backbone of rural life in Ireland. Many of these are on the margins of financial viability and are seeking ways in which they can conform to future climate challenges and emission control obligations while producing wholesome food and remaining viable (McAdam 2020).

Access to markets. People are increasingly concerned about the origins of their food and, in particular, the welfare of animals in farming systems. Growth in the organic retail sector and the increase in vegetarians and vegans are indicators of this.

4 The role of silvopasture in delivering ecosystem services

Listed, and discussed in this section, are the attributes that introducing trees onto farmland can bring to addressing some of these issues. It has been shown that there is a wide spectrum of ways trees can be integrated into farm systems and the proven benefits can be set against a range of tree-pasture scenarios. For example, these options are further explored for Ireland (McAdam 2020).

Delivering sustainable food production. The introduction of wide spaced trees in agroforestry systems can make temperate farmed landscapes more sustainable, deliver a wide range of ecosystem services and align with a sustainable agriculture and forestry land management strategy. Such policies will focus on sustainable levels of livestock output, tightened nutrient management on farms to improve water quality, healthier soil with enhanced carbon sequestration potential, increased tree cover to contribute to habitat heterogeneity, stabilisation of rural communities including the family farm structure and enhancement of biodiversity through a more complex, sustainable lower-input agriculture.

Climate resilience. Farming systems in temperate regions will have to cope with Increased temperatures, storminess and episodic rainfall severity. Trees have the potential to reduce windspeed and evapotranspiration across pasture and provide shade and shelter for stock. Agroforestry has a well proven role in riparian buffers and gives resilience to grazing during extreme rainfall through improved soil permeability (Carrol et al. 2004; Marshall et al. 2014; McAdam 2018; Cole et al. 2020).

Water quality. Trees planted in an agroforestry situation improve soil structure which allows the soil to soak and filter water at much larger volumes (Cole et al. 2020). They have root systems which extend well below the understorey root network and can absorb any excess nutrients which might escape into the lower soil horizons and ultimately end up in the groundwater pool. There are additional benefits from root differentiation, a reduction in leaching losses of nutrients, faster nutrient cycling in the presence of grazing animals and reduced soil erosion.

Biodiversity. Introducing trees into farming systems creates spatial heterogeneity and soil enrichment which greatly enhances the opportunity for a much wider range of biodiversity (Torralba et al. 2016; Valdés et al. 2020), including greater numbers and range of pollinators (Kay et al. 2024; Varah et al. 2020) than would be found under crop or livestock production alone.

GHG emissions and climate change mitigation obligations. Research in Northern Ireland has shown that agroforestry systems established with wide-spaced ash in grazed pasture can sequester up to 3.2 t C/ha/year (Olave et al.

2016). If the sequestration potential of hedgerows is added to this, silvopastoral systems can support carbon-neutral beef production at approximately 2 LU/ha. Hence, there is huge opportunity to offset these by increasing the carbon storage potential from farmland through soils, crops and trees (Kay et al. 2019). This will help meet climate change directives and implement mitigation and resilience strategies.

Tree cover and forestry. Outside the definition of commercial forestry plantations, there are existing trees, scrub and hedgerows which are an important part of the rural landscape. Agroforestry will facilitate tree planting while still allowing productive farming. Planting should allow a greater diversity of species, uses of trees (e.g. fruit/nut), intra-plot variation and ecosystem services delivery. There are opportunities for higher broadleaf targets to be met if that is the underlying objective, and together with skilled silvicultural management, agroforestry is well placed to deliver high quality timber. This has the potential to strengthen local economies and will increase positivity towards trees and instilling a greater forest and woodland culture into communities.

Air quality. Most (92%) of ammonia emitted by livestock is released while they are housed and producing slurry. Agroforestry has been shown to extend the length of the grazing season by up to 15 weeks through improved carrying capacity of the soil and better grassland utilisation (McAdam 2018). This along with other benefits of agroforestry (Bealey et al. 2014; McAdam 2020) and trees in general (Asman 2006) offers nature-based solutions to making a significant impact on ammonia emissions. Research is currently underway to quantify the amount of nitrogen trees can absorb from the atmosphere through their foliage when slurry is spread and when stock graze below them, emitting methane and ammonia.

Soil health. The living roots of healthy plants are key to soil vitality and nutrient cycling. By maintaining optimum pH and increasing legumes, herbs and a diversity of grass species, this broadens the range and depth of rooting zones being utilised. Plant diversity builds soil organic matter and increases microbial diversity which sequesters and retains more carbon in the ground (Olave et al. 2016; Mosquera-Losada 2019). Trees and woody species add another layer of complexity and root depth to the system, cycling nutrients and building fertility lower in the soil horizon. Also, many agricultural soils lack mycorrhizal and other beneficial fungi. Through symbiosis, mycorrhizal fungi can play a huge role in providing plants with nutrients which are more difficult to access. Fungal mycelia are also vital for building soil structure and increasing water infiltration. Other fungal species specialise in decomposition, having a role in recycling nutrients from leaf fall and other organic material. Trees widely integrated into the landscape will shift the soil ecology and balance the fungi-to-bacterial ratio, leading to significant benefits in soil health.

Family farm viability. Agroforestry is a low entropy measure for restoring healthy soil function and nutrient cycling; these benefits alone will create more resilient, sustainable and profitable farms. Agroforestry systems also deliver multifunctional outputs which can create extra income on the farm through labour and direct sales. There will be opportunities for specialist tree seedling supply from local producers, and the promotion of on-farm activities such as agri-tourism and wood product processing which can help people maintain the ability to farm the land. Additionally, agroforestry can make a positive impact on sustainable landscape and rural development, because of the diversity of employment opportunities created by multi-functional systems.

Access to markets. In silvopastoral systems, animals have access to 'browse' or leaf material which broadens their range of diet. This can be tailored to address mineral deficiencies in the pasture therefore adding to the overall health of the livestock (Luske and van Eekeren 2018). Also, stock have been shown to spend significantly more time under trees in hot, sunny weather and when it is rainy or windy, than in the open. This will become increasingly important as extreme weather events are predicted to become more frequent as a result of climate change. The intimate spatial integration of trees and agriculture provides shelter that reduces wind and temperature stress for animals. This, combined with the spatial diversity of their surroundings, is evidenced as a more welfare-friendly system than when no trees are present. The sustainability of animal welfare offered by well-designed and well-managed agroforestry systems is higher compared to other livestock systems (Escribano et al. 2019). Verifiable welfare credentials and environmental benefits of agroforestry can help meet farm sustainability criteria, enhancing future food marketing and export potential and landowners' access to markets. Silvopasture lends itself very well to an organic system of production (Raskin and Osborn 2019).

5 Factors governing uptake of silvopasture

It is appropriate to consider the needs of farmers and landowners then determine what silvopasture may offer to meet these needs. Commercial farms have both short-/medium-term and long-term needs:

- Short-/medium-term needs can be realised from systems which have a reasonable cash flow profile, give a quick return on capital, will enhance the value of the property and land, give good return on investment, can exploit currently available skills and expertise on the farm, deliver products for direct sale of processing on the farm, create enough activity to generate employment and help keep people on the land, help gain access to Capital Improvement grants and allow access to EU funding through participation in agri-environment schemes.

- Long-term needs can be realised from production systems which can be seen as being viable, sustainable and yield capable, while boosting assets such as land value, providing security of investment, granting tax considerations for succession and creating a general enhancement of the landscape and the living environment of the family farm.

6 Conclusion and future trends

Silvopastoral systems can be managed for multiple output and offer flexibility as a land-use option. They satisfy strategic policy objectives to realise some of the goals of decreasing levels of livestock production, tightening nutrient management on farms, increasing tree cover, stabilising rural communities and enhancing biodiversity through a more sustainable land use. Currently, sectoral support for mainstream forestry and agriculture is steadily increasing. and the goals of agroforestry align with objectives for agri-environment measures and are likely to be supported by conservation bodies and community development groups.

7 References

Asman, W.A.H. (2006) Some aspects of the atmospheric NHx cycle. In: Cox, T., Fowler, D., Monks, P. and Borrell, P. (eds.), *Understanding and Quantifying the Atmospheric Nitrogen Cycle*. The 2nd ACCENT Barnsdale expert meeting, pp. 152–155.

Bealey, W.J., et al. (2014) 'Modelling agro-forestry scenarios for ammonia abatement in the landscape.' *Environmental Research Letters*, 9, 125001. https://doi.org/10.1088/1748-9326/9/12/125001

Carroll, Z.L., et al. (2004) Can tree shelterbelts on agricultural land reduce flood risk? *Soil Use and Management*, 20(3), 357–359.

Cole, L.J., et al. (2020) Managing riparian buffer strips to optimise ecosystem services: a review. *Agriculture, Ecosystems & Environment*, 2020, 106891.

Dupraz, C. and Newman, S.M. (1997) Temperate agroforestry: the European way In: Gordon, A.M. and Newman, S.M. (eds.), *Temperate agroforestry systems*. CAB International, Wallingford, United Kingdom, pp. 181–236.

Escribano A.J., et al. (2019) Integrated crop–livestock systems with agroforestry to improve organic animal farming. In: Vaarst, M. and Roderick, S. (eds.), *Improving organic animal farming*. Burleigh Dodds Science Publishing, Cambridge, UK, pp. 1–33. www.bdspublishing.com

Gabriel, S. (2018) *Silvopasture: a guide to managing grazing animals, forage crops, and trees in a temperate farm ecosystem*. Chelsea Green Publishing, White River Junction, Vermont, p. 294.

Holegate, S., et al. (2016) *Every breath we take: the lifelong impact of air pollution*. Royal College of Physicians and Royal College of Paediatrics and Child Health, p. xiii.

Kay, S., et al. (2019) 'Agroforestry creates carbon sinks whilst enhancing the environment in agricultural landscapes in Europe'. *Land Use Policy*, 83, 581–593.

Kay, S., et al. (2024) 'Agroforestry can enhance foraging and nesting resources for pollinators with focus on solitary bees at the landscape scale'. *Agroforestry Systems*, 94(2), 379–387.

Lundgren, B.O. and Raintree, J.B. (1982) Sustained agroforestry. In: Nestel, B. (ed.), *Agricultural research for development: potentials and challenges in Asia*. INSAR, The Hague, pp. 37–49.

Luske, B. and van Eekeren, N. (2018) 'Nutritional potential of fodder trees on clay and sandy soils. *Agroforestry Systems*, 92, 975–986.

Marshall, M.R., et al. (2014) 'The impact of rural land management changes on soil hydraulic properties and runoff processes: results from experimental plots in upland UK' *Hydrological Processes*, 28(4), 2617–2629.

McAdam, J.H. (2018). 'Silvopastoral agroforestry-an option to support sustainable grassland intensification'. *Grassland Science in Europe*, 23, 628–630. In: *Sustainable meat and milk production from Grasslands.* Proceedings of the 27th General Meeting of the European Grassland Federation, Cork, Ireland.

McAdam, J.H. (2020) 'The potential for agroforestry in Ireland'. *Irish Forestry*, 77, 113–135.

McAdam, J.H., et al. (2009) 'Classification and functions of Agroforestry systems in Europe'. *Advances in Agroforestry*, 6, 21–42.

Menegat, S., et al. (2022) 'Greenhouse gas emissions from global production and use of nitrogen synthetic fertilisers in agriculture'. *Scientific Reports*, 12, 14490. https://doi.org/10.1038/s41598-022-18773-w

Merwin, M.L. (1997) *The status, opportunities and needs for agroforestry in the United States. A national report*. Association for Temperate Agroforestry, University of Missouri, Columbia, p. 41.

Mosquera-Losada, M.R., et al. (2009) 'Definitions and components of Agroforestry practices in Europe'. *Advances in Agroforestry*, 6, 3–20.

Mosquera-Losada, M.R., et al. (2019) 'Carbon sequestration in European Agroforestry systems'. In: Mohan Kumar, B. and Ramachandran Nair, P.K. (eds.), *Carbon sequestration potential of agroforestry systems. Advances in Agroforestry*, 8, 43–60.

Olave, R., et al. (2016) 'Agroforestry as a land use option to sequester carbon in a cool temperate climate'. *World Congress Silvopastoral Systems* 2016. University of Évora, Portugal, September 27–30, pp. 32–33.

Raskin, B. and Osborn, S. (2019) *The Agroforestry handbook - agroforestry for the UK*. 1st ed. Soil Association, Bristol, p. 150.

Torralba, M., et al. (2016) 'Do European agroforestry systems enhance biodiversity and ecosystem services? A meta-analysis'. *Agriculture, Ecosystems & Environment*, 230, 150–161.

Valdés, A., et al. (2020) 'High ecosystem service delivery potential of small woodlands in agricultural landscapes.' *Journal of Applied Ecology*, 57, 4–16.

Varah, A., et al. (2020) 'Temperate agroforestry systems provide greater pollination service than monoculture'. *Agriculture, Ecosystems & Environment*, 301, 107–131.

Wyer, K.E. et al. (2022) 'Ammonia emissions from agriculture and their contribution to fine particulate matter: a review of implications for human health'. *Journal of Environmental Management*, 323, 116285. https://doi.org/10.1016/j.jenvman.2022.116285

Chapter 5

Creating resilient silvopastoral systems: Mediterranean silvopastoralism as a model for agroforestry

Marina Castro and Júlio Germano de Souza, CIMO - LA SusTEC - Instituto Politécnico de Bragança, Portugal; and José Castro, Instituto Politécnico de Bragança, Portugal

1 Introduction

Spanning approximately 227 million hectares, the forests of the European Continent embody a diverse mosaic of ecosystems, characterised by coniferous, broadleaf, and mixed forests, as outlined in the State of Europe's Forests Report (FOREST EUROPE 2020). However, this distribution is far from uniform across the Continent, with considerable variations in forest cover, types, and management among European nations. Consequently, forest management strategies vary, from timber production-centric approaches in some regions to cultivating non-timber products and services in others. The overall figure does not include Other Wooded Land (OWL): land not classified as forest but which still contains a significant percentage (typically over 10%) of trees, bushes, and

http://dx.doi.org/10.19103/AS.2024.0141.10

shrubs). OWL accounts for roughly 27 million hectares, a relatively small overall area. The prevalence of OWL varies across European territories, ranging from 0.6% to 4.7%, except in Southwest Europe where OWL encompasses 14.4% of the land area (FOREST EUROPE 2020).

The dominance of broadleaf trees characterises the forests of southwestern Europe. More than 60% of the forests in this region are broadleaved, 30% are coniferous, and less than 10% are mixed forests (FOREST EUROPE 2020). The Mediterranean climate, marked by prolonged droughts and alternating wet and dry periods, provides conducive conditions for the prevalence of evergreen and sclerophyllous species (adapted to cope with periods of drought) with limited wood productivity compared to their counterparts in Northern and Central Europe. This distinction has historically shaped the utilisation of Mediterranean forests, with a rich tradition of harvesting animal and plant products over millennia (Chebli et al. 2018; Aubert 2013; Blondel 2006; Gómez Sal 2000). Today, it is estimated that silvopastoral systems cover about 6.5 million hectares in the Iberian Peninsula, representing about 30% of the agricultural area (den Herder et al., 2017).

Quezel and Medail (2003) highlighted the ecological significance of oak woodlands in the Mediterranean region (see Fig. 1, *Distribution of the most important oak species of the Iberian Peninsula in the Mediterranean area*). Due to their ecological attributes, the silvopastoral systems of these species provide essential resources such as meat, charcoal, cork, firewood, and timber and offer many ecosystem services. Combining forestry and grazing of domesticated animals on the same land unit has been a cornerstone of rural life on the Iberian Peninsula for centuries, supporting livelihoods and contributing to biodiversity conservation (Mosquera-Losada et al. 2009; Casals et al. 2009; Moreno and Rolo 2019; Castro 2009).

However, over the past century, intensifying agricultural practices to enhance food security and productivity have marginalised traditional agroforestry practices. Guerra et al. (2016) found that, over the past 35 years, cattle-related headage payments have markedly influenced tree cover density and regeneration under agricultural policy in Portugal, notably impacting the Supply of Ecosystem Services. This effect intensified after 1986, aligning with the activation of numerous Common Agricultural Policy tools to enhance farm modernisation and productivity.

This chapter investigates the complex interactions occurring on the Iberian Peninsula between key oak species – holm oak (*Quercus ilex* L.), cork oak (*Quercus suber* L.), and Pyrenean oak (*Quercus pyrenaica* Willd.) – and various livestock species, including ruminants (sheep, goats, and cattle) and monogastric animals (horses and pigs). The chapter aims to show the pivotal role of Mediterranean silvopastoral systems as resilient and sustainable agroforestry models, with a particular focus on their economic,

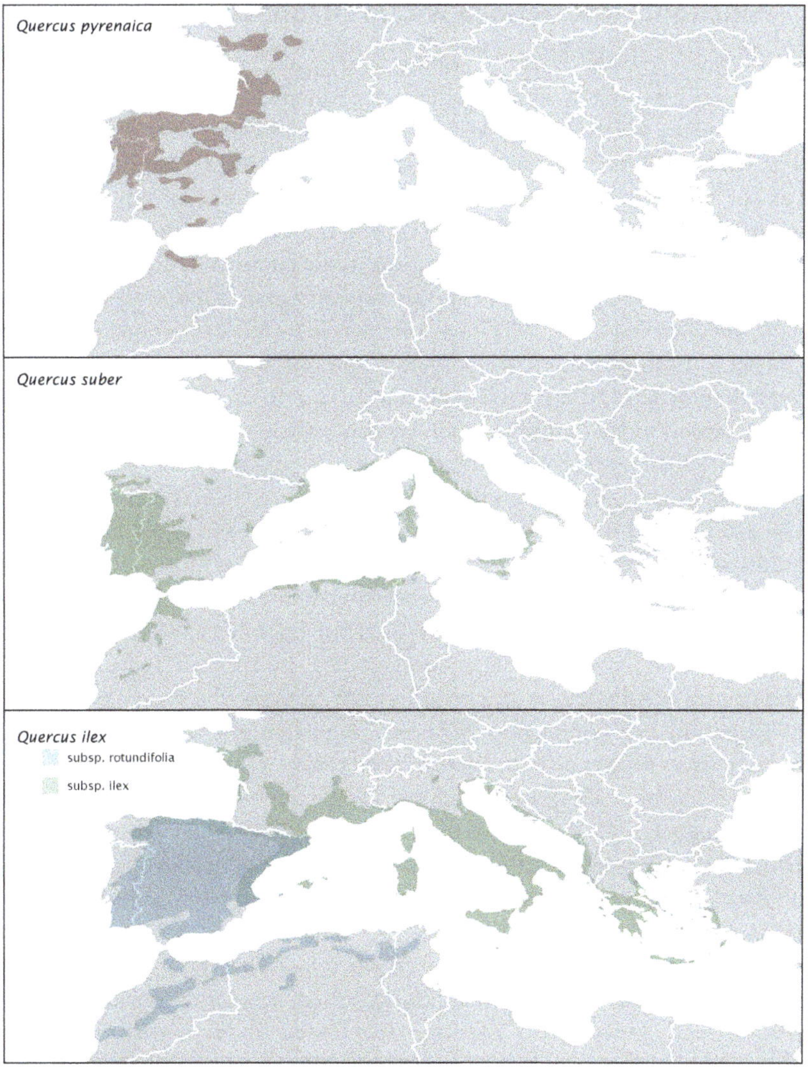

Figure 1 Distribution of the most important oak species of the Iberian Peninsula in the Mediterranean area

ecological, and cultural significance. By examining these systems, the chapter underscores the synergistic interplay between anthropogenic activities and natural ecosystems, highlighting their potential to support sustainable land management practices and biodiversity conservation within Mediterranean landscapes.

2 Functional interactions within silvopastoral systems

Forests and animals have a complex, dynamic relationship crucial for maintaining biodiversity and providing ecosystem services (Velamazán et al. 2020). Grazing domestic and wild animals interact with forest ecosystems in ways that can significantly alter their structure (Gill and Fuller 2007; Ramirez et al. 2018) and function (Beguin et al. 2016). These interactions occur through feeding habits, trampling, rubbing, dung and urine deposition, debarking, and other behaviours that affect seed dispersal, plant community composition, and soil properties (Wachiye et al. 2022; Candel-Pérez et al. 2024). Velamazán et al. (2020) and Tittonell et al. (2021) highlight the dual nature of ungulates (hoofed animals) as both providers of ecosystem services and sources of disservices (Velamazán et al. 2020; Tittonell et al. 2021).

According to Nair's agroforestry typology, silvopastoral systems involve at least two distinct components: trees and pasture/animals, so the balance between services and disservices depends on animal and plant factors (Li et al. 2021; Fan et al. 2019; Nair 1991). The supply of a specific service is rarely independent of others, and both positive (synergies) and negative (trade-offs) pairwise relationships between ecosystem services are common (Roces-Díaz et al. 2021). Sales-Baptista and Ferraz-de-Oliveira (2021) highlight that interactions can lead to different outcomes and influence ecological processes. One of the main conditions of agroforestry is reciprocal benefits between the system entities.

Exploring the connection between ungulates and the ecosystems of Mediterranean forests and scrublands requires an in-depth understanding of the unique interrelations present within these environments. The Mediterranean region, characterised by its biodiversity hotspots and distinct ecosystems, provides an illustrative example of the complex relationships between wild and domestic ungulates and their habitats. Here, pastoralism, a traditional and extensive form of livestock grazing of spontaneous fodder resources in areas commonly called rangelands, plays a crucial role in shaping these landscapes (see Figs. 2 and 3) (Dubeuf et al. 2023). While ungulates provide essential services, their impacts can become detrimental under certain conditions. Overgrazing by domestic livestock, exacerbated by insufficient management, can lead to soil degradation, loss of plant cover, and decreased biodiversity. The challenge is serious in Mediterranean woodlands, where the soil is prone to erosion and desertification.

3 Silvopastoral ecosystem services and disservices in the Mediterranean region

Silvopastoral systems embody many ecosystem services indispensable for environmental sustainability, particularly in the Mediterranean (Quintas-Soriano

Figure 2 *Preta de Montesinho* goats grazing in the oak forest

Figure 3 Sheep grazing in chestnut groves

et al. 2022; Bangash et al. 2013; Balzan et al. 2020). These systems, through the strategic integration of domestic ruminants adapted to local landscapes, contribute significantly to biodiversity conservation, soil preservation, carbon sequestration, and the mitigation of fire hazards. Each of these services reflects the sustainable and symbiotic relationship fostered between the land and livestock, showcasing the potential of these practices in enhancing the ecological integrity of agricultural landscapes.

Ungulates, including wild species like deer and boar, are instrumental in forest regeneration. By consuming fruits and traversing varied landscapes, these animals facilitate the dispersal of seeds to new locations, thereby playing a pivotal role in enhancing plant dispersal and genetic diversity. As highlighted by Janzen (1984) and Malo and Suárez (1995), this process is crucial for sustaining forest resilience and adaptability to changing environmental conditions. Moreover, in Mediterranean ecosystems, the role of ungulates extends beyond simple seed dispersal. Their movement through the woodlands aids in the propagation of numerous plant species, including those essential for maintaining the region's biodiversity, as Gómez (2003) notes. For instance, the ingestion and subsequent dispersal of seeds by these animals can lead to the regeneration of key tree species, such as oaks (*Quercus* spp.), which are pivotal in these ecosystems.

Furthermore, through selective feeding, ungulates significantly influence the composition of plant communities. This activity can enhance habitat heterogeneity and biodiversity, as Olff and Ritchie (1998) discussed. Their grazing behaviour helps control the potential dominance of certain plant species, allowing a more diverse plant community to flourish. This aspect of ungulate behaviour is particularly relevant in maintaining the open landscapes characteristic of the Mediterranean, as outlined by Seligman and Perevolotsky (1994).

Additionally, this grazing pressure aids in preventing the encroachment of shrubs, thus supporting a mosaic of habitats conducive to high biodiversity. Moreover, suppressing fire-prone species through grazing reduces the risk of wildfires, a significant concern in these regions. The movement and trampling activities of ungulates also enhance soil aeration and contribute to nutrient cycling through their manure, thereby boosting soil fertility and promoting plant growth, as detailed by Bardgett and Wardle (2003).

Despite the undeniable benefits provided by ungulates to forest ecosystems, a spectrum of adverse effects accompany their contributions. Research, including studies by Velamazán et al. (2020), has explained ungulates' significant yet complex role in ecosystems, highlighting their positive impacts on biodiversity and wildfire prevention and the negative consequences they can also cause. These include biodiversity loss in vegetation, plant recruitment failure, habitat degradation, decreased plant diversity, soil erosion, and the facilitation of invasive species spread and disease transmission, as evidenced in studies by Milchunas and Lauenroth (1993), Vitousek et al. (1997), and Gortázar et al. (2007). The complex interplay between these beneficial and detrimental effects underscores the need for informed management and conservation strategies that acknowledge and address the ecological intricacies of ungulate-forest interactions.

Mediterranean forests and scrublands serve as prime examples of these nuanced interactions, where the roles of both wild and domestic ungulates in ecological functions and biodiversity are evident yet must be mediated by careful management to mitigate potential disservices. The ecological services and disservices of silvopastoral systems, underscored by the contributions and challenges associated with ungulates, highlight the critical importance of these agricultural practices in creating resilient and sustainable landscapes (Gill and Fuller 2007; Ramirez et al. 2018).

By fostering a balanced and symbiotic relationship between livestock and their environment, silvopastoral systems also offer pragmatic solutions to contemporary environmental challenges, including climate change and wildfire management, underscoring the need for continued research, informed management, and the adoption of sustainable pastoral practices to preserve these landscapes' unique biodiversity and ecological health (Olea et al. 2006; San Miguel-Ayanz et al. 2010). By embracing sustainable pastoral practices and acknowledging the ecological roles of ungulates, it is possible to preserve the Mediterranean's unique biodiversity while supporting the livelihoods of local communities (Papanastasis et al. 2008; Rogosic et al. 2006).

4 Tree cover, forage productivity, and nutritional value

In the Mediterranean basin, characterised by fluctuating climatic conditions, the role of forest trees in supporting forage production and quality for grazing animals seems crucial (Papachristou 2016; Castro 2004). Forest trees make an essential contribution to the diet of extensive livestock production systems by diversifying the forage resources available to animals throughout the year, increasing the diversity and quality of forage, and providing a range of nutritional benefits to grazing animals, such as anthelmintic components (Torres-Fajardo et al. 2024; Torres-Acosta et al. 2016).

Climatic change is expected to have a major impact in this region due to increased temperature, reduced water availability, and inter-annual variability (IPCC 2014). The increased climatic variability will strongly influence pastoral systems, even though they have developed the capability to cope and adapt to climate uncertainty (Nardone et al. 2010). For instance, del Prado (2021) has stated that the severity of climate change threatens sheep and goat production systems in the Mediterranean area. Apart from their direct impact on animal welfare and performance, there is also an impact on the reduction of feed resources. The warmer and drier conditions forecasted for the Mediterranean region might impact net primary productivity and biomass quality (Martinez et al. 2014). Integrating the multiple contributions of forest trees to improve the production and quality of fodder resources is essential to ensure the sustainability of extensive livestock production systems and a healthy Mediterranean forest.

In silvopastoral systems, pasture yield depends on the light pastures receive and the availability of water and nutrients. The presence of the tree induces a microenvironment that is generally favourable to forage quality. Tree canopies reduce abiotic stress by moderating air and soil temperatures, reducing plant evapotranspiration, and reducing water stress in lower-lying herbaceous plants (Moreno et al. 2016; Moreno 2008; Mvondo et al. 2022). This shade is particularly beneficial during the hot and dry Mediterranean summers when the risk of drought stress is high (Canteiro et al. 2011).

Although this section describes the effects in general terms, there are significant differences between the effects produced by scattered tree systems – wood pastures – and those of dense forests in wetter mountain areas – forest grazing. This difference is because the canopy's density directly affects the radiation that reaches the lower levels, where the grass grows, and because of the availability of organic materials produced by the trees. For instance, the amount of litterfall accumulated on the soil surface is estimated at 400 in scattered tree systems and 8000 kg ha^{-1} in dense forests (Escudero et al. 1985).

Trees significantly affect the fertility of the soil, primarily by recycling leaf litter and by the turnover of nutrients that are moved through the root systems from deep in the soil (Escudero Berián 1992). This nutrient cycling is crucial for maintaining the productivity of Mediterranean pastures (Etienne 1996). Moreover, trees can facilitate the development of mycorrhizal associations and symbiotic relationships between fungal mycelium and plant roots, which improve plant access to water and soil nutrients. This symbiosis is beneficial for the growth of forage species under the nutrient-poor conditions often found in Mediterranean environments (Smith and Read 2008). In addition, trees modify soil physical properties, increasing soil water-holding capacity, macroporosity, and infiltration rates compared to open areas (Joffre and Rambal 1993), mainly explained by the increase in soil organic matter and the decreased bulk density near the trees enhancing capacity to store water in the soil.

In addition, the tree's presence reduces thermal extremes in its immediate area. Temperatures are never as high as they are beyond the tree, nor are they as low in winter, allowing pasture to grow longer in summer (Moreno et al. 2016). In dense forest systems, similar effects are also reported during the summer: temperature and light levels decrease, whereas the relative air humidity increases from the edge towards the forest interior (Meeussen et al. 2021).

Regarding forage quality, canopy cover has a positive effect on pasture phenology. It controls warming which otherwise shortens the period when plant biomass has optimal nutritional quality, increasing fibre and lignin content, and reducing digestibility (Dumont et al. 2020; Habermann et al. 2019). In addition, a moderate water deficit slows plant maturation, and biomass quality can be maintained or even improved because it delays stem development, reduces lignification, and increases digestibility (Hidalgo-Galvez et al. 2023).

Nevertheless, severe droughts usually reduce net primary productivity and quality.

Previous studies have reported that the net effect of tree canopy on forage production in Mediterranean oak woodlands can be positive, negative, or neutral, depending on inter- and intra-annual rainfall variations (Gea-Izquierdo et al. 2009; Rivest et al. 2013; López-Carrasco et al. 2015; López-Sánchez et al. 2021; Rolo et al. 2015). In addition, other factors such as soil type and fertility, as well as biological factors such as understorey species, oak species, canopy cover, tree age, and plant root architecture, play a crucial role in this variation (Rivest et al. 2011). Over-dense forests can lead to excessive competition for light, water, and nutrients between trees and understory vegetation, potentially reducing forage availability. Therefore, appropriate silvopastoral practices, including selective thinning and the maintenance of tree-pasture balance, are essential for maximising forage production while conserving forest ecosystems (Moreno and Pulido 2008).

The Mediterranean oak forest provides various fodder resources, including herbaceous pasture, acorns, foliage, and shrubs. The annual value of these products varies greatly depending on the location – weather conditions, soil type, tree or shrub species, type of system (scattered trees or dense forests), and stand density of tree or shrub cover – and, in the case of fruit, inter-annual variation. Furthermore, valuing the resource depends on who is valuing it. For instance, Evlagon et al. (2010) states that, in the Mediterranean shrubby range, the nutritional value of pasture for goats is typically more than twice that for sheep or cattle. In addition, there is almost no data on the forage production of forests or woodlands, i.e. considering all their components. However, Lledó et al. (1992) provides a global value of about 2700 kg ha^{-1} for holm oak woods, including the annual production of leaves, fruits, and flowers of trees and shrubs.

The published data for the productivity of shrubby ranges are highly variable. For instance, Ameray et al. (2022) give a value for available biomass of 3126.65 kg DM (dry matter) ha^{-1} for leguminous and cistaceous scrublands in the Mediterranean region of Portugal. However, according to these authors, only 1416.03 kg DM ha^{-1} has potentially consumed biomass, as animals only eat the year's shoots. Papanastasis et al. (1983) found annual DM production of Kermes oak pasture in Greece to vary between 378 and 742 kg DM ha^{-1}. Robles et al. (2009) give a 3264 kg DM ha^{-1} value for 'medium leguminous shrublands' in Andalusia, Spain. Evlagon et al. (2010) found an annual DM production of *Quercus calliprinos* and *Pistacia lentiscus* pasture in Israel between 700 kg and 1100 kg DM ha^{-1}. López-Díaz (2009) gives a value for tree fodder of cork and holm oaks of 350–550 kg DM ha^{-1} in wood pastures. Moreno et al. (2016) report values ranging from 500 kg to 1000 kg DM ha^{-1} obtained by pruning every 3 years.

Due to different situations, data on the productivity of herbaceous plants also vary widely between authors. In the case of Pyrenean oak, at densities between 400 and 1100 stems ha^{-1}, Castro (2004) found an annual herbaceous production of 570–2500 kg DM ha^{-1} $year^{-1}$ in mountainous regions of Portugal. In the case of scattered tree systems, according to Moreno et al. (2016), the leading natural pastures can be functionally divided into three groups:

- Annual production of between 1000 kg and 2700 kg DM ha^{-1} $year^{-1}$ for common annual pastures in the poorer soils;
- 3000 kg DM ha^{-1} $year^{-1}$ for 'majadal' pastures, created by the intense and continuous action of livestock; and
- 1500 and 2500 kg DM ha^{-1} $year^{-1}$; for the remaining type characterised by depressions with a little more water availability.

However, Gea-Izquierdo et al. (2009) reported a value varying between 5777.0 kg and 111.6 kg DM ha^{-1}, with an average annual production of 1901.8 kg DM ha^{-1} for scattered holm oaks wooded pastures in west-central Spain.

Different types of livestock eat the fruit of the oak – acorns – in the Iberian grazed woodlands. The best quality acorns come from holm oak, superior to Portuguese oak (*Quercus faginea* Lam.), cork oak, and Pyrenean oak. Gasmi-Boubaker et al. (2012) found that acorns, as a food source, contain low protein levels (less than 10% DM) and are high in carbohydrates, mainly starch (56.6–59.9% DM), making their composition like cereals. Due to their high carbohydrate content, which can be easily converted into fat, Escribano and Pulido (1998) noted that acorns are commonly fed to pigs during the finishing phase. Pigs are the best consumers of acorns in scattered tree systems, and the Iberian breed of pig can consume them without supplements (Rodriguez-Estévez et al. 2009). Joffre et al. (1999) reported that pigs graze during the seasonal acorn production between October and February, gaining about 60 kg in live weight over 75 days. Ruperez (1957) reported that 9 kg of holm oak acorns produces 1 kg of pork meat. Annual acorn production is highly variable in scattered tree systems, but a typical average value for holm oak is 500 kg ha^{-1}, with values of up to 800 kg ha^{-1} (Moreno et al. 2016).

5 Tree cover and livestock welfare

In the warm and often arduous Mediterranean climate, the welfare of grazing animals such as cattle, sheep, and goats is a significant concern. The challenges posed by the hot, dry summers, and variable precipitation patterns necessitate adaptive strategies to ensure the health and productivity of livestock (Giridhar and Samireddypalle 2015; Gitz 2016). Forest trees play a pivotal role in this context, offering both refuge and resources that enhance the welfare of these animals (Souza et al. 2019; Vizzotto et al. 2015; Roese et al. 2017; Oliveira et al. 2021; Smith et al. 2022).

Livestock animals seek shelter in various environmental and climatic conditions, such as hot or cold weather, heavy rain, or high insect infestation (Proops et al. 2019). In particular, Mediterranean summer conditions affect animal welfare by causing heat stress (Holinger et al. 2023). Heat stress poses an increasing risk to animal welfare, health, and productivity in pasture-based livestock systems (Arnott et al. 2017; Boyle et al. 2020). Heat stress occurs when environmental conditions, such as air temperature, relative humidity, air movement, and solar radiation, cause body temperature to be above the animal's thermo-neutral zone (Silanikove 2000), a limiting factor to livestock production (Paim et al. 2014). Gonzalez-Rivas et al. (2020) and Knight et al. (2023) associate heat stress with lower animal growth rates, reduced feed intake, poor fertility, inconsistent carcass composition and quality, increased veterinary costs, and animal welfare issues associated with compromised biological functioning.

Studies by Florindo et al. (2020), Maia et al. (2005), and Silanikove (2000) highlight the vital role of shade in reducing heat stress for grazing animals, leading to improved welfare and productivity. Vandenheede et al. (1995) further illustrate forests' protective role against adverse weather, enhancing livestock welfare. Silvopastoral systems, as discussed by Bertogna et al. (2022), Christensen et al. (2022), Leite-Moraes et al. (2023), and Santos Neto et al. (2022), offer a multifaceted solution by providing shade, improving fodder nutritional value, and supplementing diets during scarcity, showcasing their potential in improving animal welfare and productivity.

Finally, Ruiz and Sanz-Sánchez (2020) emphasise the indispensable role of forest ecosystems in water regulation and preserving water sources, a boon for the Mediterranean's water-stressed environments critical for livestock rearing. Trees modulate the hydrological cycle by capturing rainfall and managing water flow, fostering groundwater replenishment, and supporting springs and streams, as Bennett and Barton (2018) and Zema et al. (2021) have documented. This process is crucial for sustaining water supplies for grazing animals, mainly through the sweltering summer months, underscoring the findings of Belsky et al. (1989).

6 The complex interplay between grazing animals and forest trees

The interplay between grazing animals and forest ecosystems is critical in the Mediterranean basin's distinct climate and biodiversity framework. While crucial for landscape management and agricultural yields, grazing presents notable challenges for forest vitality and diversity. Key issues include tree damage through bark stripping, which impairs tree health and heightens susceptibility to environmental pressures. Soil compaction caused by animal trampling hinders root growth and water absorption, exacerbating tree stress in dry conditions. Moreover, selective grazing on young vegetation can shift forest dynamics, possibly diminishing biodiversity and the forest's natural regeneration processes, as highlighted in studies by Bullock and Gregory (2009), Milchunas and Noy-Meir (2002), and Plieninger et al. (2003).

Grazing's effects on Mediterranean forests vary, affected by factors like forest type, species diversity and grazing intensity, as pointed out by Doblas-Miranda et al. (2017), Stark et al. (2023), and Vaz et al. (2024). These variables pose challenges to the generalisation of results to all forest ecosystems. The aims of forest management and production direction significantly shape perceptions of grazing's effects. The evolution towards specialised forestry over recent centuries complicates the assessment of grazing, often aligning livestock management in contrast to timber production goals, as discussed by Kardell (2016), Nichiforel et al. (2018), Öllerer et al. (2019), and Serrano Alvarez (2014).

The assessment of grazing's effects is complex, influenced by regional disparities, forest types, and diverse forest management approaches. Concerns about grazing's impact on the regeneration of specific species, notably oaks, highlight the intricate dynamics of grazing, which can range from obstructing regeneration to potentially benefiting certain seedlings through altered soil water dynamics, as noted by Dufour-Dror (2007), Guy (2014), and Pinto-Correia and Azeda (2017). This intricate dynamics underscores the importance of distinguishing between the effects of undergrazing and overgrazing, pointing to the necessity of a nuanced comprehension of grazing's ecological functions, as discussed by Hidalgo-Galvez et al. (2023) and Seligman and Perevolotsky (1994).

Comprehensive strategies are vital for enhancing landscape management and mitigating the negative impacts of grazing. These strategies include managed grazing, young tree protection, and the promotion of silvopastoral systems that combine trees, shrubs, and grazing animals. This combination significantly improves biodiversity, soil quality, and carbon capture and storage. Such measures mitigate grazing's immediate adverse effects and reinforce the resilience and sustainability of Mediterranean forest ecosystems. Given grazing's dual nature – harmful and beneficial – it is imperative to implement balanced, sustainable management practices in the Mediterranean. These practices are crucial to preserving biodiversity, boosting ecosystem services, and sustaining agricultural and forest landscapes, as evidenced by Moreno and Pulido (2008) and Ruiz-Mirazo et al. (2011).

7 Silvopastoralism in Iberia: integrating nature, culture, and agriculture

The silvopastoral systems in the Iberian Peninsula, with origins dating back to ancient times, are integrally linked with the region's ecological, cultural, and economic frameworks. This interdependent relationship has been extensively documented, with notable contributions from Díaz-Maroto and Vila-Lameiro (2008), Roces-Díaz et al. (2021), and Santiago-Freijanes et al. (2021). Originally developed to address the challenges of the region's rugged and variable topography, these systems innovatively integrate forestry and livestock grazing. This strategy has enabled agricultural communities to optimise use of challenging landscapes, a concept thoroughly discussed by Glick (2005). The historical integration of domestic ruminants into these ecosystems has significantly shaped land use strategies over thousands of years, as shown by the research of Gómez Sal (2017) and McClure et al. (2006).

During the medieval period, the transhumance system, a seasonal movement of livestock between winter pastures (valley bottoms and southern latitudes) to summer pastures (mountains and northern latitudes), emerged and became a cornerstone of the Iberian agricultural calendar. This practice was fundamental to Iberian agriculture, an enhancement that authors such as López-Sáez et al. (2014) and Blanco-González et al. (2015) extensively documented. This process, involving the seasonal livestock migration, was instrumental in interweaving socio-economic structures and facilitated the exchange of goods, cultural traditions, and knowledge across regions. Specifically, sheep and goats were at the core of this system, serving as significant economic resources by producing various goods and playing a critical role in maintaining ecosystem health, a concept elaborated upon by Gómez Sal (2017).

Over time, researchers and policymakers have increasingly acknowledged the significance of silvopastoral systems in contributing to sustainable land management and biodiversity conservation, leading them to incorporate these systems within various legal frameworks aimed at their protection. Gómez Sal and Lorente (2004) have investigated the establishment of the 'Mesta' in Spain, highlighting it as a notable illustration of these systems' economic and political importance. However, the advent of industrial and agricultural modernisation presented significant challenges, leading to a decline in these traditional practices. Despite these obstacles, specific regions within the Iberian Peninsula continue to practice silvopastoralism, ingeniously integrating these age-old practices with contemporary requirements. This approach addresses modern needs and preserves the area's rich historical and ecological legacy, as documented by Herrera and Pulido (2021).

The revision of agricultural policy in 1992 marked a pivotal transformation in the European agricultural framework, prioritising biodiversity, water quality, carbon sequestration, and soil conservation. This shift is extensively analysed in the work of Mosquera-Losada et al. (2018), highlighting its significance in the context of agricultural practices. Silvopastoral systems have been identified as a practical paradigm for sustainable agriculture, aligning closely with current conservation goals in response to the pressing challenges imposed by climate change. Researchers recognise these methodologies for their critical role in reducing environmental impacts and promoting sustainability amid a global ecological crisis, a perspective that Mosquera-Losada et al. (2018) strongly support.

8 Vegetative components of Iberian silvopastoral systems

Iberian silvopastoral systems feature an integrated management strategy that synergistically combines woodlands (trees and shrubs) with livestock (ruminants and monogastric animals). This approach involves interaction among components on the same patch and landscape, fostering diverse entities to coexist and thrive. This integrative strategy optimises using environmental conditions and landscape features to achieve sustainable agricultural productivity. Silvopastoral systems are the most common and extended agroforestry systems in the Iberian Peninsula, with multiple management systems, practices, and outcomes combinations. According to Castro (2009), the two most representative are the forest grazing, mainly in mountainous areas, and the wood pastures in the more arid regions known as 'Montado' (Portugal) or the 'Dehesa' (Spain).

The vegetation component exhibits significant heterogeneity, a variation attributable to differences in regional climate, soil properties, and anthropogenic management practices. Commonly, these agroforestry systems incorporate indigenous oak species, beneath which a rich undergrowth comprising grasses, herbs, and shrubs proliferates. Drawing upon the genetic adaptations to the varied landscapes of the Peninsula, both native ruminant and monogastric livestock breeds exhibit robust performance. The domestic ruminants that graze these landscapes have evolved and been selectively bred over centuries to adapt to specific environmental conditions and best use the available resources. The structural diversity of oaks woodlands supports a multifunctional landscape conducive to livestock grazing, fosters wildlife biodiversity, and yields economically valuable non-timber forest products.

The genus *Quercus*, which includes diverse and ecologically-significant species of oak tree, is a pivotal group within the flora of the Northern Hemisphere. Vila-Viçosa et al. (2022) and Vitelli et al. (2017) highlight that these trees play crucial roles in economic and environmental contexts, including as grazing oak woods. Oaks thrive across a remarkable range of environments, from arid landscapes to the moist conditions of swamp forests and from lowland areas to high altitudes of up to 4500 metres above sea level, with Denk et al. (2023) providing insight into their adaptability.

The Iberian Peninsula is a hub for oak diversity in Eurasia, hosting over nine unique oak species that enrich its forests and woodlands. Castro and Fernandez-Nuñez (2014) and Ruiz de la Torre (1990) underline the area's ecological importance. Notably, the Pyrenean oak (*Quercus pyrenaica* Willd), cork oak (*Quercus suber* L.), and holm oak (*Quercus ilex* L.) are prominent for their extensive distribution, economic benefits, and crucial roles in silvopastoral system development. The Pyrenean woods, favouring the cooler northern climates, contribute to the region's ecological diversity, including fire resilience and habitat for wildlife. Holm oak woods are vital for southern silvopastoral practices, offering wildlife habitats and livestock grazing, while the cork oak's regenerative bark is invaluable to industry.

The extensive coverage of the Pyrenean oak, spanning 660 000 hectares in Spain and 81 700 ha in Portugal, is highlighted by Castaño-Santamaría et al. (2013) and ICNF (2016), highlighting its prevalence at various altitudes and its adaptation to siliceous soils (De Rigo and Caudullo 2016). Despite its adaptability, the species faces challenges from climate change, with vulnerabilities such as limited migration options and susceptibility to pests and diseases, as discussed by Pérez-Luque et al. (2020) and Roces-Díaz et al. (2021). The potential decline in suitable habitats, especially in southern areas like the Sierra Nevada, is a concern raised by Dorado-Liñán et al. (2019) and Pérez Luque (2022), indicating the need for focused conservation efforts.

The resilience of the Pyrenean oak to summer droughts, a characteristic feature of the Mediterranean climate, is underscored by the work of Aranda et al. (1996), Sisó et al. (2001), and Corcuera et al. (2002), emphasising its ecological importance. This species' distribution, confined mainly to the Iberian Peninsula, southwestern France, and northern Morocco, is detailed by Lorite et al. (2008) and Souto-Herrero et al. (2018), with Spain and Portugal accounting for a significant portion of its habitat, as noted by Luis Calabuig et al. (2000).

Cork and holm oaks stand out in the Mediterranean basin for their ecological and economic significance and role in the singular 'Dehesa' and 'Montado' agroforestry systems. The spatial distribution of cork oaks and holm oaks across varying moisture regimes – where cork oaks predominantly occupy coastal regions experiencing oceanic influences, and holm oaks are more prevalent in the arid inland areas – demonstrates the impact of the Mediterranean climate on vegetation patterns. This adaptation is further explored by Aronson et al. (2012) and Eriksson et al. (2017), who note the cork oak's symbolic representation of the Mediterranean's ecological and cultural heritage.

Various studies detail species' distribution, their resilience to climate extremes, and the specific environmental conditions they require, including

Oliveira et al. (1994), Eriksson et al. (2017), and Houston Durrant et al. (2016). The holm oak's adaptability across a wide range of Mediterranean climates, from semi-arid to very humid conditions, and its tolerance to temperature extremes are discussed by De Rigo and Caudullo (2016), with additional insights on its soil preferences and climatic requirements provided by Aronson et al. (2012) and Camponi et al. (2023).

9 Domestic ruminant species in silvopastoral systems

The integration of domesticated breeds into local rangelands exemplifies a remarkable adaptation to the Iberian Peninsula's unique environmental conditions. The extensive variety of indigenous livestock breeds further demonstrates this diversity. According to the national catalogues of Portugal and Spain, it includes 50 breeds of cattle, 48 breeds of sheep, 23 breeds of goats, 11 breeds of pigs, and 22 breeds of horses and donkeys, showcasing the rich biodiversity present within these silvopastoral systems and landscapes.

The co-evolution of livestock and landscape underscores the importance of traditional knowledge and selective breeding in maintaining the sustainability and productivity of these systems. Practices such as rotational grazing, controlled burning, and pruning of trees are employed to maintain the ecosystem's health, ensuring it continues to provide for the needs of the domestic ruminants while preserving biodiversity. The sustainable management of silvopastoral systems requires a nuanced understanding of the dynamic interactions between natural vegetation growth cycles, patterns, and livestock grazing practices. Joffre et al. (1988) highlight the intricate ecological dynamics between livestock and woodland habitats within the 'dehesas' and 'montados' of the Iberian Peninsula, emphasising the role of these practices in maintaining these unique agroforestry systems.

The native domestic ruminants of Iberia have developed remarkable adaptations, enabling them to thrive in challenging environments. Managing their silvopastoral grazing patterns optimises using natural resources and mitigates the impact of grazing on the environment, allowing for vegetation recovery and maintaining landscape diversity. These breeds exemplify the incredible adaptability of livestock to diverse and challenging environments as a testament to centuries of evolution and selective breeding, underlining the importance of these practices in maintaining the ecological and economic sustainability of rural mountain communities.

9.1 Sheep

Adapted across varied ecosystems within the Iberian Peninsula, ovine species significantly contribute to ecological equilibrium. In the mountain areas, sheep foraging behaviour is instrumental in sustaining open wooded landscapes by consuming understorey shrubs and herbs. Grazing practices in wood pastures are pivotal in enhancing species biodiversity and health. Pinto-Correia and Mascarenhas (1999) have explored the integration of agriculture and nature conservation, particularly in the context of traditional landscapes like the Portuguese 'montado,' underscoring the vital role of these livestock practices in promoting soil health and biodiversity.

Notably, indigenous sheep breeds such as the Merina and Churra, embedded within the transhumance tradition, exhibit remarkable resilience and are adept at exploiting the heterogeneous forage available within silvopastoral ecosystems. Historically, the commerce of Merina wool has been a cornerstone of economic development, emphasising the economic significance of these breeds. Castro et al. (2021) show the versatility of these species in terrain adaptability, including their pivotal role in forest fire prevention and landscape stewardship, mainly through their dietary preferences that curb the proliferation of shrubs, thereby reducing combustible material and lessening the likelihood of forest fires. They underscore the integral role of sheep breeds in the conservation of open woodlands and rangelands, highlighting their contribution to environmental conservation and management.

9.2 Goats

Goats also play an indispensable role, favouring rugged and mountainous woodlands in the Iberian landscapes. These animals thrive by browsing rather than grazing, focused on shrubs and trees. This specific dietary preference is crucial for preventing overgrowth and maintaining landscape diversity. Manzano and Malo (2006) describe how extensive grazing by goats in mountainous regions contributes to balancing biodiversity conservation with livestock production. Their unique browsing habits are instrumental in controlling bush encroachment.

The ecological management of marginal woodlands, often deemed unsuitable for conventional forestry, is markedly improved by the presence of goats (Rosa García et al. 2012). By consuming plant species that other livestock typically find unpalatable, goats help in the health and regeneration of woodlands. This is particularly true for breeds such as the Murciano-Granadina and the Spanish Ibex. Although the latter is not a domestic breed and thus not traditionally managed for commercial purposes, both species share similar habitat preferences and contribute positively to their environments (Delgado et al. 2017; Velamazán et al. 2020). These breeds have superbly adapted to the challenges presented by mountainous terrains, demonstrating an exceptional ability to utilise the sparse vegetation found under forest canopies (Acevedo et al. 2008; El Aich and Waterhouse 1999; Ferreira et al. 2013; Robles et al. 2009; Sanjuán et al. 2018).

The versatility of goats renders them invaluable for managing understorey and significantly reduces the risk of fire, an increasingly important issue in the face of climate change. Sanjuán et al. (2018) and Ferreira et al. (2013) have shed light on the significance of goat grazing as a strategic tool in forest management and bush encroachment control. Their research underscores how goats, through their natural behaviours and dietary preferences, serve as critical agents in the ecological stewardship of the Iberian Peninsula, enhancing biodiversity and mitigating risks associated with dense undergrowth and fire. Castro et al. (2021) provide a comprehensive understanding of how traditional and indigenous livestock breeds, through their unique behaviours and dietary preferences, serve as critical agents in the ecological stewardship of the Iberian Peninsula, enhancing biodiversity and mitigating risks associated with dense undergrowth and fire. Rigueiro-Rodríguez et al. (2009) explore the productive, ecological, and social perspectives of agroforestry systems in Europe, illustrating how these practices integrate livestock like goats into woodland ecosystems, enhancing their utility in landscape and environmental management.

9.3 Cattle

The milder climates and abundant suitable meadows make the open woodlands of the north and centre of the Iberian Peninsula ideal for cattle grazing (Gómez Sal 2001). Each breed exhibits adaptations to the unique landscape features, ranging from the rugged terrains of the Pyrenees and Cantabrian Range to the expansive plains of Castile and León. Certain breeds, such as the Sayaguesa and Maronesa, have evolved with a particular affinity for woodland environments. These breeds graze in the more humid, densely-wooded areas, benefiting from a diet of mixed vegetation. The symbiotic relationship between cattle and these woodland ecosystems is crucial for maintaining open spaces within forests, thereby promoting biodiversity by creating a variety of habitats for different species (Velado-Alonso et al. 2022).

The impact of cattle on the landscape extends beyond mere grazing. Breeds like the Cachena and Rubia Gallega serve to transform diverse vegetation into valuable resources like meat and milk (Sánchez 2005; Caballero et al. 2009). These Iberian domestic ruminants are especially appreciated for their grazing efficiency in lands deemed unsuitable for other forms of agriculture, playing an essential role in preserving open landscapes and preventing forest encroachment. López (2002) discusses the transition from traditional agricultural systems to more industrialised livestock-raising practices, the effects of rural emigration, and the subsequent impact on landscape management and biodiversity. He highlights the importance of maintaining traditional cattle grazing systems and suggests integrating them with modern agricultural practices to preserve these regions' environmental and cultural heritage.

9.4 Pigs

In the southwestern wood pastures, traditional Iberian pig rearing occurs within the 'dehesa' system, a distinctive agroforestry system in the Iberian Peninsula that merges oak woodlands with grazing livestock to enhance biodiversity and ecosystem health. Free-grazing pigs, feeding on acorns, contribute to maintaining this landscape while improving meat quality, producing premium Iberian ham. This practice, known as 'montanera,' showcases a sustainable method combining environmental conservation with economic gain, preserving the unique 'dehesa' ecosystem while ensuring the production of high-quality ham. Rodriguez-Estévez et al. (2012, 2009) emphasise the environmental advantages of integrating livestock grazing with woodland management, highlighting the symbiotic relationship between traditional rearing practices and conservation efforts in supporting the region's unique ham production and ecological balance.

9.5 Donkeys and horses

Donkeys and horses, traditionally employed as working animals, similarly graze in woodland areas, aiding in maintaining these habitats, albeit with less browsing pressure than goats. This lower grazing pressure helps to conserve woodland areas without compromising their ecological integrity. Velado-Alonso et al. (2022) emphasise the importance of native domestic grazers in bolstering the conservation value of semi-natural pastures in the Iberian Peninsula, showcasing how these animals, including donkeys, play a crucial role in the preservation of semi-natural habitats. The study by Segarra et al. (2023) also highlights the importance of donkeys in landscape maintenance in the Iberian Peninsula, particularly in less intensively managed woodland areas, shedding light on these animals' pivotal role in the ecological stewardship and biodiversity conservation of their habitats. Bartolomé et al. (2020) highlight donkeys and goats' differential grazing behaviours in holm oak habitats, emphasising their collective utility in fire prevention. Donkeys primarily target more easily digestible species, while goats consume flammable woody plants, suggesting their symbiotic grazing can enhance fire risk management and species conservation in Iberian ecosystems.

10 Conclusion: challenges facing Iberian silvopastoral systems

Agroforestry, which intertwines woody vegetation with agricultural production, can be seen as a symbiotic paradigm of land management. It an integral part of the agricultural practices of the Iberian Peninsula. Categorised as 'silvopastoral' systems, these practices serve as quintessential models of sustainable land use, embodying the intricate balance between forestry, agriculture, and grazing (Moreno and Rolo 2019; Santiago-Freijanes et al. 2021). They help create unique agroecosystems across more than 6.5 million hectares across the Iberian landscape (den Herder et al. 2017). These ecosystems are pivotal for their contributions to biodiversity conservation, carbon sequestration, and soil stabilisation, necessitating focused conservation efforts to safeguard these critical habitats (Campos et al. 2020; Laporta et al. 2021; Marta-Pedroso et al. 2018).

Beyond their ecological footprint, silvopastoral systems underpin the socio-economic fabric and cultural heritage of rural communities within the Iberian Peninsula. The production of distinctive, high-quality local products, such as artisanal cheeses and the renowned Jamón Ibérico ham, not only fuels the local economy but also acts as a bastion for traditional practices and breeds, thus reinforcing regional cultural identity and heritage (Haddad et al. 2022; Castro 2009). At the nucleus of these systems lies the institution of family farming, encapsulating a generational legacy of agrarian practice deeply intertwined with the land.

Small-scale, familial enterprises are essential for the preservation and operational efficiency of silvopastoral systems, given their adaptive management capacity and intrinsic understanding of the multifaceted nature of the landscapes in which they live (Mosquera-Losada et al. 2020; Rois-Díaz et al. 2018). Moreover, silvopastoral systems advocate for sustainable consumption models by providing locally-sourced, ecologically-sourced products. This paradigm minimises carbon footprints and fosters a robust linkage between producers and consumers, enhancing societal awareness of food provenance, and the environmental and community impact of consumption behaviours (Pantera et al. 2021).

Contemporary challenges such as climate change, land-use conflicts, and globalisation pressures in agricultural markets actively threaten the sustainability of silvopastoral systems. These threaten the sustainability of these systems and, by extension, the livelihoods they support. Addressing these challenges requires a multifaceted approach, incorporating policy interventions, advancements in sustainable agronomic practices, and the strategic promotion of local products globally to ensure the resilience and sustainability of these agroecological systems (Haddad et al. 2022). This holistic strategy is paramount for safeguarding the legacy and viability of these systems for future generations, thus reinforcing their role as a linchpin in the socio-economic and ecological landscape of the Iberian Peninsula (Campos et al. 2020; Laporta et al. 2021; Marta-Pedroso et al. 2018).

Adopting integrated land use policies is crucial for the sustainable management of Iberian silvopastoral systems (Moreno et al. 2014; Varela et al.

2022). By recognising their multifunctional nature, these policies can enhance biodiversity and ecosystem services (Mastrangelo et al. 2014). Coupled with financial incentives, such as subsidies or payments for ecosystem services, they motivate farmers to adopt sustainable practices (Piñeiro et al. 2020). This dual approach ensures that silvopastoral systems remain economically viable while conserving the environment, thus fostering a balance between agricultural productivity and ecological preservation (Fernández-Núñez and Castro 2016).

The future of Iberian silvopastoral systems hinges on innovation and market integration. Advanced agroforestry techniques, such as drought-resistant species or improved grazing management, can significantly enhance sustainability and productivity. Diversifying value chains and developing niche markets for organic and sustainable products can enhance economic outcomes. Embracing digital technologies to access broader markets also plays a pivotal role, allowing small-scale producers to connect with consumers directly, fostering transparency, and boosting profitability through premium pricing on sustainable products.

Enhancing social cohesion and economic resilience is also vital for the sustained success of silvopastoral systems. Community-led initiatives, education, and participatory management involve stakeholders in sustainable resource management, ensuring knowledge transfer and innovation. Focusing on resilience planning against climate change and market volatilities, these strategies aim to diversify income sources and enhance adaptive capacities. Ultimately, fostering a community-centric approach with robust support mechanisms ensures the longevity and prosperity of silvopastoral practices, securing both ecological and economic benefits for future generations.

Addressing the challenges faced by Iberian silvopastoral systems requires a multifaceted approach that integrates sustainable land management practices, market innovations, and social engagement. Strategic imperatives focused on policy support, innovation, and enhancing community cohesion are crucial for the long-term sustainability and resilience of these systems. By implementing these strategies, it is possible to preserve the ecological, economic, and cultural values of silvopastoral systems in the Iberian Peninsula.

11 Where to look for further information

- Journal of Mountain Research and Development https://www.mrd-journal.org
- Pastoralism, Research, Policy and Practice https://pastoralismjournal.springeropen.com/
- Agroforestry Systems (Springer) https://link.springer.com/journal/10457

- Rangeland Ecology & Management (Elsevier / Society for Range Management) https://www.journals.elsevier.com/rangeland-ecology-and-management
- International Year of Rangelands and Pastoralists (IYRP 2026) https://iyrp.info
- FAO Pastoralist Knowledge Hub https://www.fao.org/pastoralist-knowledge-hub
- L'Association Française de Pastoralisme: L'AFP http://www.pastoralisme.net/

12 References

Acevedo, P., et al., 2008. The Iberian ibex is under an expansion trend but displaced to suboptimal habitats by the presence of extensive goat livestock in central Spain, in: Hawksworth, D.L. and Bull, A.T. (Eds.), *Biodiversity and Conservation in Europe*. Springer Netherlands, Dordrecht, pp. 119–134. https://doi.org/10.1007/978-1-4020-6865-2_9

Ameray, A., et al., 2022. Potential greenhouse gas emissions mitigation through increased grazing pressure: a case study in North Portugal. *Carbon Manag* 13, 142–153. https://doi.org/10.1080/17583004.2022.2029575

Aranda, I., et al., 1996. Seasonal water relations of three broadleaved species (*Fagus sylvatica* L., *Quercus petraea* (Matt.) Liebl. and *Quercus pyrenaica* Willd.) in a mixed stand in the centre of the Iberian Peninsula. *For Ecol Manage* 84, 219–229. https://doi.org/10.1016/0378-1127(96)03729-2

Arnott, G., et al., 2017. Review: welfare of dairy cows in continuously housed and pasture-based production systems. *Animal* 11, 261–273. https://doi.org/10.1017/S1751731116001336

Aronson, J., et al. (Eds.), 2012. *Cork Oak Woodlands on the Edge: Ecology, Adaptive Management, and Restoration*. Island Press, Washington, DC.

Aubert, P.M., 2013. Les évolutions de la politique forestière au Maroc: entre réappropriation du modèle forestier français et idéalisation de la tribu. *Revue forestière française* 65, 305–316.

Balzan, M.V., et al., 2020. Assessing ecosystem services supplied by agroecosystems in mediterranean Europe: a literature review. *Land (Basel)* 9, 245. https://doi.org/10.3390/land9080245

Bangash, R.F., et al., 2013. Ecosystem services in Mediterranean river basin: Climate change impact on water provisioning and erosion control. *Sci Total Environ* 458–460, 246–255. https://doi.org/10.1016/J.SCITOTENV.2013.04.025

Bardgett, R.D. and Wardle, D.A., 2003. Herbivore-mediated linkages between aboveground and belowground communities. *Ecology* 84(9), 2258–2268. https://doi.org/10.1890/02-0274

Bartolomé, J., et al., 2020. Preference by Donkeys and Goats among five mediterranean forest species: implications for reducing fire hazard. *Animals* 10, 1302. https://doi.org/10.3390/ani10081302

Beguin, J., et al., 2016. Management of forest regeneration in boreal and temperate deer–forest systems: challenges, guidelines, and research gaps. *Ecosphere* 7, e01488. https://doi.org/10.1002/ecs2.1488
Belsky, A.J., et al., 1989. The effects of trees on their physical, chemical and biological environments in a semi-arid savanna in Kenya. *J Appl Ecol* 26, 1005–1024. https://doi.org/10.2307/2403708
Bennett, B.M. and Barton, G.A., 2018. The enduring link between forest cover and rainfall: a historical perspective on science and policy discussions. *For Ecosyst* 5, 5. https://doi.org/10.1186/s40663-017-0124-9
Bertogna, K.F., et al., 2022. Parameters associated with sexual precocity of Nellore heifers in integrated systems. *Agrofor Syst* 96, 669–679. https://doi.org/10.1007/s10457-021-00702-x
Blanco-González, A., et al., 2015. Medieval landscapes in the Spanish Central System (450-1350): a palaeoenvironmental and historical perspective. *J Mediev Iber Stud* 7, 1–17. https://doi.org/10.1080/17546559.2014.925135
Blondel, J., 2006. The 'design' of mediterranean landscapes: a millennial story of humans and ecological systems during the historic period. *Hum Ecol* 34, 713–729. https://doi.org/10.1007/s10745-006-9030-4
Boyle, W.A., et al., 2020. Hygric niches for tropical endotherms. *Trends Ecol Evol* 35, 938–952. https://doi.org/10.1016/j.tree.2020.06.011
Bullock, P. and Gregory, P.J., 2009. Soils: a neglected resource in Urban Areas, Bullock, P. and Gregory, P.J. (Eds.), in: *Soils in the Urban Environment*. Wiley Online Library. pp. 1–4. https://doi.org/10.1002/9781444310603.ch1
Caballero, R., et al., 2009. Grazing systems and biodiversity in Mediterranean areas: Spain, Italy and Greece. *Pastos* 39, 9–154.
Camponi, L., et al., 2023. Holm oak (Quercus ilex L.) cover: a key soil-forming force in controlling C and nutrient stocks in long-time coppice-managed forests. *J Environ Manage* 330, 117181. https://doi.org/10.1016/J.JENVMAN.2022.117181
Campos, P., et al., 2020. Environmental incomes: refined standard and extended accounts applied to cork oak open woodlands in Andalusia, Spain. *Ecol Indic* 117, 106551. https://doi.org/10.1016/j.ecolind.2020.106551
Candel-Pérez, D., et al., 2024. Effects of grazing on soil properties in mediterranean forests (Central-Eastern Spain). *J Environ Manage* 354, 120316. https://doi.org/10.1016/j.jenvman.2024.120316
Canteiro, C., et al., 2011. Conservation of Mediterranean oak woodlands: understorey dynamics under different shrub management. *Agrofor Syst* 82. https://doi.org/10.1007/s10457-011-9375-6
Casals, P., et al., 2009. Silvopastoral systems in the Northeastern Iberian Peninsula: a multifunctional perspective, in: *Agroforestry in Europe*. Current Status and Future Prospects, Springer, Dordrecht, pp. 161–181. https://doi.org/10.1007/978-1-4020-8272-6_8
Castaño-Santamaría, J., et al., 2013. Regional-scale stand density management diagrams for Pyrenean oak (Quercus pyrenaica Willd.) stands in north-west Spain. *IForest* 6, 113–122. https://doi.org/10.3832/ifor0880-006
Castro, J., et al., 2021. Changes on the climatic edge: adaptation of and challenges to pastoralism in Montesinho (Northern Portugal). *Mt Res Dev* 41, R29–R37. https://doi.org/10.1659/MRD-JOURNAL-D-21-00010.1

Castro, M., 2004. Análisis de la interacción vegetación-herbivoro en sistemas silvopastorales basados en Quercus pyrenaica. Análisis de la interacción vegetación-herbivoro en sistemas silvopastorales basados en Quercus pyrenaica.

Castro, M., 2009. Silvopastoral systems in Portugal: current status and future prospects, in: *Agroforestry in Europe*. Current Status and Future Prospects, Springer, Dordrecht, pp. 111–126. https://doi.org/10.1007/978-1-4020-8272-6_6

Castro, M. and Fernandez-Nuñez, E., 2014. Soil properties and understory herbaceous biomass in forests of three species of Quercus in Northeast Portugal. *For Syst* 23, 425–437. https://doi.org/10.5424/FS/2014233-05086

Chebli, Y., et al., 2018. Forest and silvopastoral cover changes and its drivers in northern Morocco. *Appl Geogr* 101, 23–35. https://doi.org/10.1016/j.apgeog.2018.10.006

Christensen, J.W., et al., 2022. Shelter use by horses during summer in relation to weather conditions and horsefly (Tabanidae) prevalence. *Appl Anim Behav Sci* 253, 105676. https://doi.org/10.1016/j.applanim.2022.105676

Corcuera, L., et al., 2002. Functional groups in Quercus species derived from the analysis of pressure–volume curves. *Trees* 16, 465–472. https://doi.org/10.1007/s00468-002-0187-1

De Rigo, D. and Caudullo, G., 2016. Quercus ilex in Europe: Distribution, habitat, usage and threats, in: San-Miguel-Ayanz, J., et al. (Eds.), *European Atlas of Forest Tree Species*. Publ. Off. EU, Luxembourg, pp. 152–153.

del Prado, A., 2021. Climate change impacts on Mediterranean small ruminant production systems and climate change mitigation options, in: López-Francos, A., et al. (Eds.), *Efficiency and Resilience of Forage Resources and Small Ruminant Production to Cope with Global Challenges in Mediterranean Areas*. (Options Méditerranéennes : Série A. Séminaires Méditerranéens; n. 125. CIHEAM, Zaragoza, pp. 162–163.

Delgado, J.V., et al., 2017. Murciano-Granadina goat: a Spanish local breed ready for the challenges of the twenty-first century, in: Simões, J. and Gutiérrez, C. (Eds.), *Sustainable Goat Production in Adverse Environments: Volume II: Local Goat Breeds*. Springer International Publishing, Cham, pp. 205–219. https://doi.org/10.1007/978-3-319-71294-9_15

den Herder, M., et al., 2017. Current extent and stratification of agroforestry in the European Union. *Agric Ecosyst Environ* 241. https://doi.org/10.1016/j.agee.2017.03.005

Denk, T., et al., 2023. Niche evolution in a northern temperate tree lineage: biogeographical legacies in cork oaks (Quercus Cerris). *Ann Bot* 131, 769–787. https://doi.org/10.1093/aob/mcad032

Díaz-Maroto, I.J. and Vila-Lameiro, P., 2008. Historical evolution and land-use changes in natural broadleaved forests in the north-west Iberian Peninsula. *Scand J For Res* 23, 371–379. https://doi.org/10.1080/02827580802203552

Doblas-Miranda, E., et al., 2017. A review of the combination among global change factors in forests, shrublands and pastures of the Mediterranean Region: Beyond drought effects. *Glob Planet Change* 148, 42–54. https://doi.org/10.1016/J.GLOPLACHA.2016.11.012

Dorado-Liñán, I., et al., 2019. Geographical adaptation prevails over species-specific determinism in trees' vulnerability to climate change at Mediterranean rear-edge forests. *Glob Chang Biol* 25, 1296–1314. https://doi.org/10.1111/gcb.14544

Dubeuf, J.-P., et al., 2023. Assessing the operational perspectives of agrosilvopastoralism in the Mediterranean Region: learnings from observations in Corsica. *Small Rumin Res* 229, 107131. https://doi.org/10.1016/j.smallrumres.2023.107131

Dufour-Dror, J.M., 2007. Influence of cattle grazing on the density of oak seedlings and saplings in a Tabor oak forest in Israel. *Acta Oecologica* 31, 223–228. https://doi.org/10.1016/J.ACTAO.2006.11.003

Dumont, B., et al., 2020. Incorporating diversity into animal production systems can increase their performance and strengthen their resilience. *Front Sustain Food Syst* 4. https://doi.org/10.3389/fsufs.2020.00109

El Aich, A. and Waterhouse, A., 1999. Small ruminants in environmental conservation. *Small Rumin Res* 34, 271–287. https://doi.org/10.1016/S0921-4488(99)00079-6

Eriksson, G., et al., 2017. Genetic conservation and management of Quercus suber. Technical Bulletin. European Forest Genetic Resources Programme (EUFORGEN), Bioversity International, Rome, Italy.

Escribano, M. and Pulido, F., 1998. La dehesa en Extremadura. Estructura económica y recursos naturales. Colección Monograf\'\ias. Servicio de Investigación y Desarrollo Tecnológico. Junta de Extremadura.

Escudero, A., et al., 1985. The nutrient cycling in Quercus rotundifolia and Quercus pyrenaica ecosystems ('dehesas') of Spain. *Acta Oecologica-oecologia Plantarum* 6, 73–86.

Escudero Berián, A., 1992. Intervención del arbolado en los ciclos de los nutrientes, in: *El Libro de Las Dehesas Salmantinas*. Junta de Castilla y León, pp. 241–260.

Etienne, M., 1996. *Western European Silvopastoral Systems*. INRA, Paris.

Evlagon, D., et al., 2010. How much browse is available for goats that graze Mediterranean woodlands? *Small Rumin Res* 94, 103–108. https://doi.org/10.1016/j.smallrumres.2010.07.008

Fan, F., et al., 2019. Effects and relationships of grazing intensity on multiple ecosystem services in the Inner Mongolian steppe. *Sci Total Environ* 675, 642–650. https://doi.org/10.1016/j.scitotenv.2019.04.279

Fernández-Núñez, E. and Castro, M., 2016. *Management of Agroforestry Systems: Ecological, Social and Economic Approaches*. Instituto Politécnico de Bragança, País.

Ferreira, L.M.M., et al., 2013. Foraging behaviour of domestic herbivore species grazing on heathlands associated with improved pasture areas. *Livest Sci* 155, 373–383. https://doi.org/10.1016/j.livsci.2013.05.007

Florindo, T.J., et al., 2020. Multicriteria decision-making and probabilistic weighing applied to sustainable assessment of beef life cycle. *J Clean Prod* 242, 118362. https://doi.org/10.1016/j.jclepro.2019.118362

FOREST EUROPE, 2020. State of Europe's Forests 2020.

Gasmi-Boubaker, A., et al., 2012. Importance of Mediterranean forest products as food resource of domestic herbivores: the case of oak acorn, in: Bouche, R., et al. (Eds.), *New Trends for Innovation in the Mediterranean Animal Production*. Wageningen Academic Publishers, Wageningen, pp. 123–126. https://doi.org/10.3920/978-90-8686-726-4_16

Gea-Izquierdo, G., et al., 2009. Changes in limiting resources determine spatio-temporal variability in tree–grass interactions. *Agrofor Syst* 76, 375–387. https://doi.org/10.1007/s10457-009-9211-4

Gill, R.M.A. and Fuller, R.J., 2007. The effects of deer browsing on woodland structure and songbirds in lowland Britain. *Ibis* 149, 119–127. https://doi.org/10.1111/j.1474-919X.2007.00731.x

Giridhar, K. and Samireddypalle, A., 2015. Impact of climate change on forage availability for livestock, in: *Climate Change Impact on Livestock: Adaptation and Mitigation*.

Springer India, New Delhi, pp. 97–112. https://doi.org/10.1007/978-81-322-2265-1_7
Gitz, V., 2016. Climate change and food security: risks and responses. *Watch Letter* n°37.
Glick, T.F., 2005. Agriculture, settlement, and the moving frontier, in: *Islamic and Christian Spain in the Early Middle Ages*. Brill, Leiden, pp. 42–112. https://doi.org/10.1163/9789047415589_005
Gómez, J.M., 2003. Spatial patterns in long-distance dispersal of Quercus ilex acorns by jays in a heterogeneous landscape. *Ecography* 26, 573–584. https://doi.org/10.1034/j.1600-0587.2003.03586.x
Gómez Sal, A., 2017. *Patterns of Vegetation Cover Shaping the Cultural Landscapes in the Iberian Peninsula*. pp. 459–497. https://doi.org/10.1007/978-3-319-54867-8_10
Gómez Sal, A., 2000. The variability of Mediterranean climate as an ecological condition of livestock production systems, in: *Livestock Production and Climatic Uncertainty in the Mediterranean*. Proceedings of the Joint ANPA-EAAP-CIHEAM-FAO Symposium, Agadir, Morocco, 22–24 October 1998. pp. 3–11.
Gómez Sal, A. and Lorente, I., 2004. The present status and ecological consequences of transhumance in Spain, in: *Transhumance and Biodiversity in European Mountains*. Alterra, Wageningen, pp. 233–248.
Gómez Sal, A., 2001. The ecological rationale and nature conservation value of extensive livestock systems in the Iberian Peninsula. Examples of European agri-environment schemes and livestock systems and their influence on Spanish cultural landscapes 103.
Gonzalez-Rivas, P.A., et al., 2020. Effects of heat stress on animal physiology, metabolism, and meat quality: a review. *Meat Sci* 162, 108025. https://doi.org/10.1016/J.MEATSCI.2019.108025
Gortázar, C., et al., 2007. Diseases shared between wildlife and livestock: a European perspective. *Eur J Wildl Res*. https://doi.org/10.1007/s10344-007-0098-y
Guerra, C.A., et al., 2016. Policy impacts on regulating ecosystem services: looking at the implications of 60 years of landscape change on soil erosion prevention in a Mediterranean silvo-pastoral system. *Landsc Ecol* 31, 271–290. https://doi.org/10.1007/s10980-015-0241-1
Guy, B., 2014. Wood-pastures and the common agricultural policy: rhetoric and reality, in: *European Wood-Pastures in Transition*. Routledge, Abingdon, UK, pp. 273–281.
Habermann, E., et al., 2019. Warming and water deficit impact leaf photosynthesis and decrease forage quality and digestibility of a C4 tropical grass. *Physiol Plant* 165, 383–402. https://doi.org/10.1111/ppl.12891
Haddad, F.F., et al., 2022. *Grazing with Trees: A Silvopastoral Approach to Managing and Restoring Drylands*. Food & Agriculture Org, Rome.
Herrera, P.M. and Pulido, F., 2021. Conflict and the governance of Iberian silvopastoral systems: conflict and interaction among different actors and levels of governance today, in: *Governance for Mediterranean Silvopastoral Systems*. Routledge, Abingdon, UK, pp. 191–206.
Hidalgo-Galvez, M.D., et al., 2023. Impact of climate change on pasture quality in Mediterranean dehesas subjected to different grazing histories. *Plant Soil* 488, 465–483. https://doi.org/10.1007/S11104-023-05986-9/METRICS
Holinger, M., et al., 2023. Behavioural changes to moderate heat load in grazing dairy cows under on-farm conditions. https://doi.org/10.1016/j.livsci.2023.105376

Houston Durrant, T., et al., 2016. Quercus suber in Europe: distribution, habitat, usage and threats, in: San-Miguel-Ayanz, J., et al. (Eds.), *European Atlas of Forest Tree Species*. Publ. Off. EU, Luxembourg, pp. 164–165.

ICNF, I. da C. da N. e das F., 2016. Inventário Florestal Nacional [WWW Document].

IPCC, 2014. Fifth Assessment Report–Synthesis Report, IPPC Rome, Italy.

Janzen, D.H., 1984. Dispersal of small seeds by big herbivores: foliage is the fruit. *Am Nat* 123. https://doi.org/10.1086/284208

Joffre, R. and Rambal, S., 1993. How tree cover influences the water balance of mediterranean rangelands. *Ecology* 74, 570–582. https://doi.org/10.2307/1939317

Joffre, R., et al., 1999. The dehesa system of southern Spain and Portugal as a natural ecosystem mimic. *Agrofor Syst* 45, 57–79. https://doi.org/10.1023/A:1006259402496

Joffre, R., et al., 1988. The dehesa: an agrosilvopastoral system of the Mediterranean region with special reference to the Sierra Morena area of Spain. *Agrofor Syst* 6. https://doi.org/10.1007/BF02344747

Kardell, Ö., 2016. Swedish forestry, forest pasture grazing by livestock, and game browsing pressure since 1900. *Environ Hist Camb* 22, 561–587. https://doi.org/10.3197/096734016X14727286515817

Knight, M.I., et al., 2023. The effect of shade on sheep grazing pasture during summer conditions. *J Vet Behav* 64–65, 16–24. https://doi.org/10.1016/j.jveb.2023.05.005

Laporta, L., et al., 2021. It's a keeper: valuing the carbon storage service of Agroforestry ecosystems in the context of CAP Eco-Schemes. *Land Use Policy* 109, 105712. https://doi.org/10.1016/j.landusepol.2021.105712

Leite-Moraes, A.E., et al., 2023. Environmental impacts in integrated production systems: an overview. *J Clean Prod* 420, 138400. https://doi.org/10.1016/j.jclepro.2023.138400

Li, M., et al., 2021. Ecosystem services under different grazing intensities in typical grasslands in Inner Mongolia and their relationships. https://doi.org/10.1016/j.gecco.2021.e01526

Lledó, M.J., et al., 1992. Structure, biomass and production of a resprouted holm-oak (Quercus ilex L.) forest in NE Spain. *Vegetatio* 99, 51–59. https://doi.org/10.1007/BF00118209

López, M.M., 2002. Landscapes of northern Spain and pastoral systems, in: Redecker, B. (Ed.), *Pasture Landscapes and Nature Conservation*. Springer, New York, NY, pp. 67–86.

López-Carrasco, C., et al., 2015. The effect of tree cover on the biomass and diversity of the herbaceous layer in a Mediterranean dehesa. *Grass Forage Sci* 70, 639–650. https://doi.org/10.1111/GFS.12161

López-Díaz, M. L. 2009. Nutritional resources in the dehesa. Improved pastures. In: *Agroforestry systems as a technique for sustainable territorial management,* Mosquera-Losada, M. R. Fernández-Lorenzo, J. L. Rigueiro-Rodríguez, A. (Eds.). Ministerio de Asuntos Exteriores-AECID. Madrid. 185–196.

López-Sáez, J.A., et al., 2014. Vegetation history, climate and human impact in the Spanish Central System over the last 9000 years. *Quat Int* 353, 98–122. https://doi.org/10.1016/j.quaint.2013.06.034

López-Sánchez, A., et al., 2021. Effects of declining oak vitality on ecosystem multifunctionality: Lessons from a Spanish oak woodland. *For Ecol Manage* 484, 118927. https://doi.org/10.1016/j.foreco.2021.118927
Lorite, J., et al., 2008. Phytosociological review on the forests of Quercus pyrenaica Willd. *Acta Botanica Gallica* 155, 219–233. https://doi.org/10.1080/12538078.2008.10516105
Luis Calabuig, E., et al., 2000. History of landscape changes in northwest Spain according to land use and management, in: Louis Trabaud (Ed.), *Life and Environment in the Mediterranean*. Wit Press, Southampton, pp. 43–86.
Maia, A.S.C., et al., 2005. Sensible and latent heat loss from the body surface of Holstein cows in a tropical environment. *Int J Biometeorol* 50, 17–22. https://doi.org/10.1007/s00484-005-0267-1
Malo, J.E. and Suárez, F., 1995. Herbivorous mammals as seed dispersers in a Mediterranean dehesa. *Oecologia* 104, 246–255. https://doi.org/10.1007/BF00328589
Manzano, P. and Malo, J.E., 2006. Extreme long-distance seed dispersal via sheep. *Front Ecol Environ* 4, 244–248. https://doi.org/https://doi.org/10.1890/1540-9295(2006)004[0244:ELSDVS]2.0.CO;2
Marta-Pedroso, C., et al., 2018. Economic valuation and mapping of ecosystem services in the context of protected area management (Natural Park of Serra de São Mamede, Portugal). *One Ecosystem* 3, e26722. https://doi.org/10.3897/ONEECO.3.E26722
Martinez, C.A., et al., 2014. Moderate warming increases PSII performance, antioxidant scavenging systems and biomass production in Stylosanthes capitata Vogel. *Environ Exp Bot* 102, 58–67. https://doi.org/10.1016/j.envexpbot.2014.02.001
Mastrangelo, M.E., et al., 2014. Concepts and methods for landscape multifunctionality and a unifying framework based on ecosystem services. *Landsc Ecol* 29, 345–358. https://doi.org/10.1007/s10980-013-9959-9
McClure, S.B., et al., 2006. Human behavioral ecology, domestic animals, and land use during the transition to agriculture in Valencia, eastern Spain, in: Kennett, D. J. and Winterhalder, B. (Eds.), *Behavioral Ecology and the Transition to Agriculture*, University of California Press, Berkeley, CA, pp. 197–216.
Meeussen, C., et al., 2021. Microclimatic edge-to-interior gradients of European deciduous forests. *Agric For Meteorol* 311, 108699. https://doi.org/10.1016/j.agrformet.2021.108699
Milchunas, D.G. and Lauenroth, W.K., 1993. Quantitative effects of grazing on vegetation and soils over a global range of environments. *Ecol Monogr* 63, 327–366. https://doi.org/10.2307/2937150
Milchunas, D.G. and Noy-Meir, I., 2002. Grazing refuges, external avoidance of herbivory and plant diversity. *Oikos* 99, 113–130. https://doi.org/10.1034/j.1600-0706.2002.990112.x
Moreno, G., 2008. Response of understorey forage to multiple tree effects in Iberian dehesas. *Agric Ecosyst Environ* 123, 239–244. https://doi.org/10.1016/j.agee.2007.04.006
Moreno, G., et al., 2016. System report: Iberian Dehesas, Spain. AGFORWARD Project Report.
Moreno, G., et al., 2014. Multifunctionality and dynamics of silvopastoral systems. https://www.researchgate.net/publication/263905628_Multifunctionality_and_dynamics_of_silvopastoral_systems

Moreno, G. and Pulido, F.J., 2008. The Functioning, Management and Persistence of Dehesas, in: *Agroforestry in Europe*. Current Status and Future Prospects, Springer, Dordrecht. https://doi.org/10.1007/978-1-4020-8272-6_7
Moreno, G. and Rolo, V., 2019. Agroforestry practices: silvopastoralism, in: *Agroforestry for Sustainable Agriculture*. Burleigh Dodds Science Publishing, Cambridge, UK, pp. 119–164.
Mosquera-Losada, M.R., et al., 2009. Definitions and Components of Agroforestry Practices in Europe, in: *Agroforestry in Europe*. Springer Netherlands, Dordrecht, pp. 3–19. https://doi.org/10.1007/978-1-4020-8272-6_1
Mosquera-Losada, M.R., et al., 2020. *The Importance of Agroforestry Systems in Supporting Biodiversity Conservation and Agricultural Production: A European Perspective*. pp. 243–258. https://doi.org/10.19103/AS.2020.0071.14
Mosquera-Losada, M.R., et al., 2018. Agroforestry in Europe: a land management policy tool to combat climate change. https://doi.org/10.1016/j.landusepol.2018.06.052
Mvondo, E.A., et al., 2022. Tree diversity and shade rate in complex cocoa-based agroforests affect citrus foot rot disease. *Basic Appl Ecol* 64, 134–146. https://doi.org/10.1016/j.baae.2022.08.003
Nair, P.K.R., 1991. State-of-the-art of agroforestry systems. *For Ecol Manage* 45, 5–29. https://doi.org/10.1016/0378-1127(91)90203-8
Nardone, A., et al., 2010. Effects of climate changes on animal production and sustainability of livestock systems. *Livest Sci* 130, 57–69. https://doi.org/10.1016/J.LIVSCI.2010.02.011
Nichiforel, L., et al., 2018. How private are Europe's private forests? A comparative property rights analysis. *Land Use Policy* 76, 535–552. https://doi.org/10.1016/J.LANDUSEPOL.2018.02.034
Olea, L., et al., 2006. The Spanish dehesa. A traditional Mediterranean silvopastoral system linking production and nature conservation, in: *Sustainable Grassland Productivity: Proceedings of the 21st General Meeting of the European Grassland Federation*, Badajoz, Spain, 3–6 April, 2006, pp. 3–13.
Olff, H. and Ritchie, M.E., 1998. Effects of herbivores on grassland plant diversity. *Trends Ecol Evol* 13(7), 261–265. https://doi.org/10.1016/S0169-5347(98)01364-0
Oliveira, C.C., et al., 2021. Daytime ingestive behaviour of grazing heifers under tropical silvopastoral systems: responses to shade and grazing management. *Appl Anim Behav Sci* 240, 105360. https://doi.org/10.1016/j.applanim.2021.105360
Oliveira, G., et al., 1994. Phenological and growth patterns of the Mediterranean oak Quercus suber L. *Trees* 9. https://doi.org/10.1007/BF00197868
Öllerer, K., et al., 2019. Beyond the obvious impact of domestic livestock grazing on temperate forest vegetation – A global review. *Biol Conserv* 237, 209–219. https://doi.org/10.1016/J.BIOCON.2019.07.007
Paim, T. do P., et al., 2014. Thermal comfort index and infrared temperatures for lambs subjected to different environmental conditions. *Sci Agric* 71, 356–361. https://doi.org/10.1590/0103-9016-2013-0191
Pantera, A., et al., 2021. Agroforestry and the environment. *Agrofor Syst* 95, 767–774. https://doi.org/10.1007/S10457-021-00640-8/METRICS
Papachristou, T.G., 2016. Greece's grazing/forage resources for livestock production, in: *Options Méditerranéennes. Series A: Mediterranean Seminars*, pp. 29–42.

Papanastasis, V.P., et al., 1983. Productivity and management of kermes oak brushlands for goats, in: Le Houerou, H.N. (Ed.), *Browse in Africa, the Current State of Knowledge*, ILCA, Ethiopia, pp. 375–381.

Papanastasis, V.P., et al., 2008. Integrating woody species into livestock feeding in the Mediterranean areas of Europe. *Anim Feed Sci Technol* 140, 1–17. https://doi.org/10.1016/j.anifeedsci.2007.03.012

Pérez Luque, A.J., 2022. Análisis de la dinámica de los robledales (Quercus pyrenaica Willd.) frente al cambio global en el límite sur de su distribución (Sierra Nevada).

Pérez-Luque, A.J., et al., 2020. Ecological diversity within rear-edge: a case study from mediterranean quercus pyrenaica willd. *Forests* 12, 10. https://doi.org/10.3390/f12010010

Piñeiro, V., et al., 2020. A scoping review on incentives for adoption of sustainable agricultural practices and their outcomes. *Nat Sustain* 3, 809–820. https://doi.org/10.1038/s41893-020-00617-y

Pinto-Correia, T. and Azeda, C., 2017. Public policies creating tensions in Montado management models: Insights from farmers' representations. *Land Use Policy* 64, 76–82. https://doi.org/10.1016/J.LANDUSEPOL.2017.02.029

Pinto-Correia, T. and Mascarenhas, J., 1999. Contribution to the extensification/intensification debate: new trends in the Portuguese montado. *Landsc Urban Plan* 46, 125–131. https://doi.org/10.1016/S0169-2046(99)00036-5

Plieninger, T., et al., 2003. Effects of land-use history on size structure of holm oak stands in Spanish dehesas: implications for conservation and restoration. *Environ Conserv* 30. https://doi.org/10.1017/S0376892903000055

Proops, L., et al., 2019. Shelter-seeking behavior of donkeys and horses in a temperate climate. *J Vet Behav* 32, 16–23. https://doi.org/10.1016/j.jveb.2019.03.008

Quezel, P. and Medail, F., 2003. Que faut-il entendre par "forêts méditerranéennes" ? *Forêt Méditerranéenne* XXIV, 11–31.

Quintas-Soriano, C., et al., 2022. Effects of land abandonment on nature contributions to people and good quality of life components in the Mediterranean region: a review. *Land Use Policy* 116, 106053. https://doi.org/10.1016/j.landusepol.2022.106053

Ramirez, J.I., et al., 2018. Effects of wild ungulates on the regeneration, structure and functioning of temperate forests: a semi-quantitative review. *For Ecol Manage* 424, 406–419. https://doi.org/10.1016/J.FORECO.2018.05.016

Rigueiro-Rodríguez, A., et al., 2009. Agroforestry in Europe: current status and future prospects, Rigueiro-Rodríguez, A., McAdan, J., Mosquera-Losada, M.R (Eds.), in: *Agroforestry Systems in Europe: Productive, Ecological, and Social Perspective*. . Springer Science, pp. 43–66.

Rivest, D., et al., 2013. A meta-analysis reveals mostly neutral influence of scattered trees on pasture yield along with some contrasted effects depending on functional groups and rainfall conditions. *Agric Ecosyst Environ* 165, 74–79. https://doi.org/10.1016/J.AGEE.2012.12.010

Rivest, D., et al., 2011. Shrub encroachment in Mediterranean silvopastoral systems: Retama sphaerocarpa and Cistus ladanifer induce contrasting effects on pasture and Quercus ilex production. *Agric Ecosyst Environ* 141, 447–454. https://doi.org/10.1016/j.agee.2011.04.018

Robles, A.B., et al., 2009. Role of grazing livestock in sustainable use, fire prevention and naturalization of marginal ecosystems of southeastern Spain, in: *Agroforestry* in *Europe*. Current Status and Future Prospects. Springer, Dordrecht, Netherlands, pp. 211–231.

Roces-Díaz, J.V., et al., 2021. Temporal changes in Mediterranean forest ecosystem services are driven by stand development, rather than by climate-related disturbances. *For Ecol Manage* 480, 118623. https://doi.org/10.1016/J.FORECO.2020.118623

Rodriguez-Estévez, V., et al., 2009. Foraging of Iberian fattening pigs grazing natural pasture in the dehesa. *Livest Sci* 120, 135–143.

Rodriguez-Estevez, V., et al., 2012. Consumption of Acorns by Finishing Iberian Pigs and Their Function in the Conservation of the Dehesa Agroecosystem. Agroforestry for Biodiversity and Ecosystem Services - Science and Practice. InTech, pp. 1–22.

Roese, A.D., et al., 2017. Microclimate in agrosilvopastoral system enhances powdery mildew severity compared to agropastoral and non-integrated crop. Trop Plant Pathol 42, 382–390. https://doi.org/10.1007/s40858-017-0162-4

Rogosic, J., et al., 2006. Sheep and goat preference for and nutritional value of Mediterranean maquis shrubs. *Small Rumin Res* 64, 169–179. https://doi.org/10.1016/J.SMALLRUMRES.2005.04.017

Rois-Díaz, M., et al., 2018. Farmers' reasoning behind the uptake of agroforestry practices: evidence from multiple case-studies across Europe. *Agrofor Syst* 92, 811–828. https://doi.org/10.1007/S10457-017-0139-9/TABLES/5

Rolo, V., et al., 2015. Microhabitat effects on herbaceous nutrient concentrations at the community and species level in Mediterranean open woodlands: the role of species composition. *Grass Forage Science* 70, 219–228. https://doi.org/10.1111/GFS.12110

Rosa García, R., et al., 2012. Goat grazing, its interactions with other herbivores and biodiversity conservation issues. *Small Rumin Res* 107, 49–64. https://doi.org/10.1016/j.smallrumres.2012.03.021

Ruiz de la Torre, J., 1990. Mapa forestal de España: escala 1: 200.000. Memoria general. Instituto Nacional para la Conservación de la Naturaleza.

Ruiz, I. and Sanz-Sánchez, M.J., 2020. Effects of historical land-use change in the Mediterranean environment. *Sci Total Environ* 732, 139315. https://doi.org/10.1016/j.scitotenv.2020.139315

Ruiz-Mirazo, J., et al., 2011. Two-year evaluation of fuelbreaks grazed by livestock in the wildfire prevention program in Andalusia (Spain). *Agric Ecosyst Environ* 141. https://doi.org/10.1016/j.agee.2011.02.002

Ruperez, A., 1957. La encina y sus tratamientos. Ediciones Selv\'\icolas, Madrid.

Sales-Baptista, E. and Ferraz-de-Oliveira, M.I., 2021. Grazing in silvopastoral systems: multiple solutions for diversified benefits. *Agrofor Syst* 95, 1–6. https://doi.org/10.1007/s10457-020-00581-8

San Miguel-Ayanz, A., et al., 2010. Wild ungulates vs extensive livestock. Looking back to face the future. *Options Meditérranneenes* 92, 27–34.

Sánchez, L., 2005. Indigenous breeds and silvopastoral systems, in: *Silvopastoralism and Sustainable Land Management. Proceedings of an International Congress on Silvopastoralism and Sustainable Management Held in Lugo*, Spain, April 2004. CABI Publishing, UK, pp. 231–235. https://doi.org/10.1079/9781845930011.0231

Sanjuán, Y., et al., 2018. Woody plant encroachment following grazing abandonment in the subalpine belt: a case study in northern Spain. *Reg Environ Change* 18, 1103–1115. https://doi.org/10.1007/s10113-017-1245-y

Santiago-Freijanes, J.J., et al., 2021. Global and European policies to foster agricultural sustainability: agroforestry. *Agrofor Syst* 95, 775–790. https://doi.org/10.1007/S10457-018-0215-9/TABLES/2

Santos Neto, C.F., et al., 2022. Microclimate and animal thermal comfort indexes in different silvopastoral system arrangements in Caatinga. *Int J Biometeorol* 66, 449–456. https://doi.org/10.1007/s00484-021-02182-1

Segarra, J., et al., 2023. Managing abandoned Mediterranean mountain landscapes: the effects of donkey grazing on biomass control and floral diversity in pastures. *Catena (Amst)* 233, 107503. https://doi.org/10.1016/J.CATENA.2023.107503

Seligman, N.G. and Perevolotsky, A., 1994. *Has Intensive Grazing by Domestic Livestock Degraded Mediterranean Basin Rangelands?* pp. 93–103. https://doi.org/10.1007/978-94-011-0908-6_9

Serrano Alvarez, J.A., 2014. When the enemy is the state: common lands management in northwest Spain (1850–1936). *Int J Commons* 8, 107. https://doi.org/10.18352/ijc.389

Silanikove, N., 2000. Effects of heat stress on the welfare of extensively managed domestic ruminants. *Livest Prod Sci* 67, 1–18. https://doi.org/10.1016/S0301-6226(00)00162-7

Sisó, S., et al., 2001. Relationship between hydraulic resistance and leaf morphology in broadleaf Quercus species: a new interpretation of leaf lobation. *Trees* 15, 341–345. https://doi.org/10.1007/s004680100110

Smith, M.M., et al., 2022. Silvopasture in the USA: a systematic review of natural resource professional and producer-reported benefits, challenges, and management activities. *Agric Ecosyst Environ* 326, 107818. https://doi.org/10.1016/j.agee.2021.107818

Smith, S.E. and Read, D.J., 2008. *Mycorrhizal Symbiosis*. Academic Press, New York, NY.

Souto-Herrero, M., et al., 2018. Earlywood vessels and latewood width explain the role of climate on wood formation of Quercus pyrenaica Willd. across the Atlantic-Mediterranean boundary in NW Iberia. *For Ecol Manage* 425, 126–137. https://doi.org/10.1016/J.FORECO.2018.05.042

Souza, E.C., et al., 2019. Thermal comfort and grazing behavior of Girolando heifers in integrated crop-livestock (ICL) and crop-livestock-forest (ICLF) systems. *Acta Sci* 41, 46483. https://doi.org/10.4025/actascianimsci.v41i1.46483

Stark, S., et al., 2023. The ecosystem effects of reindeer (Rangifer tarandus) in northern Fennoscandia: Past, present and future. *Perspect Plant Ecol Evol Syst* 58, 125716. https://doi.org/10.1016/J.PPEES.2022.125716

Tittonell, P., et al., 2021. Ecosystem services and disservices associated with pastoral systems from Patagonia, Argentina: a review. *Cahiers Agricultures* 30, 43. https://doi.org/10.1051/cagri/2021029

Torres-Acosta, J.F.J., et al., 2016. Revalorizing the tropical deciduous forest for ruminant production. *Trop Subtrop Agroecosyst* 19, 73–80.

Torres-Fajardo, R.A., et al., 2024. The complexity of goats' feeding behaviour: an overview of the research in the tropical low deciduous forest. *Small Rumin Res* 231, 107199. https://doi.org/10.1016/j.smallrumres.2024.107199

Vandenheede, M., et al., 1995. Use of a shelter by grazing fattening bulls: effect of climatic factors. *Anim Sci* 60, 81–85. https://doi.org/10.1017/S135772980000816X

Varela, E., et al., 2022. Unravelling opportunities, synergies, and barriers for enhancing silvopastoralism in the Mediterranean. *Land Use Policy* 118, 106140. https://doi.org/10.1016/j.landusepol.2022.106140

Vaz, P.G., et al., 2024. Grazing hinders seed dispersal during crop failure in a declining oak woodland. *Sci Total Environ* 907, 167835. https://doi.org/10.1016/J.SCITOTENV.2023.167835

Velado-Alonso, E., et al., 2022. The landscapes of livestock diversity: grazing local breeds as a proxy for domesticated species adaptation to the environment. *Landsc Ecol* 37, 1035–1048. https://doi.org/10.1007/S10980-022-01429-5/FIGURES/4

Velamazán, M., et al., 2020. Ungulates and ecosystem services in Mediterranean woody systems: a semi-quantitative review. *J Nat Conserv* 55, 125837. https://doi.org/10.1016/j.jnc.2020.125837

Vila-Viçosa, C.M., et al., 2022. New annotated checklist of the Portuguese oaks (Quercus L., Fagaceae). *Mediterranean Botany Online First*, 1–46. https://doi.org/10.5209/mbot.79286

Vitelli, M., et al., 2017. Phylogeographic structuring of plastome diversity in Mediterranean oaks (Quercus Group Ilex, Fagaceae). *Tree Genet Genomes* 13, 3. https://doi.org/10.1007/s11295-016-1086-8

Vitousek, P.M., et al., 1997. Introduced species: a significant component of human-caused global change. *N Z J Ecol* 21, 1–6.

Vizzotto, E.F., et al., 2015. Access to shade changes behavioral and physiological attributes of dairy cows during the hot season in the subtropics. *Animal* 9, 1559–1566. https://doi.org/10.1017/S1751731115000877

Wachiye, S., et al., 2022. Effects of livestock and wildlife grazing intensity on soil carbon dioxide flux in the savanna grassland of Kenya. *Agric Ecosyst Environ* 325, 107713. https://doi.org/10.1016/j.agee.2021.107713

Zema, D.A., et al., 2021. Influence of forest stand age on soil water repellency and hydraulic conductivity in the Mediterranean environment. *Sci Total Environ* 753, 142006. https://doi.org/10.1016/j.scitotenv.2020.142006

Chapter 6

Integrating tree crops and livestock in silvopastoral systems

Adolfo Rosati, Research Centre for Olive, Fruit and Citrus Crops – Council for Agricultural Research and Economics (CREA), Italy; Mariano Pauselli, University of Perugia, Italy; and Alberto Mantino, University of Pisa, Italy

1 Introduction

About 45% (48 million km^2) of the planet's inhabitable land is utilized for agriculture, while about 38% is covered by forest, with the remainder being shrubland, water bodies, urban, and built-up land (FAO, 2024). About 80% of agricultural land (38 million km^2) is utilized for livestock production:

- 6 million km^2 of cropland for feed production; and
- 32 million km^2 of grazing land.

http://dx.doi.org/10.19103/AS.2024.0141.11

Only 16% of agricultural land is used for food crops, and 4% for non-food crops. While occupying 80% of all agricultural land, livestock production produces only 17% of the global calorie supply and 38% of the global protein supply. Global population is increasing (Bodirsky et al., 2015) and so is the per capita consumption of livestock products. This has resulted in a more than proportional increase in the demand for livestock products, making food security a concern (Smith and Glauber, 2019). This increasing demand can be matched in different possible ways, including increasing land for livestock production, increasing feed conversion efficiency (i.e. producing more livestock products with the same feed-crop area and pasture area), or increasing land productivity (i.e. more feed and forage with the same land). Alternatively, livestock product consumption per capita could be reduced, either by reducing actual intake or by reducing food waste (or both).

While all these options are likely to be adopted eventually, the most likely in the short term is increasing the land used for livestock (particularly for feed) production, as this option is easiest to adopt, while the others require more research and social change to implement. Additional land for livestock production can be obtained from conversion of forest into agricultural land (i.e. deforestation) or from grazing in the forest. Deforestation has already occurred to a very large extent and is still ongoing; further continuation of this land use change is clearly undesirable. Grazing in the forest could have a large potential, but it is difficult to achieve and could also damage the forest ecosystem if not well managed, and for these reasons it is often subject to legal restrictions.

An alternative possibility to obtain additional land for livestock production is to expand grazing and feed production into existing cropland, without reducing crop production (or reducing it less than the increase in forage, thus increasing overall land productivity). This approach is referred to as crop–livestock integration or integrated crop livestock systems (ICLS). When livestock is integrated into crop areas that include both herbaceous crops and trees, the resulting system is called integrated crop–livestock–forestry systems (ICLFSs), or agrosilvopastoral systems, a particular case of agroforestry combining all three elements (trees, crops, and livestock). Livestock integration in wooded areas is referred to as silvopasture, and grazing in fruit tree crops is a particular case of silvopasture, often referred to as orchard grazing. The latter is the silvopasture part of fruit-tree-based agroforestry systems (FTB-AFS), which include orchard intercropping as the silvoarable part.

In addition to introducing grazing and forage production into existing tree crops, another way to obtain tree crop–livestock integration is to plant trees on pastureland without trees, to a density that does not reduce, but may even increase, forage production, thus increasing overall land productivity in a similar way as the opposite approach (i.e. introducing livestock in existing permanent crops). In fact, introducing fruit trees into grazing and agricultural land is the primary driver of modern agroforestry adoption worldwide (Wolz and Delucia, 2018) as fruit trees provide income much earlier than forest trees.

Worldwide, there are about 189 million ha of permanent crops, including trees, bushes, and vines (FAO, 2024). Although it is difficult to find precise statistics, most of this land is not grazed nor used to produce forage. More precise statistics are available for Europe, where agroforestry is practiced on about 15.4 million ha (8.8% of the agricultural area) if considering only tree-based systems (Den Herder et al., 2015) or 19.77 million ha (5% of the EU land area) if including shrubs and home gardens (Mosquera-Losada et al., 2018). Most (85.6%) of this EU agroforestry area is silvopasture (grazing on wooded areas), but only 11.26% is FT-AFS (Lauri et al., 2019) and only 4.3% (0.85 MHa) is grazed orchards (Mosquera-Losada et al., 2018).

Among the fruit tree species grown in FTB-AFS, olive is by far the most represented species (60% of the total FTB-AFS area), followed by apple (15%) and pear (5%), while all other species represent less than 20% of the FTB-AFS area (Lauri et al., 2019). Pooling together all FTB-AFS in the EU, fruit–livestock systems cover about 68% of the total area, followed by woody polycultures (20%) and fruit-crops polycultures (12%) (Lauri et al., 2019). From these data, it is evident that grazing in the olive orchard is the most widespread FTB-AFS in the EU. Yet, even for this species, most of the acreage (about 90%, Lauri et al., 2019) is not grazed nor intercropped (i.e. monoculture). The situation is even more dramatic for other fruit tree species, with 94% of the area in monoculture (Lauri et al., 2019).

In conclusion, most of the area cultivated with permanent crops is not currently integrated with livestock production. This provides ample opportunities for expansion of agroforestry, by converting fruit tree monocultures into FTB-AFS, in addition to the possibility of planting fruit trees on treeless pastures. Exploiting the forage potential of permanent crops, by integrating them with livestock, can contribute to decreasing grazing land requirements, deforestation and pesticide use, while increasing land productivity (ecological intensification) and farm resiliency. For instance, incorporating livestock into existing oil palm plantations is a recommended climate mitigation strategies that could reduce deforestation by 1–2% on a worldwide scale and by 10–23% in Eastern (Peninsular) and Western (Borneo) Malaysia (Azhar et al., 2021; Grinnell et al., 2022). Extending this practice to all permanent crops worldwide would have a much greater impact.

This chapter deals with the integration of livestock (or forage) production with tree crops, particularly fruit trees and vines. After a brief section on the current extent of FTB-AFSs, the chapter highlights the opportunities this integration offers, at a global and local scale. These opportunities relate to increasing overall land productivity, without competition between forage and food production, while reducing inputs and environmental impact. The chapter then discusses in detail the many challenges this integration entails.

2 Trends in orchard management

Fruit orchards have been grazed and intercropped since time immemorial (Sestini, 1963; Nair, 1993; Lelle and Gold, 1994; Herzog, 1998; Eichhorn et al., 2006; Vannucci, 2009; Rosati et al., 2021), forming systems known as:

- Joualles, Verger haute tige, and Pré-verger in France (Laugénie, 1965; Bérion and Pourcelot, 2007; Dupraz and Liagre, 2011);
- Dehesa, Pomadaras, and Bocage in Spain (Guichard, 1975);
- Piantate and Seminativi arborati (arable land with trees) in Italy (Dupraz et al., 2018); and
- Streuobst in Germany, Austria and Switzerland (Herzog, 1998).

Even more diversified systems were called (Lauri et al., 2019):

- 'coltura promiscua' in Italy;
- 'huertas' in Spain; and
- 'Oasis' in North Africa and the Eastern Mediterranean.

These systems provided diverse goods and greater overall land productivity, making it possible to satisfy different needs of the farming family (Herzog, 1998).

However, crop specialization and intensification in the last century greatly reduced the extent of FTB-AFS (Eichhorn et al., 2006; Mosquera-Losada et al., 2012; Dupraz et al., 2018). Specialization allowed mechanization, economies of scale, increasing labor use efficiency, higher productivity, and crop yields. However, this specialization also oversimplified the system into a monoculture typically characterized by continuous tillage, inorganic fertilization, pest and weed control. This led to soil structural degradation, erosion, environmental pollution, and biodiversity loss, including serious erosion in orchards cultivated on slopes (Abid and Lal, 2009; Montanaro et al., 2017; Novara et al., 2018; Fleskens and Stroosnijder, 2007; Romero et al., 2007).

Soil erosion by water and wind reduces soil fertility, removing the upper and more fertile soil layer. Soil organic matter content is reduced due to continuous tillage and the replacement of manure by chemical fertilizers, further decreasing soil fertility (Palese et al., 2014). Poor soil quality means that fertilizers and other chemicals applied to the orchards are lost by leaching, runoff, and soil erosion, which is detrimental to both the orchard and the environment. The production, transportation and application of fertilizers and other chemicals, and soil tillage with machinery, all contribute greenhouse gas emissions. Decreasing soil organic matter contributes to carbon emissions in the atmosphere and reduces soil carbon stocks (Vignozzi et al., 2019).

To reverse these negative trends, green mulching with either temporary or permanent soil cover has been increasingly adopted (Lauri et al., 2019), because of the opportunities it offers in terms of increasing orchard sustainability (Fig. 1) (Granaststein and Sanchez, 2009). Green mulching protects soil structure, improves water infiltration, promotes biodiversity, prevents erosion, and increases (or at least reduces the loss of) soil organic matter and thus enhances soil fertility, biological activity, and carbon sequestration in the orchard (Hernández et al., 2005, Zuazo et al., 2009; Ramos et al., 2011; Gucci et al., 2012; Palese et al., 2014; Rodrigues et al., 2015; Correia et al., 2015; Alvarez et al., 2017; Garcia et al., 2018; Repullo-Ruibérriz de Torres et al., 2018; Vignozzi et al., 2019). This improves water and nutrient retention, reduces environmental pollution, and can help break down pollutants. Living covers also protect the soil surface from the negative effects of rainfall on soil aggregates, preventing crusting and erosion, and reducing soil compaction due to field traffic (Pardini et al., 2002; Gucci et al., 2012). Green mulching improves field access to machinery and increases agrobiodiversity, with potential benefits for pest control (Pardini et al., 2002; Pantera et al., 2018), although this is an area of research that needs further studying. Given their biological nitrogen fixation properties, legume species are particularly useful in replacing chemical fertilizers, as was done before the introduction of synthetic fertilizers in the last century (Smil, 2001). This practice has become attractive again given the

Figure 1 Cover crop in an olive orchard.

increasing economic and environmental costs of synthetic fertilizers, especially given that nitrogen (N) fertilizers are the largest single source of emissions from agriculture (IPCC, 2007). In the short term, living covers can compete with trees for water and nutrients, especially with young trees, but in the long run, this is usually not the case (Gucci et al., 2012; Vignozzi et al., 2019; Fang et al., 2022). Instead, residues from living covers and other organic materials (e.g. pruning materials) left on the ground as mulch can increase water storage, reduce soil evaporation, and increase soil biological activity (Palese et al., 2014; Neilsen et al., 2017).

While the agronomic benefits of living covers are well known, the economic benefits are less clear (Gómez et al., 2018; Kavvadis and Koubouris, 2019). These benefits can be substantial in the long run, in terms of enhancing the soil capital but, in the short term, the costs might outweigh the economic benefits, explaining why green mulching is not always adopted (Sastre et al., 2016; Gómez et al., 2018). Nonetheless, this technique has been increasingly adopted, especially where there is sufficient rainfall and/or irrigation. Green mulching adoption in orchards is now fostered by the Common Agricultural Policy (CAP) of the EU, which (as of 2023) subsidizes it within the so-called 'ecoschemes', which have been implemented by several member states.

Despite the increasing adoption, there is little use of the biomass produced by naturally occurring green mulches or sown cover crops (Lauri et al., 2019). If this vegetation is now (as in the past) desirable for its services to the orchard and the environment, why then not use it as an additional source of income (as in the past), rather than it representing just a cost? Why not green mulch the orchard with profitable crops if this can be done without detriment to orchard yield? A good living cover should combine environmental services (including agronomical services to the orchard) with economic profitability, increasing overall income (Rosati, 2011; Rosati et al., 2009; Lauri et al., 2019). Surprisingly, this topic has received relatively little attention in the literature.

3 Growing forage crops in orchards

What can be grown under orchards? Fruit trees can be combined with many kinds of crops, depending on orchard design (i.e. species, distance between trees) and pedoclimatic conditions, etc. Modern FTB-AFS require that the crops/livestock cultivated/reared in the orchard be compatible with modern cultivation techniques (see more details in Rosati et al., 2021). High value-added crops can provide better potential income. However, they can be more difficult to grow successfully in an orchard, and without affecting tree yield, than forage crops which are comparatively easier to grow.

Many different species that can grow in an orchard can be grazed or used for forage. This larger choice of forages makes it easier to choose species and cultivars compatible with orchard trees, i.e. those that can be grown either without reducing tree yield or which may even enhance yields. Additionally, weeds that grow within the forage crop can form part of the forage, reducing the need for weed control. In fact, spontaneously growing forage (weeds) can have a similar forage value to cultivated forages (Genty et al., 2023). Furthermore, the whole plant is harvested as forage, so that all the biomass obtained is useful. This contrasts with other crops where e.g. only the fruit may have a value which may be reduced if fruit ripening and yields are reduced more than total biomass by factors such as tree shade. Grassland species are less affected by shade than food crops (Pardini et al., 2010), making them more productive under trees. Forage crops can also be harvested and terminated at different times, with more flexibility compared to a crop, which cannot be harvested until the harvestable part is mature.

These characteristics make it possible to harvest or terminate the forage crop in a flexible manner, adjusting to seasonal conditions. For instance, in a dry spring, the forage crop can be terminated early, leaving more water in the soil for the main (i.e. tree) crop: the forage yield will be lower, but not null (as it would be for a seed or fruit crop, if harvested before seed/fruit formation), and of better quality. On the other hand, in a wet spring, the forage can be grown for longer, favoring orchard accessibility to machinery, and water infiltration and storage in the soil, while increasing forage yield with no risk of competition for water. This larger flexibility of forages, in terms of wider choice of species and management practices, makes it easier to better design and manage the system, in order to maximize complementarity and minimize competition between trees and forages. This makes it possible to produce additional forage from the orchard, while maintaining or even enhancing fruit tree yield, via optimized ecosystem services, such as biological nitrogen fixation, weed suppression, enhanced water infiltration and storage, nutrient cycling, reduced soil erosion, etc. Finally, while perennial forage crops tend to improve soil fertility and properties (see below), cultivating annual crops can both reduce soil organic matter and increase erosion (Caravaca et al., 2002). For these reasons, it may be concluded that, although many other crops are also possible, forages are among the easiest and most compatible crops to be grown as understory crops in orchards.

However, there is currently insufficient literature to guide the best choice of forages for each situation (Ahmad et al., 2017) that can maximize forage production while maintaining or enhancing tree yield. Pardini et al. (2002) reviewed the literature on species and cultivars of living covers most suitable for olive orchards and vineyards, concluding that much work remains to be done. A particular case of forage potential is melliferous plants, which could entail an additional ecosystem service for pollination and a potential income source if exploited for honey production. Planting of melliferous plants in orchards is another subsidy which falls within CAP ecoschemes that EU member states can adopt, which will foster greater adoption of this practice in the EU. Melliferous plants include many culinary herbs, which have an additional value in the herb market.

In conclusion, re-thinking orchard agrosystems to consider their feed potential is probably one of the most promising avenues to sustainably increase and exploit their potential productivity, as has long been recommended by the FAO (Sansoucy et al., 1985). Combining fruit growing with livestock rearing, whether on the same farm or in separate farms, would significantly increase overall productivity while reducing costs and environmental impact in both the fruit and the livestock sectors, promoting a circular economy (Paolotti et al., 2016; Berbel and Posadillo, 2018; Rocchi et al., 2019; Quevedo et al., 2024). Integrating crops and/or livestock in the orchard is not the prerogative of small family farms or marginal systems. On the contrary, for any production system, and especially for livestock rearing, it is necessary to reach a sufficient farm size to make the necessary scale economies for a profitable business. The largest farms often have the best conditions to achieve sufficient scale economies for each of the different combined types of productions and the ones that can invest in the necessary skills and equipment to develop innovative techniques and products and explore new markets.

4 Forage production in fruit orchards: benefits and potential

About 50% of olive cultivation area in Europe is managed with green mulching (Sastre et al., 2016; Gómez et al., 2018), mostly spontaneous vegetation with occasional sown covers. Green mulching can be permanent or temporary (during the wet season) and can be on the whole area or only in alleys (strips) between trees, keeping the tree line weed free. This green mulch is not used as forage in most instances. In fact, while green mulching has been adopted on about 50% of the olive cultivation area in the EU, grazing occurs on less than 10% of the area (see above). The numbers are similar for other fruit tree species (Lauri et al., 2019). This represents a waste of production potential and a missed opportunity to obtain livestock production without competing for land with food production. It is also a missed opportunity to reduce off-farm inputs and costs and to obtain additional ecosystem services (Wilson and Hardesty, 2006; Atucha et al., 2011). Grazing animals can also produce better quality products than intensive production systems without access to grazing, including poultry products (Spencer, 2013; Dal Bosco et al., 2016; Cartoni Mancinelli et al., 2019).

The productivity of the orchard cover varies with a number of factors, including pedoclimatic conditions and management, but can be quite high. For instance, grass–legume mixtures produced up to 48 (Misri, 1988) and 32

(Ahmad et al., 2018) Mg ha^{-1} of green forage in Kashmir apple orchards, up to 25 Mg ha^{-1} in an almond orchard (Makaya and Gangoo, 1995), and about 8–10 Mg ha^{-1} of dry matter in an olive orchard in the Lazio region in central Italy (Campiglia et al., 2005). Cocksfoot (or orchard grass: *Dactylis glomerata* L.) grown under apple trees in the Loess Plateau of China produced 2–6 Mg ha^{-1} of dry biomass (Cao et al., 2021). Growing perennial forages in mango orchards reduced fodder deficiency from 50.6% to 9% (Biradar et al., 2019).

The forage production potential of an orchard does not end with the biomass produced by the green mulch. Fallen fruits before and during harvest can be consumed by livestock (with due attention to avoid choking hazards). This may also contribute to pest control, as fallen fruits are often those more infested with pathogens or pests and consumption by livestock destroys the pests still present in the fruit (see Section 5.5). Additionally, processing of fruit (e.g. for juicing, oil extraction, etc.) produces by-products such as pomace, peel, pulp, etc. These by-products often have good forage value. For instance, 1 ha of olive orchard produces about 1–2 Mg of dry matter in de-pitted olive pomace or cake, a valuable and abundant material, rich in sugars and phenolic compounds which can be used as a feed source, which can be stored dry or ensiled and thus used year-round. This can not only be used as a component of feed without detriment to livestock performance, but it can also improve livestock health and the quality of the livestock products, with health benefits for the consumers (e.g. increased antioxidants in milk, cheese, eggs, etc.), while also reducing livestock emissions (reviewed in Rosati et al., 2021).

Summer pruning materials, as well as winter pruning for evergreen species, also represent a potential source of forage. In olive, for instance, with annual pruning, 1–3 Mg of fresh material is removed, 50% of which is palatable by most livestock and available for a long time during winter, when other fresh forages are scarce (Fig. 2). Deciduous fruit trees will drop their leaves in autumn, and this material also has good forage value (Temel and Pehluvan, 2015). Pruned oil palm fronds are also used as fodder (Thohiran et al., 2023). In summary, using all these mostly unexploited forage resources from the orchard (i.e. green mulch, fallen fruits, processing by-products, pruning materials, and fallen leaves) increases the overall yield of the orchard (i.e. ecological intensification) while allowing product diversification, thus increasing farm resiliency. At the same time this approach also has potential benefits for livestock health and welfare, product quality, consumers' health, and reduced environmental impact. These feeds produced in the orchards can be cut and carried to the animals, avoiding animals' damage to the trees. Alternatively, livestock can graze directly in the orchard, saving costs for mowing, transportation and storage, while avoiding problems related to using machinery in orchards (Burgess et al., 2017; Rosati et al., 2021). Additionally, replacing mowing with grazing prevents the loss of N, in the form of ammonia, from decaying mowed material (Phelan et al., 2015),

Figure 2 Cows and goats grazing olive pruning materials.

which can otherwise be up to almost 50% of the total (Ferreira et al., 2015). However, orchard direct grazing entails other challenges (see below).

The benefits of orchard–livestock integration do not end with the feed potential of the orchard. The trees provide shelter and shade to the grazing animals, which are particularly important in cold and hot weather. Trees can also protect some livestock from predators, like poultry from raptors. They can also create a more natural environment which promotes natural livestock behavior, improving animals' welfare and increasing forage intake, grazing range and livestock performance.

An increasingly important issue is heat stress which already causes significant production losses. According to Thornton et al. (2022), global economic losses in cattle milk and meat production are estimated at $14.65 billion annually currently compared to 2005 levels and could rise to as much as $39.94 billion under the worst-case scenario of increasing greenhouse gas emissions by the end of the century. Integrating livestock with tree crops is a promising approach that can contribute to mitigating these losses. Other benefits of fruit trees on livestock welfare and product quality are reviewed in Lauri et al. (2019) and Rosati et al. (2023).

5 Integrating livestock into tree orchards: benefits and potential

After analyzing the benefits of fruit tree–livestock integration for the animals, let us now analyze the benefits of this integration for the orchard. Among the services that livestock can provide to the orchard there are mowing and weed control, leaf thinning to improve aeration, sucker (water sprout) removal and pruning, soil tillage, fertilizing, and pest control.

5.1 Mowing and weed control

Orchards were traditionally grazed to provide both pasture for the animals and weed control for the orchard. Orchards used to have large spacing between large trees, allowing for grazing of large animals without much damage to the trees, although animals still had to be managed carefully to avoid potential damage. Most modern orchards are now managed with low-vigor or dwarf trees, making it more difficult to avoid damage from large animals. This is probably one of the reasons why grazing was gradually abandoned in the last century, although another reason was farm specialization, with farmers preferring to orient their farms toward a single or few products, resulting in the progressive separation of crop and livestock production.

However, with the current need to increase the economic and environmental sustainability of agriculture, via ecosystem diversification and ecological intensification, grazing in orchards is gaining new interest. Examples abound, with sheep, geese, chickens, and other poultry species grazing in apple, walnut, chestnut, olive, and other orchards and vineyards (Fig. 3) (Reviewed in Niles et al., 2018; Rosati et al., 2021; Paut et al., 2021). Cows are also employed in olive orchards (Iacurto et al., 2021), vineyards, and oil palm plantations (Tohiran et al., 2019; Grinnell et al., 2022). Pigs have been used to weed orchards for centuries as well as currently (Fig. 4) (reviewed in Buehrer and Grieshop, 2014). Rabbits, hares, and other species could also be used.

Figure 3 Sheep grazing in a vineyard.

Figure 4 Pigs grazing in an olive orchard.

Mixing livestock species has been found to be more effective for weed control, promoting a better sward quality (Roepstorff et al., 2000; Sehested et al., 2004). This avoids the frequent situation where the weeds unpalatable to a single livestock species employed become dominant, reducing both weeding efficacy and the forage potential. Previous studies have shown that sequential grazing with cattle followed by sheep offers several benefits. Sheep benefit from mixed grazing with cattle on taller swards, as this arrangement enhances complementary grazing behavior and helps reduce the presence of intestinal parasites (Fraser et al., 2007). Similarly, co-grazing pigs and cows reduces the parasite load of the cows (Roepstorff et al., 2000; Sehested et al., 2004).

Chickens and other birds, as well as sheep and other livestock, can control weeds and fertilize fruit trees, as well as understory crops. Chickens can e.g. weed in between wild asparagus or artichoke crops in orchards (Rosati et al., 2023). Since mature wild asparagus plants are prickly and artichoke plants are bitter to the taste, chickens graze beneath them while destroying weeds, both fertilizing and weeding both the trees and the crops (Fig. 5).

Whatever the species used, grazing must be managed. It is a mistake to fence the orchard and keep the livestock there indefinitely. This will create overgrazing, decreasing forage production, increasing parasite load and inducing soil compaction, and tree damage. Grazing should be done according to forage allowance (Sollenberger et al., 2005) and should be intense (to obtain a mowing effect, with consumption of all species and not just the most palatable ones) and rotational (to allow for forage re-growth and avoiding unnecessary soil compaction). It may be desirable to overgraze the orchard

Figure 5 Chickens grazing in an olive orchard.

before the summer drought to reduce competition for water with the trees. Overgrazing when the soil is dry has lower impact on soil structure than when the soil is moist.

5.2 Leaf thinning to improve aeration, sucker (water sprout) removal, and pruning

Sheep have been employed in New Zealand to pluck leaves near unripe grape bunches, to promote aeration and light availability to berries, improving quality and decreasing the incidence of disease (Hawkes Bay Winegrowers Association, 2010; Niles et al., 2018). Sheep grazing in the orchard also consume lower shoots, pruning the canopy to about 1 m. Whilst this can be damaging in some low-stem orchards, it is acceptable or even desirable in mid- to high-stem orchards (Fig. 6) (Burrough et al., 2010). This can be useful in high-stem orchards and in orchards where trunk shakers are used for harvesting, needing a clean trunk up to a certain height (e.g. most olive orchards). Livestock, including sheep as well as chickens and other poultry, can destroy suckers when they appear in Spring and are tender, saving the need for manual, mechanical, or chemical removal.

5.3 Soil tillage

Chickens can not only graze (i.e. mow the weeds) but till (and fertilize) the soil superficially, thus preparing it for sowing cover or other crops. Similarly, pigs

Figure 6 Sheep grazing in a fig orchard.

can till the soil, uprooting weeds with features such as stolons and rhizomes and preparing the soil for sowing (Andresen et al., 2001; Hermansen et al., 2004).

5.4 Soil fertilization

All animals fertilize the soil while grazing by depositing manure. Fertilization input varies according to management. If animals spend 100% of their time in the orchard and consume 100% of their feed from the orchard, the net input would be zero. However, transforming forage into manure is likely to be beneficial in terms of nutrient cycling. Yet, in most cases, in addition to grazing, animals are fed additional forages grown outside the orchard. If animals spend sufficient time in the orchard, then the external feed becomes an input fertilizer for the orchard. This is particularly the case with poultry, which are usually kept in the orchard 100% of the time, while most of the feed consumed is grown outside the orchard (grazing replaces external feed only minimally for chickens) and is deposited on the orchard floor in the form of chicken manure.

Keeping 120 geese per ha in a vineyard has been reported to contribute 69 kg per ha of nitrogen (N), 82 kg per ha of phosphorus (P), and 44 kg per ha of potassium (K) (Fig. 7) (Palliotti et al., 2020). Sheep grazing in an apple orchard has improved apple leaf K and P content (Corroyer, 2017). Sheep grazing increases soil carbon content compared to chicken manure applications in chestnut orchards (Landi et al., 2016). Grazing had the greatest positive effects on soil properties, including increased soil moisture, compared to cover crops alone or tillage, probably because it combines cover crop benefits with

Figure 7 Geese grazing in a vineyard.

manuring and more frequent vegetation control, compared to mowing, which reduced competition for water and nutrients (Ramos et al., 2010). Three cycles of broilers with 1000 birds per ha in an olive orchard have been estimated to contribute 180 kg per ha of N, 234 kg per ha of P_2O5, and 126 kg per ha of K_2O (Paolotti et al., 2016). These authors also used life cycle assessment (LCA) to assess the environmental impact of integrating chickens in an olive orchard. Results indicated that, when chickens provide fertilization and weeding, the environmental impact of olive cultivation is dramatically reduced, as these are the two operations with the highest environmental impact. The impact of free-range poultry rearing is also reduced, as the need for grazing land is eliminated when using an existing orchard.

5.5 Pest control

Livestock can consume fallen fruit, destroying pests present within the fruit, as is the case with pigs grazing unharvested chestnuts (Mosquera-Losada et al., 2017) and apples (Nunn et al., 2007; Buehrer and Grieshop, 2014). Sheep also consume chestnuts infested with pests such as tortricids (*Cydia* spp.) and curculionids (*Curculio* spp.) that fall earlier, reducing infestations (Landi et al., 2016). Chickens consume fallen olives containing the olive fly larva (*Bactrocera oleae*) (Rosati et al., 2021). Both pigs and chickens disturb the soil, potentially disrupting the pest cycle for those species that overwinter in the topsoil, like plum curculio, oriental fruit moth, codling moth, and olive fly. Geese can also disturb pest life cycles (Clark et al., 1996). Chickens used to be allowed to graze orchards in part to improve insect pest control (Bertoni, 1906; Donno, 1937). Chickens (like ducks) have been used to control asparagus beetles (Chittenden, 1917), which can be useful in fruit-tree–asparagus polycultures (Rosati et al., 2021). Chickens can also help in the control of the olive weevil (*Otiorhynchus cribricollis*) and potentially in the control of spittlebugs, including *Philaenus spumarius*, the main vector of *Xylella fastidiosa*, the bacterium causing the olive quick decline syndrome (OQDS), but research is currently lacking. Sheep can consume fallen apples, destroying codling moth and sawfly present in the fruit (Corroyer, 2017), but this can be extended to all other livestock species consuming fallen apples. Poultry grazing in pear and apple orchards reduced pear midges (*Contarinia pyrivora*) and sawflies (*Hoplocampa testudinea*), respectively (Pedersen et al., 2002). Pigs have been found to consume 96% (Nunn et al., 2007) or 100% of fallen fruits and provide partial control of plum curculio and codling moth (reviewed in Buehrer and Grieshop, 2014).

Livestock readily consume fallen leaves from fruit trees. This can eliminate inoculum of leaf diseases and pests, reducing incidence in subsequent years (Burgess et al., 2016; Corroyer, 2017; Bestman and Timmermans, 2017). Grazing has been also found to reduce voles, which can damage tree roots (Corroyer, 2017), probably because the animals damage vole tunnels by trampling. Several other examples of pest control with livestock are reported in Buehrer and Grieshop (2014). Cover crops can influence insect populations by harboring and sheltering predators as well as pests. Growing grasses instead of legumes e.g. in vineyards can reduce vigor and available N, thus reducing pests that feed on plants, while improving wine quality (see Muscas et al., 2017 and references therein).

Because of the large variety of services that livestock and forage covers can provide to the orchard, some farmers find it convenient to keep some livestock as 'free' workers, independent of their potential direct productivity in terms of livestock products. In conclusion, integrating livestock and/or forage production and permanent crops has many potential benefits for both the trees and the forage/livestock. However, this integration also poses many challenges, which will be explored below.

6 Challenges to livestock-orchard integration: competition for water, nutrients, and light

Given all the potential benefits of livestock–orchard integration, it seems surprising that this integration is not adopted on a larger scale. As mentioned above, a small percentage of the area of permanent crops is grazed or used for forage production. This is probably because this integration entails some challenges which need to be addressed in order to find solutions. These challenges include the following, among others:

- competition for water, nutrients, and light;
- damage to trees and/or to the green cover;
- contamination of fruit;
- risks to animals;
- management challenges related to the complexity of livestock integration; and
- sociocultural and policy challenges, including regulatory burdens.

These are discussed in more detail below.

6.1 Competition for water and nutrients

Fruit trees have lower root density than most herbaceous vegetation (Merwin, 2003), making them poor competitors to other plants. However, research shows that growing a green mulch or cover crop, including a forage crop, in an orchard (or other permanent crop system) can both increase or decrease soil water and nutrient availability for the trees. For instance, growing herbaceous crops under trees can promote deeper tree roots, favoring root compartmentalization in the soil and better exploitation of water and nutrients (Yao et al., 2009; Cardinael et al., 2015; Wei et al., 2024). Cover crops can promote water infiltration and improve water storage in the soil (Hernàndez et al., 2005; Palese et al., 2014). Growing ryegrass with apples increased soil water availability (Wei et al., 2024). However, growing cocksfoot (orchard grass: *Dactylis glomerata* L.) in apple orchards decreased soil water availability, although increasing mowing frequency reduced this effect, and apple yield was not affected (Cao et al., 2021). In fact, increasing mowing frequency decreases cover crop evapotranspiration (Centinari et al., 2013).

A cover of legumes increased soil organic C, total N, and potentially available N for apple trees (Hoagland et al., 2008). Clover or clover–grass mixtures improved soil chemical and physical parameters in an apple orchard (Ahmad et al., 2018). Cover crops increased soil chemical, physical, and biological health but reduced soil water availability and orchard development and yield in an almond orchard (Ramos et al., 2010) and an apple orchard (Fang et al., 2016). In a jujube orchard, however, soil water was decreased in the alley but was increased under the tree row (Ling at al., 2017).

Research also shows that tree yields can be increased or decreased with green covers and grazing. Clover or clover–grass mixtures e.g. increased apple yield (Kuhn and Pedersen, 2009; Ahmad et al., 2018) while annual ryegrass increased pear yield (Fu et al., 2023). Geese grazing in a vineyard increased grape yield (and grape and soil quality) (Palliotti et al., 2020). Sheep grazing in walnut plantations increased tree growth compared to mowing, while slightly improving soil moisture and C and N cycling (Lopez and Moreno, 2017).

The variability in increasing/decreasing tree yield with green covers is related to many factors, but the competition for water and nutrients is probably the main driver (Fang et al., 2022). Legume species are more likely to increase soil N availability, favoring N uptake by the tree, while grasses and other non-N fixing species are more likely to compete with the tree for N availability. For instance, growing cereals under olive trees decreased olive yield, while growing legume crops did not (Daoui and Fatemi, 2014; Amassaghrou et al., 2021, 2023). In other cases, while cereals decreased olive yield, fava beans increased it (Razouk et al., 2016). Olive yield was also increased with annual legumes compared to tillage and natural vegetation (Correia et al., 2015; Chehab et al., 2019). Apple yield was increased by clover or clover–grass mixtures but decreased with grasses alone or cultivated control (Ahmad et al., 2018) although it was not significantly affected by cocksfoot (Cao et al., 2021). For a comprehensive review on the effect of the cover on orchard yield, see the meta-analysis of Fang et al. (2022). An interesting case is that of Fu et al. (2023) who reported increased pear yield with a grass cover (annual ryegrass): the yield increase was attributed to ryegrass removing excess nutrients from over fertilization. While excessive fertilization should be avoided for obvious economic and environmental reasons, these results suggest that the more aggressive competition of grasses might be due to nutrient competition, which can be compensated for with increased fertilization (Pardini et al., 2002, and reference therein). This suggests that, while legume covers can reduce fertilization needs, maintaining or increasing orchard yield, grasses could also be used (producing more biomass/forage) without necessarily reducing orchard yield, if fertilization is appropriately increased. This hypothesis should be investigated in future research.

Herbaceous species with deep roots are likely to compete more for water, while shallow-rooted ones may die back when top-soil water is depleted, leaving a dead mulch which can preserve water deep in the soil, where trees can access it. The presence of a cover increases soil water infiltration and storage into the deep soil, where it will be accessible by the trees during droughts. Mowing or grazing frequency and intensity may also affect competition: frequent mowing/grazing and intense grazing reduce the green biomass and its transpiration, as well as root growth, reducing water and nutrient uptake, while returning biomass to the soil, with a mulching and fertilizing effect. Termination time, i.e. the time at which the cover is killed (by mechanical means like tilling or mowing for non-resprouting species, or by chemical means), is also an important determinant of the onset of competition: early termination reduces competition (Granatstein and Sánchez, 2009) although it reduces the benefits, like the increase in organic matter, which would favor water and nutrient availability to the tree in the long run.

The variable results, in terms of water and nutrient competition and tree yield, imply that, with the right design and management of cover forages, in terms of species (and even cultivar), mowing/grazing and termination management, it should be possible to minimize competition or even increase water and nutrient availability for the tree. The right design depends also on the goal of the farmer. If the tree crop is the primary focus, the cover forage can be designed and managed to minimize the competition or even favor the trees. However, if the forage production is not secondary, it may be desirable to allow some competition and tolerate some reduction in tree yield, if this is more than compensated by an increase in forage production. For instance, in a year of failure of the tree crop (for instance due to unfavorable weather conditions during flowering or to an off year in alternate bearing species), it can be appropriate to allow for maximum forage production. In this case, competition might be even useful to contain tree vigor, which would otherwise be excessive in the absence of fruit (Rosati et al., 2018a, 2018b).

Despite the variable results reported, the literature appears to converge on one aspect: young fruit trees suffer from competition with green covers. This is probably because young trees have shallow roots, occupying the same soil compartment as the herbaceous vegetation (Zhou et al., 2018), while older trees have deeper roots and can uptake water and nutrients from deeper soil levels (Suo et al., 2019). Reduced growth/yield with cover crops in young trees was found, for instance, in olive (Gucci et al., 2012), peach (Welker and Glenn, 1989), and pecan (Carroll, 2016). Similarly, young trees were found to be more sensitive to competition than older trees in the meta-analyses of Fang et al. (2022) even if yield did not decrease. However, the data were pooled across different climatic zones (tropical, subtropical, and temperate). It is likely that, for temperate zones alone, yield would have been found to be reduced for young trees, as the cover crop benefit on yield was generally lower for temperate climate compared to the other climates. It is therefore advisable to leave a weed-free area near young trees. Even better is to apply a plastic, straw, or wood chip mulch near the trees to prevent any weed growth and maintain soil moisture during the first few years of establishment of trees. Fang et al. (2022) found mulch most effective at increasing tree yield.

6.2 Competition for light

A forage or other cover crop does not compete for light with trees. However, trees will shade the cover which can greatly reduce the cover's development and yield (Dufour et al., 2013). This may be positive if it means better weed control.

However, in most instances heavy shade will greatly reduce cover biomass, and thus forage production and grazing allowance. Biomass production does not decrease in proportion to the reduction of photosynthetically active radiation (PAR) since plant radiation use efficiency (RUE) increases with shade (Rosati et al., 2004), including for legume cover crop species (Mauro et al., 2014). This makes forage production in an orchard more efficient in the use of light than separate monocultures. Different species respond differently to tree shade. For instance, legume crops, like fava bean and lentil, experience lower yield reductions, when grown under olive trees, than cereals like wheat and barley (Panozzo et al., 2022; Amassaghrou et al., 2021, 2023). Crops can adapt to moderate shade by morphological or physiological adaptations (Rosati et al., 2018c; Arenas-Corraliza et al., 2018) partly or totally compensating potential yield loss from shade reduction.

Tree shade may even increase biomass production under conditions where high temperature and drought, rather than light, are the limiting factors (Arenas-Corraliza et al., 2018; Hernandez-Esteban et al., 2018; Pitchers et al., 2021), conditions that are increasingly experienced with climate change (Ahmed et al., 2013; Iglesias et al., 2011). Under more arid conditions, tree shade can improve crop microclimate (Lott et al., 2009; Mugunga et al., 2017), buffering extreme temperatures (Peng et al., 2015), increasing air humidity (Campi et al., 2009), and reducing evapotranspiration (Coussement et al., 2018). Increased biomass production under tree shade could also derive from increased soil fertility under trees when soil fertility, rather than light, is the limiting factor. These reasons might explain why forage production under tree shade is at times increased rather than decreased (e.g. Kyriazopoulos et al., 1999).

Choosing species of cover plants that are particularly tolerant to shade provides one solution. Most mature orchards intercept about 50–75% of incident PAR when in leaf. Olive, the most cultivated tree species in the Mediterranean, is still mostly grown in traditional orchards, usually having less than 30% canopy cover (Miranda-Fuentes et al., 2015). Balancing higher RUE with lower PAR means that forage production can be reduced less than the reduction in PAR. More importantly, most orchards are deciduous and don't have leaves for about half of the year. This allows for greater light availability when cultivating winter crops under deciduous trees (Dufour et al., 2013),although trees usually leaf out in spring before the winter crop cycle is over which can then decrease crop performance (Friday and Fownes 2002; Jose et al., 2004; Zamora et al., 2009). Nonetheless, on an annual basis, PAR interception by deciduous trees is much less than 50–75%. Most PAR is available for cover crops in the autumn-spring period, when water is usually not limiting (at least in Mediterranean and temperate climates), and nutrient competition is also not a problem, since trees are dormant. In most orchards, light transmission is also not uniform and there are positions under the trees where light is much higher than 25–50%. For instance, in the alleys between olive hedgerow systems daily PAR levels can be greater than 60% (Rosati et al., 2021), which could be enough to produce as much biomass as in full sun for some species. One more factor to consider is that orchards need some time from planting before they reach full canopy development. During these initial years, the PAR transmitted and available for understory covers is much higher and could be exploited to produce biomass such as forage, while preventing weeds.

In conclusion, considering the whole orchard cycle, there is abundant light availability for green covers, which, combined with greater RUE at lower PAR, provide large opportunities to produce cover biomass such as forages. Additionally, orchards can produce additional forage sources such as fallen fruits and leaves, pruning materials and processing by-products. Combining all of these, the orchard might have a forage potential that is not much reduced compared to a pure pasture, while maintaining tree yield. Additionally, while usually reducing quantity, shade also affects forage/crop quality: protein content is often higher both in grains (e.g. Panozzo et al., 2019) and in forages (e.g. Figueiredo et al., 2023; Mantino et al., 2021). Fiber content can also be affected with variable results, probably depending on the combined effect of shade on both the proportion of leaves and stems and on their tissue composition. Shade also affects the concentration of secondary metabolites (Re et al., 2019). All these aspects need further studying.

7 Challenges to livestock-orchard integration: damage and contamination risks

7.1 Damage to trees and/or to the green cover

Direct grazing in an orchard might be compromised by potential damage to trees by livestock. Damage includes browsing leaves or branches, debarking and rubbing. Animals do consume bark which might have an attractive taste or contain secondary metabolites of nutritional and therapeutic value. Debarking tends to happen particularly at certain times of the year, either because the bark is more palatable or because animals seek out bark more in those times. Certain trees are more targeted than others. Adult grapevines seem less vulnerable to debarking than most trees, at least in the trunk, but very vulnerable to browsing with summer grazing (Schoof et al., 2021; authors' personal observation), including browsing of 1-year-old shoots. Young vines need protection and, if summer grazing is implemented, the training system can be adjusted to move the fruit and young wood further up, for instance raising the first wire or using hanging canes as occurs in top-wire cordon training systems (Schoof et al., 2021).

Walnut trees appear to be unpalatable to sheep during the first years of growth, but trampling may damage the trees during the first 5–6 years (Lopez and Moreno, 2017). Debarking seems to occur more frequently on already damaged trees, whether because it is easier for the animal to continue debarking on the same spot or because they develop a habit to debark the same trees (Lopez-Sanchez et al., 2020). Different livestock species have different inclination to browse and debark trees and, within species, there is large variation between breeds and even more between individuals. Lowland sheep breeds tend to be more sedentary while upland breeds browse more intensively, often standing on their hind legs (McAdam, 2017). Individuals that initiate debarking will quickly teach others and should be promptly removed. Donkeys and goats are among the most destructive to trees. Cows and horses can do more debarking damage than sheep (Lopez-Sanchez et al., 2020). Cows are active browsers and debarkers, especially at certain times of the year. Sheep are better but, if there is not enough good ground forage to consume, they will learn quickly to brows. Some breeds are more suitable due to their low tendency to browse, like Shropshire sheep (Corroyer, 2017; Burgess et al., 2017). Sheep breeds with smaller/shorter body size are preferable to taller ones in low-stem orchards. Shropshire sheep, especially the shorter-legged Danish type and Southdown (or Babydoll) sheep, show a widespread inability to stand on two legs, making them suitable even for vineyard grazing (Conrad et al., 2022; Niles et al., 2018). An Austrian winery has been cross-breeding sheep explicitly to develop animals better suited for vineyard use (Schoof et al., 2021).

Geese are even less prone to debark, but will damage herbaceous shoots that they can reach although their small stature prevents most damage except when the tree canopy is very low. Chickens will peck at almost any vegetation but will not debark trees, and their short stature prevents them from damaging most trees. They can damage small-fruits bushes (currents and other crops with low-hanging fruits). They will even learn to jump to reach the fruit but won't reach more than up to 1 meter or so. Ducks, like geese, will not scratch the soil, having webbed feet, and will reach lower than geese, so their potential damage is very limited. Additionally, they eat insects, snails, and slugs, unlike geese which are vegetarian, which could be useful for pest control. Pigs can uproot very young trees or debark and break them but usually do not damage older trees. They will browse and eat low hanging fruit but will not reach very high. Horses have quite variable behaviors. Individuals that are used to grazing in wooded areas will browse and debark trees with different intensities, depending on season and availability of other forages. However, individuals that don't have browsing experience (i.e. indoor animals, or those kept in environments with no trees) and used to consume hay and/or graze only herbaceous vegetation are much less likely to damage trees and vines. They are therefore often allowed to graze in vineyards and orchards, particularly olive orchards. In fact, among farmers, it is rather common knowledge that horses do not browse olive trees. This is mostly true, but with exceptions, so that monitoring and prompt removal of individuals that browse is always necessary, as for other species (McAdam, 2017). Large animals can compact the soil and should only graze for short periods, possibly when soil moisture is low, and must be removed promptly when the forage has been consumed. This is in fact a general principle for all grazing livestock although smaller animals are generally safer in terms of soil compaction. Chickens and other scratching birds (e.g. turkeys, guinea hens) are very effective weeders: even when they do not consume the forage entirely (as they are mainly granivorous and cannot consume large, fibrous parts), they destroy it by scratching. For this reason, their action can lead to complete destruction of any vegetation and thus to soil compaction, especially near housing and feeding areas. Rotational grazing is clearly essential (Dal Bosco et al., 2014; Westway, 2017).

There are several strategies to control tree damage by livestock. As modern orchards tend to be made with smaller trees, one solution is to use smaller animals. In fact, while cow grazing in the orchard has virtually disappeared (except in countries where traditional agricultural systems still survive), using chickens is gaining interest. However large animals can still be used by choosing species, breeds, and individuals that do not damage trees and vines, and using them at the right time e.g. using horses and cows in vineyards during the dormant season (Fig. 8). There are reports of training sheep, using lithium chloride, to avoid browsing during the vegetative season (Wilson and Hardesty, 2006; Meadows, 2008; Albanel et al., 2017). This compound gives the animals temporary nausea. Administering this compound soon after the animals have consumed tree leaves will induce the animals to stay away from such feed for a while. The animals can then safely graze the orchard/vineyard, with no interest in browsing. Another solution, particularly suited to evergreen trees, like olive, is to feed pruning materials to the animals daily, before allowing them into the orchard: once they have consumed enough material, they are unlikely interested in consuming more for that day, and can safely graze the orchard (Rosati et al., 2021). Another strategy is to use fencing, particularly portable electric fences. This works better in orchards with large alleys, where less fencing per unit of grazable land is necessary. Tree shelters can be used to protect individual trees, allowing grazing between trees. Shelters can be used to protect young trees but will not protect low-stem fruit trees from browsing by large animals. As shelters are frequently used in silvopasture and are not specific to fruit trees, this chapter will not discuss this topic.

Figure 8 Horses grazing in a vineyard.

7.2 Contamination risks for fruit and animals

One of the main reasons why many orchards are not grazed is the risk, or perceived risk, of contamination of fruit and tree products by livestock pathogens. Livestock can carry zoonotic diseases that are transmissible to humans like *Salmonella*, *Campylobacter*, *Listeria,* etc. (Salaheen et al., 2015, and references therein). In some countries there are rules about livestock grazing in orchards and cropland in general. In the US, for instance, the USDA National Organic Program (NOP) and the Good Agricultural Practices (GAP) policies restrict manure application and impose a livestock withdrawal for some months before crop harvest or even for the whole crop cycle. In the UK, some cider manufacturers require that the sheep are removed from apple orchards at least 56 days before harvest (Burgess et al., 2016). Small farms that sell their products directly to the final consumers are exempt from some of these regulations in the US. Independent of the legal frame, it is advisable to respect some withdrawal time on a voluntary basis (Savanna Institute, 2021) although more research is needed to establish the correct withdrawal duration. Existing research findings are insufficient and contradictory, with studies reporting both increased, decreased, or similar risks with livestock integration (Salaheen et al., 2015, and reference therein). The literature is even more scarce for livestock–orchard integration. Contamination of the orchard can also occur with pathogen transport from neighboring livestock facilities (Theofel et al., 2020). Postharvest grazing is one way to comply with these regulations while still providing some pest control (Buehrer and Grieshop, 2014).

Contamination goes also the opposite way: livestock can be at risk from chemicals applied to the crops (Matei et al., 2023), including contamination from persistent chemicals applied to orchards many years before, like DDT and other organochlorine compounds, or heavy metals (Cu, Hg, Zn, Pb, Mn), as widely documented (e.g. Durst and Fluharty, 1988; Willett et al., 1993; Morgan and Taylor, 2004; Genova et al., 2022). Although livestock are most at risk from contaminated forage (Abdulrahman et al., 2023), grazing in orchards is a particular risk, given that orchards tend to be subject to more intense spraying with pesticides which contaminate not only the tree vegetation and the fruit but also ground vegetation and soil. While there are well-established withholding periods (i.e. the time interval after spraying before a crop can be consumed) which are included in pesticide labeling and guidance on use, there is usually no information on when livestock can reenter the orchard and graze. Product labels setting out a re-entry period after spraying refer only to the safety of humans re-entering the orchard to execute field operations and do not cover animals consuming contaminated vegetation. There is currently a lack of information and guidance in this area.

Some studies report the levels of contamination for different chemicals used in orchards. For instance, geese grazing in a vineyard were reported to have increase copper levels in the liver, compared to indoor, non-grazing animals (151 vs. 92 ppm) although both values were below toxicity thresholds (Palliotti et al., 2020). One problematic aspect is that the fate of a chemical, once sprayed into the environment, is not necessarily known, making it difficult to measure the level of unknown metabolites or their impact on animal and human health.

Risks to animals are reduced in organic farms (Penagos-Tabares et al., 2023) where the chemicals allowed are limited and often have shorter or even absent withholding periods. Clearly, a reduction in the use of pesticide chemicals in orchards would be desirable, as it would not only reduce the environmental impact of orchard production, but it would also help livestock integration in orchards. High-stem cider apple orchards appear particularly suitable for sheep grazing, since the tall trees do not suffer from browsing, while spraying is reduced, since fruit appearance is less important for cider making (Burgess et al., 2017; McAdam, 2017), and traditional varieties are often more resistant to pests and diseases. For these reasons, sheep grazing is still practiced in a considerable proportion of traditional orchards in the UK (Burrough et al., 2010). Similarly, organic bush apple orchards have reduced contamination problems with grazing chickens (Bestman and Timmermans, 2017). The current trend toward integrated pest management, reduced chemical inputs, and diversification of farm production creates new opportunities for tree crop–livestock integration (Wilson and Hardesty, 2006).

8 Challenges to livestock–orchard integration: management and policy issues

8.1 Management challenges related to the complexity of livestock integration

While livestock integration into the orchard makes ecological sense, getting the right combination of trees and livestock, and managing them properly, is essential for its success. Managing an orchard is already challenging, and so is managing livestock. Combining them not only means addressing both sets of challenges but also creates new ones related to the interaction between the two elements. The right combination of trees and forages/animals depends on the goal of the farmer. If fruit production is the main goal, then forages and animals must provide as many services as possible to the main crop, while not competing with or damaging it. If the farmer is interested in the additional crops that can be obtained, then some competition between forage and tree crops can be tolerated if overall production is increased. This section briefly mentions some technical aspects to be considered.

Choosing the right forage species is essential. Spontaneous vegetation is the least expensive green cover option as it requires no seed purchase and no sowing operation. Spontaneous covers can also be quite productive. However, many species within a spontaneous cover are well adapted in collective use of natural resources and are thus highly competitive with trees. Spontaneous covers are usually dominated by grasses which are more competitive with trees than legumes.

Using a sown cover can be preferable, as it is then possible to select species that best meet the objectives of the cover, including competing less with trees, by exploiting the potential complementary use of natural resources between the cover and the tree. Legume species fix N and can provide some or all the N fertilization for the tree crop. Providing N in this way can be attractive, depending on the cost of fertilizers. With the recent increase in the cost of fertilizers, biological N fixation is gaining increasing interest as it is becoming more competitive. However, fertilizing with the cover biomass (rather than chemical fertilization) has many other well-known advantages (discussed above) that should be considered.

Grasses tend to be more competitive with the orchard but a mixed grass–legume sward is usually more productive (Tramacere et al., 2024) and provides more balanced nutrition for livestock. Grasses are also more resistant to trampling by livestock. Perennial ryegrass in particular appears to be one of the most resistant whilst, among legumes, white clover is one of the best (Phelan et al., 2015, and reference therein). Most legumes species present a high risk of bloating in ruminants when directly grazed, with exceptions like birds' foot trefoil (*Lotus corniculatus*), berseem clover (*Trifolium alexandrinum*), crown vetch (*Coronilla varia*), sulla (*Hedysarum coronarium*), and sainfoin (*Onobrychis viciifolia*). Mixing high-risk legumes with grasses can be safer if the plan is to have the livestock graze directly in the orchard.

Sown forages can be annual or perennial. Annuals have the advantage that they die naturally: this can create the opportunity for providing a dead mulch, which reduces weeds and soil evaporation, while not competing for water and nutrients in the summer. However, annuals have greater seed and sowing costs than perennials, unless the annuals self-reseed, as it is the case, for instance, with subterranean clover (*Trifolium subterraneum*). Perennials can last a few to several years, reducing establishment costs, but might compete more for water and nutrients with the trees, unless they have a dormant season during the time when the tree crop is active. Nutrient competition is less important as additional nutrients can be applied. Water competition can be tolerated if there is enough rainfall for the tree species or if there is irrigation.

Depending on the shade level in the orchard, shade tolerance of the forage species can be important to increase forage yield and quality (Ahmad et al., 2017). Shade-tolerant species include sulla, several clovers and medics, among leguminous species, and several grasses like orchard grass (*Dactylis glomerata*) and tall fescue (*Festuca arundinacea*) (Campiglia et al., 2005; Borek and Mosquera-Losada, 2019; Hernandez-Esteban and Moreno, 2017, Franca et al., 2017; Tramacere et al., 2025). However, there is intraspecific variation in shade tolerance and more research is needed to select species and cultivars especially suited for shade tolerance and orchard (and silvopasture in general) intercropping. In fact, it is often difficult to even find sufficient species and cultivars to choose from that are suitable for local conditions (Pardini et al., 2002). Most seeds are marketed by a few large, international seed companies that focus on species and cultivars selected for certain areas specialized for seed production, like Oregon and New Zealand, which might not perform well in different climates, or in the orchard, rather than in the open. There is an urgent need for selecting species and cultivars for agroforestry in different pedoclimatic conditions.

Once the right species and cultivar, or mixture of them, has been selected, the right management is necessary for best results. Frequency of mowing/grazing is important as overgrazing and frequent mowing can destroy the forage crop, which might be desirable before a period of drought but might unnecessarily reduce forage yield or compromise the persistence of perennial species or the reseeding of annual ones. If the forage is used on a cut-and-carry basis (i.e. no direct grazing) it may be desirable to let it develop more before cutting and reduce the number of cuttings. However, with direct grazing it might be preferable to graze often to keep the sward small to reduce competition.

It is often not desirable to till the soil. This can be for different reasons, including not exposing the soil to erosion (especially for orchards on slopes), or because tilling promotes undesirable weeds, is expensive or the right machinery is not present, or the soil is too wet and inaccessible. In other cases, the farmer may want to retain an existing cover. In some cases, subsidy schemes might not allow tilling, as is the case with some good agricultural and environmental conditions (GAECs) or ecoschemes within CAP. In these cases, sod seeding might be a solution, but sod seeders are both rare and expensive. Alternative sod seeding methods with commonly available machinery might be possible, but there is lack of knowledge on this practice. The authors have had some success with broadcasting seed into the existing sod, then harrowing with a tooth harrow, and then flail mowing the existing cover very low to the ground to set back (but not kill) the existing vegetation, thus reducing its competition with the emerging sown species. However, seeds of different size and germination ability might require different sod seeding approaches and more knowledge is required for best management.

8.2 Sociocultural and policy challenges

Contemporary farmers often prefer a specialized production of crops or livestock and are not often interested in diversifying their activities. Integrating trees in a pasture, or livestock in an orchard, can be overwhelming as it entails an additional set of skills. Furthermore, livestock requires a daily commitment, unlike fruit trees, and there may be no time/will in the farm for this more intense commitment of time. Having both trees and livestock requires marketing more products and acquiring more facilities and infrastructure, and it may be too expensive and demanding in terms of skills and workload. The biggest obstacle, however, is probably the bureaucratic burden related to each type of product. A large farm with a single monoculture has much less bureaucracy than a small farm producing many different products, making it very difficult for the small farm to diversify. This is possibly the biggest hindrance to diversification and thus to sustainability (Reckling et al., 2023). Many examples could be cited, including the mutual exclusion of permitted CAP actions.

A specific example of policies that challenge orchard–livestock integration is the 'ecoscheme' of the CAP of the EU, which subsidizes green mulching of permanent crops but, in some of the countries adopting this measure, grazing is not allowed. This represents a wasted opportunity to promote ecological intensification and to increase farm productivity, diversification, resilience, and income, while reducing the use of external inputs and the environmental impact of agriculture. Why pay the farmers to mow the green mulch with tractors and gasoline, when livestock could mow it for free (in economic and ecological terms) while producing additional income and sparing agricultural land for grazing elsewhere? This policy should be changed to allow and even subsidise grazing in orchards, given the environmental benefits. In fact, given that the environmental costs of agricultural practices are often not a direct cost to the farmer, policies should favor orchard (and crop in general)–livestock integration to benefit society (Schoof et al., 2021, and references therein). Modifying existing policies, i.e. reducing the bureaucratic burdens to be proportional to the production level (i.e. simplifying the commitments for small productions and/or exempting farmers from those commitments under certain production thresholds), would stimulate farm diversification, promoting resilience and sustainability, and would make it possible for small farmers to make a living, rather than seek other jobs or unemployment subsidies.

Despite these hindrances, livestock integration into orchards and vineyards is growing slowly, often thanks to collaboration between livestock farmers and orchard managers, as reported e.g. for New Zealand (Niles et al., 2018), France (Paut et al., 2021), the US (Wilson and Hardesty, 2006), and Germany (Schoof et al., 2021). Orchard managers are finding it convenient to allow livestock farmers to bring their animals into the orchard during the winter rest, without having to care for the animals personally. The livestock farmers gain free pasture, and orchard managers gain free weeding and fertilization.

9 Future research needs

As mentioned above, shade-adapted species and cultivars should be developed as understory crops (i.e. forages) for agroforestry systems. This would enhance productivity since understory species and cultivars have been selected under full sun conditions and their performance is not necessarily the best possible in the shade (Rosati et al., 2018c; Arenas-Corraliza et al., 2018; Panozzo et al., 2020; Mantino et al., 2021). However, light under the orchard is not uniform. Where the illumination pattern varies in a regular/predictable way, as in hedgerow systems, an interesting possibility is to design, test, and develop multiple understory crop systems, with different species having different shade tolerance/requirements placed according to the illumination available: a sort of precision agroforestry. Although there is some discussion about this in the literature, this topic is still mostly unexplored.

Given that avoiding excessive competition with the trees for water and nutrients is important, selecting earlier maturing species and varieties of cover crops is an important research priority. Earlier crops can be harvested/terminated earlier, before summer drought or even before the onset of fast tree growth, thus avoiding competition. One interesting aspect is the selection of early seed crops, particularly protein-rich legume crops that could be 'harvested' directly by grazing animals. Due to smaller tree sizes in modern orchards, poultry are becoming an interesting option for grazing in orchards, providing effective weeding and fertilization without damaging the trees (Rosati et al., 2021). However, unlike ruminants, which were more common orchard grazers in the past, poultry are not herbivores. Despite consuming the green cover, they still need external feed in similar quantities as non-grazing animals. If the orchard cover could produce seeds that the poultry could 'self-harvest,' this would provide a saving in feed which represents 70% of the cost of chicken production.

Current species/cultivars of seed crops are designed to optimize yield in monocultures, taking advantage of the whole growing season and the water available and produce too late to avoid competition with trees, besides being selected for full-sun exposure. Earlier, shade-tolerant species/cultivars, with lower vegetative development, although less productive than in full sun, would be more compatible (i.e. less competitive for water and nutrients) with orchard management, while still producing enough to substantially reduce feed cost. The benefits would be even greater if N-fixing legume cover crops were used, because of the N fixation and because legumes are higher than cereals in proteins, which are the costliest part of feeds, often imported from distant countries.

Another aspect that would be interesting to explore in future research is the selection of species/cultivars that grow faster during winter (cold tolerance). Winter is the period when most fruit trees are dormant, thus there is low or no risk of competition for water and nutrients, while light is also available. This is an ideal time to produce forages with no detriment to (and many advantages for) trees, but low temperatures limit this potential production. With climate change, however, winter temperatures are becoming much milder in many regions, and this might provide an opportunity to increase winter forage production. There are very strong differences among species/cultivars in terms of winter growth, with some growing early, and others waiting for warmer temperatures in Spring. Some grass species grow very fast during the coldest months. These species might be too competitive if left until late Spring but might provide abundant forage during Winter and can be terminated in early Spring before the onset of competition.

Whether using new or old genotypes, management practices must be appropriate. There is abundant literature on grazing and mowing techniques to optimize forage production in monocultures but little information on forage production in an orchard. Termination time, termination techniques (mowing, flail mowing, roller-crimping, tilling), grazing timing (continuous, rotational), and intensity, all affect the level of competition with trees, forage production, cover duration, and the self-seeding or re-sprouting ability of the cover. The combination of the wide range of possible cover choices and possible cover management techniques with the local pedoclimatic conditions creates an infinite number of possibilities. This is positive as it creates more opportunities to achieve the best design for maximum tree and forage yield. However, it also requires the need for local experimentation and exchange of knowledge on good practices. In this regard, it is worth mentioning that peer-review publications, where anecdotal experiences are rarely reported, are less likely to foster adoption of diversification approaches in the orchard. Guided peer-to-peer farm experience, where farmers can see other farms where diversification is successfully practiced, may have a greater impact.

As mentioned above, annual seeding is expensive and self-reseeding annuals might combine economy (no need to buy and sow seeds annually) with reduced competition (they dye and create a dead mulch during the summer). However, self-reseeding is rare and improving this characteristic would be useful (Mauro et al., 2014). Selecting perennials with a dormant phase during the tree vegetative season would also be useful, achieving a similar goal as self-reseeding annuals.

Since bloating is a problem with grazing animals, selecting species/cultivar with low bloating risk could also be of interest. Bloating is known to be limited by the presence of tannins in the forage. Tannins also appear to reduce methane emissions from ruminants and to improve protein use efficiency and reduce parasitic load (Phelan et al., 2015). High-tannin forages that could combine all three positive aspects could be a research goal. Since achieving the highest forage yield is not the goal (and perhaps not even possible) under orchard conditions, the priority being to avoid competition and maximize fruit production, then the goal of forage production in the orchard could be to produce small quantities (i.e. avoiding competition) of high-quality forage, which could be used to complement lower-quality forages produced in more abundance elsewhere. Higher quality could entail higher protein content (e.g. legumes instead of grasses) but also higher tannin content and diverse species for bloat prevention, parasite control, and diet diversification.

Modern orchards use dwarfing or strongly vigor-reducing rootstocks to reduce tree size and induce earlier fruiting. The right rootstock is usually selected considering an orchard monoculture, with no competition from covers (i.e. tilled soil, at least during the tree growing season). Introducing a competitive cover into an orchard can decrease tree growth and yield. An interesting research topic would be whether using slightly more vigorous rootstocks could compensate for the growth and yield reduction with a competitive cover (Granatstein and Sánchez, 2009), making it possible to obtain the same tree size, growth and yield as in monoculture, but with additional forage production. To the best of our knowledge, no previous studies have investigated this possibility.

New machinery, like less expensive sod seeders, machinery for haymaking in orchard alleys, for collecting and transporting pruning materials out of the orchard and to the animals consuming them, or for storing/drying/ensiling fruit by-products, might greatly facilitate exploiting the orchard forage potential.

10 Conclusion

Up to the middle of the last century livestock grazing in orchards was common. This has been reduced due to a number of factors, including farm specialization into either crop or livestock production only. Increasing use of chemicals for weed and pest control also made it difficult to keep livestock in orchards. However, removal of livestock from permanent crops resulted in problems such as lack of manure and consequent decrease in soil fertility and soil carbon stocks (thus contributing to global warming), increasing costs for weed control, and reduced overall productivity of the land (at least without an increasing amount of external inputs). More sustainable and profitable agricultural approaches are now needed. Changed socio-economic conditions are also favoring more sustainable products and shorter supply chains, including direct sales, creating the conditions for reconsidering polycultures in a modern context. Addressing climate change also calls for reconsidering the mitigation and adaptation potential of polycultures. Most permanent crops do not intercept all the available PAR, because of the necessary space between individual plants or plant rows (due to the plant's physiology or the cultivation/harvesting technique), and/or because the trees are deciduous, intercepting little PAR during the dormant season. The PAR not absorbed by the permanent crop promotes weed growth, which is controlled mechanically or chemically, representing a cost, both economically and environmentally, and a wasted opportunity to produce forage.

Exploiting this biomass potential with suitable understory crops and management, thus avoiding reductions in tree yield, would increase the overall productivity of the permanent cropland, while providing other ecosystem services (i.e. ecological intensification), including those that support orchard yield (improvement of soil fertility, water retention, reduced erosion, N fixation, etc.) and orchard management (e.g. orchard accessibility to machinery). Legume covers are to be preferred when orchard yield is the priority, as they improve soil fertility and are less likely to compete for water and nutrients. Legume–grass mixtures can be preferable when overall (orchard + forage) yield is the goal, as they produce more forage while still not competing severely with the trees. Grasses alone are less desirable, unless forage production is the main goal, or if yield and vigor reduction are actually desirable to improve fruit quality, as found in grapes (Muscas et al., 2017). The shade of the permanent crops could even be beneficial for some crops, especially with climate change and for shade-adapted crops (Arenas-Corraliza et al., 2018). Finally, trees themselves can produce forages in addition to the tree crop for which they are cultivated, like fallen (or harvested) leaves and fruit, and processing by-products. This ecological intensification would provide opportunities for increasing crop diversification and farm resilience, productivity and profits, while reducing the use of external inputs, thus reducing costs and the environmental impact of agriculture.

Considering that permanent crops are cultivated over about 189 million ha worldwide and that the majority of this area is currently not exploited for its forage potential with livestock integration, there is a huge potential for increasing forage production on already cultivated land. This can greatly contribute to meet the increasing food demand of the growing population, particularly livestock products, without further deforestation, while also making agriculture more sustainable and resilient. Agriculture is responsible for about one-quarter of anthropic GHG emissions, most of which are related to producing feed for livestock. At the same time, 80% of all agricultural land worldwide is used for livestock production. Exploiting the already existing forage potential of permanent cropland would help reducing both the need for additional agricultural land for forage production and the related emissions.

However, growing understory crops in an orchard presents many challenges, although forages are less challenging than food crops. Challenges include technical ones, related to field management of tree/forage crops and livestock, choice and availability of suitable species and cultivars/breeds, sociocultural and policy challenges, including bureaucratic burdens. In fact, any farm diversification approach entails increasing the number of products. Exploiting the orchard forage potential might entail raising livestock directly, or allowing a neighbor to graze the orchard, or gathering the forage and selling it outside the farm. In all cases, there would be more products to sell and more bureaucracy to deal with. Much too often, it is this bureaucratic burdens that discourage small farmers in particular. Researchers and policy makers should strive to provide technical solutions and to remove obstacles, including policy and bureaucracy barriers, in order to facilitate orchard–livestock integration (and other sustainable practices like crop diversification), allowing it to unleash its potential in terms of productivity and sustainability.

11 Where to look for further information

- Rosati, A., et al. (2012c). Olivo Asparago selvatico, Pollo rustico. Video on the olive, wild asparagus and free-range chicken system. https://www.youtube.com/watch?v=ALw73WwUr2o (Last accessed on 2 July 2025).

- Further information on fruit trees in temperate agroforestry systems can be found in the recently published second edition of the book Temperate Agroforestry Systems (Gordon, A.M., Newman, S.M. and Coleman, B., Eds), CABI, 2018 and in the book chapters listed in the reference list as Lauri et al., 2019 and Rosati et al., 2023).
- The European Agroforestry Federation website (EURAF; http://www.eurafagroforestry.eu/welcome) provides much information on agroforestry, some of which on fruit trees, including projects results, documents and conference presentations, including the seven European agroforestry conferences organized by EURAF.
- The website of the AGFORWARD (AGroFORestry that Will Advance Rural Development) project, the largest European agroforestry project (https://www.agforward.eu), contains documents and material on different innovative and best agroforestry practices, including on fruit trees.
- The journal Agroforestry Systems (https://link.springer.com/journal/10457) publishes agroforestry articles, including on fruit tree-based agroforestry systems.

12 References

Abdulrahman, N. M., et al. (2023). Effects of pesticide residues in animal by-products relating to public health. *Journal of Applied Veterinary Sciences*, 8(4), 95–103.

Abid, M. and Lal, R. (2009). Tillage and drainage impact on soil quality: II. Tensile strength of aggregates, moisture retention and water infiltration. *Soil and Tillage Research*, 103(2), 364–372.

Ahmad, S., et al. (2017). Scope and potential of hortipastoral systems for enhancing livestock productivity in Jammu and Kashmir. *Indian Journal of Agroforestry*, 19(1), 48–56.

Ahmad, S., et al. (2018). Forage production and orchard floor management through grass/legume intercropping in apple based agroforestry systems. *International Journal of Chemical Studies*, 6(1), 953–958.

Ahmed, M., et al. (2013). Could agricultural system be adapted to climate change? A review. *Australian Journal of Crop Science*, 7, 1642–1653.

Albanell, E., et al. (2017). Using long-term averted goats for selective grazing in olive groves. *Animal*, 11(10), 1832–1838.

Alvarez, R., et al. (2017). Cover crop effects on soil and subsequent crops in the pampas: a meta-analysis. *Soil and Tillage Research*, 170, 53–65. https://doi.org/10.1016/j.still.2017.03.005.

Amassaghrou, A., et al. (2021). Productivité et efficience des systèmes agroforestiers à base d'oliviers au Maroc: cas de Moulay Driss Zerhoun. *Cahiers Agricultures*, 30, 2. https://doi.org/10.1051/cagri/2020041.

Amassaghrou, A., et al. (2023). Impact of legumes and cereals on olive productivity in the South Mediterranean. *Malaysian Journal of Sustainable Agriculture*, 7(1), 6–13.

Andresen, N., et al. (2001). Pigs on grassland–animal growth rate, tillage work and effects in the following winter wheat crop. *Biological agriculture & Horticulture*, 18, 327–343.

Arenas-Corraliza, M. G., et al. (2018). Winter cereal production in a Mediterranean silvoarable walnut system in the face of climate change. *Agriculture, Ecosystems & Environment*, 264, 111–118. https://doi.org/10.1016/j.agee.2018.05.024.

Atucha, A., et al. (2011). Long-term effects of four groundcover management systems in an apple orchard. *HortScience*, 46(8), 1176–1183.

Azhar, B., et al. (2021). Mitigating the risks of indirect land use change (ILUC) related deforestation from industrial palm oil expansion by sharing land access with displaced crop and cattle farmers. *Land Use Policy*, 107, 105498.

Berbel, J. and Posadillo, A. (2018). Review and analysis of alternatives for the valorisation of agro-industrial olive oil by-products. *Sustainability*, 10(1), 237.

Bérion, P. and Pourcelot, J. (2007). Approche géographique du terroir du Kirsh de Fougerolles. *Territoire*, 36, 6–9.

Bestman, M. and Timmersmans, B. (2017). Commercial apple orchards in poultry free-range areas. Agroforestry innovation leaflet produced within the AGFORWARD project. https://www.agforward.eu/documents/leaflets/39_Commercial_apple_orchards_in_poultry_free-range_areas.pdf. Accessed 7 October 2024.

Bertoni, G., 1906. Chickens and Poultry production (Original title: "Polli e pollicoltura"). Il coltivatore. G. Agric. Prat. 33, 200e203.

Biradar, N., et al. (2019). Effect of participatory dissemination of fodder innovations in mango orchards of livestock farmers on fodder availability. *Range Management and Agroforestry*, 40(1), 156–161.

Bodirsky, B.L., et al. (2015). Global food demand scenarios for the 21st century. *PLoS One*, 10, 1–27. https://doi.org/10.1371/journal.pone.0139201.

Borek, R. and Mosquera-Losada, M. R. (2019). Shade tolerant grasses and legumes. https://euraf.isa.utl.pt/files/pub/20191006_factsheet_43_en_web.pdf. Accessed 20 January 2025.

Buehrer, K. A. and Grieshop, M. J. (2014). Postharvest grazing of hogs in organic fruit orchards for weed, fruit, and insect pest management. *Organic Agriculture*, 4, 223–232.

Burrough, A. E., et al. (2010). *Traditional Orchard Project in England – The Creation of an Inventory to Support the UK Habitat Action Plan*, Natural England, London, UK.

Burgess, P., et al. (2016). System report: grazed orchards in England and Wales. https://www.agforward.eu/documents/WP3_UK_Grazed_Orchards_system_description.pdf. Accessed 7 October 2024.

Burgess, P., et al. (2017). Economic benefits of grazed apple orchards in England. Agroforestry innovation leaflet produced within the AGFORWARD project. https://www.agforward.eu/documents/leaflets/24_Economic_benefits_of_grazed_apple_orchards_in_England.pdf. Accessed 7 October 2024.

Campi, P., et al. (2009). Effects of tree windbreak on microclimate and wheat productivity in a Mediterranean environment. *European Journal of Agronomy*, 30, 220–227. https://doi.org/10.1016/j.eja.2008.10.004.

Campiglia, E., et al. (2005). *L'inerbimento dell'oliveto con leguminose annuali autoriseminanti*. ARSIAL (AGENZIA REGIONALE PER LO SVILUPPO). E L'INNOVAZIONE DELL'AGRICOLTURA DEL LAZIO).

Cao, Q., et al. (2021). The effects of cocksfoot cover crop on soil water balance, evapotranspiration partitioning, and system production in an apple orchard on the Loess Plateau of China. *Soil and Tillage Research*, 205, 104788.

Caravaca, F., et al. (2002). Land use in relation to soil chemical and biochemical properties in a semiarid Mediterranean environment. *Soil and Tillage Research*, 68(1), 23–30.

Cardinael, R., et al. (2015). Competition with winter crops induces deeper rooting of walnut trees in a Mediterranean alley cropping agroforestry system. *Plant and Soil*, 391(1), 219–235.

Carroll, B. (2016). *Use of Legumes in Pecan Orchards*. Oklahoma Cooperative Extension Service.

Cartoni Mancinelli, A., et al. (2019). Rearing Romagnola geese in vineyard: pasture and antioxidant intake, performance, carcass and meat quality. *Italian Journal of Animal Science*, 18(1), 372–380.

Centinari, M., et al. (2013). Cover crop water use in relation to vineyard floor management practices. *American Journal of Enology and Viticulture*, 64, 522–526. https://doi.org/10.5344/ajev.2013.13025.

Chehab, H., et al. (2019). Effects of compost, olive mill wastewater and legume cover cropson soil characteristics, tree performance and oil quality of olive trees cv. Chemlali grown under organic farming system. *Scientia Horticulturae*, 253, 163–171.

Chittenden, F. H. (1917). *The asparagus beetles and their control* (No. 837). US Department of Agriculture.

Conrad, L., et al. (2022). Sheep in the Vineyard: suitability of different breeds and potential breeding objectives. *Animals*, 12(19), 2575.

Correia, C. M., et al. (2015). Leguminous cover crops improve the profitability and the sustainability of rainfed olive (Olea europaea L.) orchards: from soil biology to physiology of yield determination. *Procedia Environmental Sciences*, 29, 282–283.

Corroyer, N. (2017). Apple orchards grazed in France. Agroforestry innovation leaflet produced within the AGFORWARD project. https://www.agforward.eu/documents/leaflets/23_Apple_orchards_grazed_in_France.pdf. Accessed 7 October 2024.

Coussement, T., et al. (2018). A tree-bordered field as a surrogate for agroforestry in temperate regions: where does the water go? *Agricultural Water Management*, 210, 198–207.

Clark, M. S., and Gage, S. H. (1996). Effects of free-range chickens and geese on insect pests and weeds in an agroecosystem. *American Journal of Alternative Agriculture*, 11(1), 39–47.

Dal Bosco, A., et al. (2014). Effect of range enrichment on performance, behaviour and forage intake of freerange chickens. *Journal of Applied Poultry Research*, 23, 137–45.

Dal Bosco, A., et al. (2016). Transfer of bioactive compounds from pasture to meat in organic free-range chickens. *Poultry Science*, 95, 2464–2471.

Daoui, K. and Fatemi, Z. (2014). Agroforestry systems in Morocco: the case of olive tree and annual crops association in Saïs Region. In: Behnassi, M., et al. (eds) *Science, Policy and Politics of Modern Agricultural System*. Springer. https://doi.org/10.1007/978–94-007-7957-0_19.

Den Herder, M., et al. (2017). Current extent and stratification of agroforestry in the European Union. *Agriculture, Ecosystems & Environment*, 241, 121–132.

Donno, G., 1937. Pollicoltura e Frutticoltura – Rivista di Avicoltura 10, 281–284.

Dufour, L., et al. (2013). Assessing light competition for cereal production in temperate agroforestry systems using experimentation and crop modelling. *Journal of Agronomy and Crop Science*, 199, 217–227.

Dupraz, C. and Liagre, F. (2011). *Agroforesterie, des arbres et des cultures*. Éditions France-Agricole.

Dupraz, C., et al. (2018). Temperate agroforestry: the European way. In: Gordon, A., et al. (eds) *Temperate agroforestry systems*. CABI.

Durst, H. I. and Fluharty, L. B. (1988). Pesticide-laden orchard soils used as pasture for dairy cows. *Journal of Dairy Science*, 71(Suppl. 1), 149.

Eichhorn, M. P., et al. (2006) Silvoarable systems in europe - past, present and future prospects. *Agroforestry Systems*, 67, 29–50.

Fang, K. K., et al. (2016). Comparative analysis on spatial variability of soil moisture under different land use types in orchard. *Scientia Horticulturae*, 207, 65–72. https://doi.org/10.1016/j.scienta.2016.05.017.

Fang, L., et al. (2022). The effects of ground cover management on fruit yield and quality: a meta-analysis. *Archives of Agronomy and Soil Science*, 68(13), 1890–1902.

FAO. (2024).Faostat, FAO Statistics Division. http://faostat3.fao.org Accessed 26 July 2024.

Ferreira, I. Q., et al. (2015). Management of nitrogen-rich legume cover crops as mulch in traditional olive orchards. *Communications in Soil Science and Plant Analysis*, 46(15), 1881–1894.

Figueiredo, C. C. D., et al. (2023). Nitrogen use efficiency in an agrisilviculture system with gliricidia sepium in the Cerrado Region. *Plants*, 12(8), 1647.

Fleskens, L. and Stroosnijder, L. (2007). Is soil erosion in olive groves as bad as often claimed? *Geoderma*, 141(3–4), 260–271.

Franca, A., et al. (2017). Shade tolerant legumes. https://www.agforward.eu/documents/leaflets/08_Shade_tolerant_legumes.pdf. Accessed 20 January 2025.

Fraser, M. D., et al. (2007). Effects on animal performance and sward composition of mixed and sequential grazing of permanent pasture by cattle and sheep. *Livestock Science*, 110(3), 251–266.

Friday, J. B. and Fownes, J. H. (2002). Competition for light between hedgerows and maize in an alley cropping system in Hawaii, USA. *Agroforestry Systems*, 55, 125–137. https://doi.org/10.1023/A:1020598110484.

Fu, H., et al. (2023). Effect of planting and mowing cover crops as livestock feed on soil quality and pear production. *Frontiers in Plant Science*, 13, 1105308.

Garcia, L., et al. (2018). Management of service crops for the provision of ecosystem services in vineyards: a review. *Agriculture, Ecosystems & Environment*, 251, 158–170. https://doi.org/10.1016/j.agee.2017.09.030.

Genova, G., et al. (2022). Copper and zinc as a window to past agricultural land-use. *Journal of Hazardous Materials*, 424, 126631.

Genty, L., et al. (2023). Promising weeds forage potential in perennial Mediterranean agroecosystems. *Agriculture, Ecosystems & Environment*, 347, 108388.

Gómez, J. A., et al. (2018). Soil erosion control, plant diversity, and arthropod communities under heterogeneous cover crops in an olive orchard. *Environmental Science and Pollution Research*, 25(2), 977–989.

Granatstein, D. and Sánchez, E. (2009). Research knowledge and needs for orchard floor management in organic tree fruit systems. *International Journal of Fruit Science*, 9(3), 257–281.

Grinnell, N. A., et al. (2022). Cattle-oil palm integration–a viable strategy to increase Malaysian beef self-sufficiency and palm oil sustainability. *Livestock Science*, 259, 104902.

Gucci, R., et al. (2012). Changes of soil properties and tree performance induced by soil management in a high-density olive orchard. *European Journal of Agronomy*, 41, 18.

Guichard, M. F. (1975). La région du VinhoVerde (Portugal). *Revue géographique des Pyrénées et du Sud-Ouest*, 46(4), 403–429.

Hawkes Bay Winegrowers Association. (2010). Guide to using sheep for leaf-plucking in the vineyard. https://www.premier1supplies.com/img/newsletter/09-05-13-sheep/sheep-forleaf-plucking-booklet.pdf. Accessed 13 September 2024.

Hermansen, J. E., et al. (2004). Integration of organic animal production into land use with special reference to swine and poultry. *Livestock Production Science*, 90(1), 11–26.

Hernández, A. J., et al. (2005). Effects of different management practices on soil conservation and soil water in a rainfed olive orchard. *Agricultural Water Management*, 77(1–3), 232–248.

Hernández-Esteban, A., et al. (2019). Are sown legume-rich pastures effective allies for the profitability and sustainability of Mediterranean dehesas? *Agroforestry Systems*, 93, 2047–2065.

Hernandez-Esteban, A. and Moreno, G. (2017). Establishing pastures rich in legumes. https://www.agforward.eu/documents/leaflets/01_Establishing_pastures_rich_in_legumes.pdf. Accessed 20 January 2025.

Herzog, F. (1998). Streuobst: a traditional agroforestry system as a model for agroforestry development in temperate Europe. *Agroforestry Systems*, 42, 61–80.

Hoagland, L. O. R. I., et al. (2008). Orchard floor management effects on nitrogen fertility and soil biological activity in a newly established organic apple orchard. *Biology and Fertility of Soils*, 45, 11–18.

Iacurto, M., et al. (2021). Interaction between beef herd and olive grove in Lazio (Italy) organic farm. In: Spano, D., et al. (eds) *Book of abstracts of the 5th EUROPEAN AGROFORESTRY CONFERENCE*, 17–19 May 2021 Italy, pp. 439–441.

Iglesias, A., et al. (2011). Towards adaptation of agriculture to climate change in the Mediterranean. *Regional Environmental Change*, 11, 159–166. https://doi.org/10.1007/s10113-010-0187-4.

Intergovernmental Panel on Climate Change. (2007). Climate change 2007: the physical science basis. Contribution of working group I to the fourth assessment report of the intergovernmental panel on climate change. https://www.ipcc.ch/publications_and_data/publications_ipcc_fourth_assessment_report_wg1_report_the_physical_science_basis.htm.

Jose, S., et al. (2004). Interspecific interactions in temperate agroforestry. *Agroforestry Systems*, 61, 237–255.

Kavvadias, V. and Koubouris, G. (2019). Sustainable soil management practices in olive groves. In: Panpatte, D. G. and Jhala, Y. K. (eds) *Soil fertility management for sustainable development*. Springer, pp 167–188. https://doi.org/10.1007/978-981-13-5904-0_8.

Kuhn, B. F., and Pedersen, H. L. (2009). Cover crops and mulching effects on yield and fruit quality in un-sprayed organic apple production. *European Journal of Horticultural Science*, 74, 247–253.

Kyriazopoulos, A. P., et al. (1999). Effects of wild cherry (Prunus avium) and grazing on the herbaceous understorey and its crude protein content.

Landi, S., et al. (2016). Evaluation of sheep grazing effects on nematode community, insect infestation and soil fertility in sweet chestnut orchards: a case of study. *Redia: Journal of Zoology/Giornale di Zoologia*, 99, 117–126.

Laugénie, C. (1965). La culture du Noyer au Perigord. *Revue géographique des Pyrénées et du Sud-Ouest*, 36(2), 135–58.

Lauri, P. É., et al. (2019). Agroforestry for fruit trees in the temperate Europe and dry Mediterranean. In: Mosquera-Losada, M. R. and Prabhu, R. (ed.) *Agroforestry for sustainable agriculture*, Burleigh Dodds Science Publishing. pp. 385–418. ISBN: 978 1 78676 220 7; www.bdspublishing.com

Lelle, M. A. and Gold, M. A. (1994). Agroforestry systems for temperate climates: lessons from Roman Italy. Forest & Conservation History, 38, 118–126.

Ling, Q., et al. (2017). Soil water effects of agroforestry in rainfed jujube (Ziziphus jujube Mill.) orchards on loess hillslopes in Northwest China. *Agriculture, Ecosystems & Environment*, 247, 343–351. https://doi.org/10.1016/j.agee.2017.06.031.

Lopez, M. L. and Moreno, G. (2017). Grazing sheep under walnut trees. Agroforestry innovation leaflet produced within the AGFORWARD project. https://www.agforward.eu/documents/leaflets/16_Grazing_sheep_under_walnut_trees.pdf. Accessed 7 October 2024.

López-Sánchez, A., et al. (2020). Challenges on the conservation of traditional orchards: tree damage as an indicator of sustainable grazing. *Journal of Environmental Management*, 257, 110010.

Lott, J. E., et al. (2009). Agricultural and forest meteorology understorey microclimate and crop performance in a Grevillea robusta -based agroforestry system in semi-arid Kenya. *Agricultural and Forest Meteorology*, 149, 1140–1151. https://doi.org/10.1016/j.agrformet.2009.02.002.

Makaya, A. S. and Gangoo, S. A. (1995). Forage yield of pasture grasses and legumes in Kashmir valley. *Forage Research*, 21(3), 152–154.

Mantino, A., et al. (2021). Competition for light affects alfalfa biomass production more than its nutritive value in an olive-based alley-cropping system. *Forests*, 12(2), 233.

Matei, M., et al. (2023). Persistent organic pollutants (POPs): a review focused on occurrence and incidence in animal feed and cow milk. *Agriculture*, 13(4), 873.

Mauro, R. P., et al. (2014). Phenological and growth response of legume cover crops to shading. *The Journal of Agricultural Science*, 152(6), 917–931.

McAdam, J. (2017). Key challenges of orchard grazing. Agroforestry innovation leaflet produced within the AGFORWARD project. https://www.agforward.eu/documents/leaflets/25_Key_challenges_of_orchard_grazing.pdf. Accessed 7 October 2024.

Meadows, R. (2008). Trained ovines chomp on weeds, avoid vines. *California Agriculture*, 62(1), 10–10.

Merwin, I. A. (2003). Orchard-floor management systems. In: Ferreeand, D. C. and Warrington, I. J. (eds.) *Apples: botany, production, and uses*. CABI, pp. 303–318.

Miranda-Fuentes, A., et al. (2015). Towards an optimized method of olive tree crown volume measurement. *Sensors*, 15, 3671–3687. https://doi.org/10.3390/s150203671.

Montanaro, G., et al. (2017). Carbon budget in a mediterranean peach orchard under different management practices. *Agriculture, Ecosystems & Environment*, 238, 104–113. https://doi.org/10.1016/j.agee.2016.05.031.

Mosquera-Losada, M. R., et al. (2012). Past, present, and future of agroforestry systems in Europe. In: Nair, P.K. and Garrity, D. (eds) *Agroforestry - the future of global land use*. Springer, pp. 286–312.

Mosquera-Losada, M. R., et al. (2017). Protecting trees in chestnut stands grazed with Celtic pigs. Agroforestry innovation leaflet produced within the AGFORWARD project. https://www.agforward.eu/documents/leaflets/17_Protecting_trees_in_chestnut_stands_grazed_with_Celtic_pigs.pdf. Accessed 7 October 2024.

Mosquera-Losada, M. R., et al. (2018). Agroforestry in Europe: a land management policy tool to combat climate change. *Land Use Policy*, 78, 603–613.

Morgan, R. K. and Taylor, E. (2004). Copper accumulation in vineyard soils in New Zealand. *Environmental Sciences*, 1(2), 139–167.

Mugunga, C. P., et al. (2017). Tree-crop interactions in maize-eucalypt woodlot systems in Southern Rwanda. *European Journal of Agronomy*, 86, 78–86. https://doi.org/10.1016/j.eja.2017.03.004.

Muscas, E., et al. (2017). Effects of vineyard floor cover crops on grapevine vigor, yield, and fruit quality, and the development of the vine mealybug under a Mediterranean climate. *Agriculture, Ecosystems & Environment*, 237, 203–212.

Nair, P. K. R. (1993). *An introduction to agroforestry*. Kluwer.

Niles, M. T., et al. (2018). Ecological and economic benefits of integrating sheep into viticulture production. *Agronomy for Sustainable Development*, 38, 1–10.

Neilsen, G. H., et al. (2017). Advances in soil and nutrient management in apple cultivation. In Kate, E. (ed.) *Achieving sustainable cultivation of apples, burleigh dodds series in agricultural science, number 018*. Burleigh Dodds Science Publishing Limited, pp. 239–278.

Novara, A., et al. (2018). The impact of soil erosion on soil fertility and vine vigor: a multidisciplinary approach based on field, laboratory and remote sensing approaches. *Science of the Total Environment*, 622–623, 474–480. https://doi.org/10.1016/j.scitotenv.2017.11.272.

Nunn, L., et al. (2007). Rotationally grazing hogs for orchard floor management in organic apple orchards. *Acta Horticulturae*, 737, 71.

Palese, A. M., et al. (2014). Influence of soil management on soil physical characteristics and water storage in a mature rainfed olive orchard. *Soil and Tillage Research*, 144, 96–109.

Panozzo, A., et al. (2019). Durum wheat in organic olive orchard: good deal for the farmers? *Agroforestry Systems*, 94, 1707–1717.

Panozzo, A., et al. (2020). Morphology, phenology, yield, and quality of durum wheat cultivated within organic olive orchards of the mediterranean area. *Agronomy*, 10(11), 1789. https://doi.org/10.3390/agronomy10111789.

Panozzo, A., et al. (2022). Impact of olive trees on the microclimatic and edaphic environment of the understorey durum wheat in an alley orchard of the mediterranean area. *Agronomy*, 12. https://doi.org/10.3390/agronomy12020527.

Pantera, A., et al. (2018). Agroforestry for high value tree systems in Europe. *Agroforestry Systems* (in press). https://doi.org/10.1007/s10457-017-0181-7.

Paut, R., et al. (2021). Orchard grazing in France: multiple forms of fruit tree–livestock integration in line with farmers' objectives and constraints. *Forests*, 12(10), 1339.

Paolotti, L., et al. (2016). Combining livestock and tree crops to improve sustainability in agriculture: a case study using the Life Cycle Assessment (LCA) approach. *Journal of Cleaner Production*, 131, 351–363.

Palliotti, A., et al. (2020). Là dove osano le oche. Terra e Vita 35: 46–48.Terra e Vita is a national Italian journal, written in Italian.

Pardini, A., et al. (2002). Cover crop species and their management in vineyards and olive groves. *Advances in Horticultural Science*, 16, 225–234.

Pardini, A., et al. (2010). Efecto del arbolado enla producción de pasto y trigo (Triticum aestivum L.) ecológicos en la Maremma Toscana (Italia central). *Pastos*, 40(2), 221–223.

Pedersen, H. L., et al. (2002). *Combined production of broilers and fruit*. NJF Seminar No. 346, Organic Production of Fruit and Berries. Arslev, October 22, 2002.

Peng, X., et al. (2015). Photosynthetic response of soybean to microclimate in 26-year-old tree-based intercropping systems in Southern Ontario, Canada. *PLoS One*, 10, 1–10. https://doi.org/10.1371/journal.pone.0129467.

Phelan, P., et al. (2015). Forage legumes for grazing and conserving in ruminant production systems. *Critical Reviews in Plant Sciences*, 34(1–3), 281–326.

Pitchers, B., et al. (2021). Apple tree adaptation to shade in agroforestry - An architectural approach. *American Journal of Botany*, 108 (in press).

Quevedo-Cascante, M., et al. (2024). Life cycle assessment and modeling approaches in silvopastoral systems: a case study of egg production integrated in an organic apple orchard. *Journal of Environmental Management*, 372, 123377.

Ramos, M. E., et al. (2010). Cover crops under different managements vs. frequent tillage in almond orchards in semiarid conditions: effects on soil quality. *Applied Soil Ecology*, 44(1), 6–14.

Ramos, M. E., et al. (2011). Soil responses to different management practices in rainfed orchards in semiarid environments. *Soil and Tillage Research*, 112(1), 85–91.

Razouk, R., et al. (2016). Optimal distance between olive trees and annual crops in rainfed intercropping system in northern Morocco. *Journal of Crop Science Research*, 1, 23–32.

Re, G. A., et al. (2019). Polyphenolic composition and antioxidant capacity of legume-based swards are affected by light intensity in a Mediterranean agroforestry system. *Journal of the Science of Food and Agriculture*, 99(1), 191–198.

Reckling, M., et al. (2023). Diversification for sustainable and resilient agricultural landscape systems. *Agronomy for Sustainable Development*, 43(4), 44.

Repullo-Ruibérriz de Torres, M. A., et al. (2018). Efficiency of four different seeded plants and native vegetation as cover crops in the control of soil and carbon losses by water erosion in olive orchards. *Land Degradation & Development*, 29(8), 2278–2290.

Rocchi, L., et al. (2019). Assessing the sustainability of different poultry production systems: a multicriteria approach. *Journal of Cleaner Production*, 211, 103–114.

Rodrigues, M. Â., et al. (2015). Early-maturing annual legumes: an option for cover cropping in rainfed olive orchards. *Nutrient Cycling in Agroecosystems*, 103(2), 153–166.

Roepstorff, A., et al. (2000). Mixed grazing with sows and heifers: parasitological aspects. In: Hermansen, J.E., et al. (eds.), *Ecological animal husbandry in the Nordic Countries*, DARCOF Report, Tjele, vol. 2, pp. 41–44.

Romero, P., et al. (2007). Curve number values for olive orchards under different soil management. *Soil Science Society of America Journal*, 71(6), 1758–1769.

Rosati, A., et al. (2004). A simple method to estimate photosynthetic radiation use efficiency of canopies. *Annals of Botany*, 93(5), 567–574.

Rosati, A., et al. (2009). Olive, asparagus and animals: an agroforestry model for temperate climate in developed countries. Proceedings of the III OLIVEBIOTEQ -For a Renovated, Profitable and Competitive Mediterranean Olive Growing Sector, Sfax, Tunisia, 15–19 December 2009, pp. 229–33, ISBN: 978-9938-9513-0-1.

Rosati, A. (2011). Asparagi nell'oliveto e la produzione raddoppia. *Olivo e Olio*, 7/8, 20–24.
Rosati, A., et al. (2018a). Resource investments in reproductive growth proportionately limit investments in whole-tree vegetative growth in young olive trees with varying crop loads. *Tree Physiology*, 38, 1267–1277. https://doi.org/10.1093/treephys/tpy011.
Rosati, A., et al. (2018b). Fruit production and branching density affect shoot and whole-tree wood to leaf biomass ratio in olive. *Tree Physiology*, 38, 1278–1285. https://doi.org/10.1093/treephys/tpy009.
Rosati, A., et al. (2018c). Improving crop productivity in agroforestry systems: low leaf respiration is a key trait. In *Proceedings of the 4rd European Agroforestry Conference – Agroforestry as Sustainable land use*, Nijmegen, Netherland, 28–30 May 2018, pp 442–446, ISBN: 978-84-09-02384-4.
Rosati, A., et al. (2021). Intercepted photosynthetically active radiation (PAR) and spatial and temporal distribution of transmitted PAR under high-density and super high-density olive orchards. *Agriculture*, 11(4), 351.
Salaheen, S., et al. (2015). Zoonotic bacterial pathogens and mixed crop-livestock farming. *Poultry Science*, 94(6), 1398–1410.
Sansoucy, R., et al. (1985). *Olive by-products for animal feed*. Food and Agriculture Organization of the United Nation, FAO.
Sastre, B., et al. (2016). The effect of soil management on olive yield and VOO quality in a rainfed olive grove of Central Spain. *Journal of Chemistry*. https://doi.org/10.1155/2016/4974609.
Savanna Institute. (2021). Managing food safety liability risks when integrating livestock with specialty crops. https://www.savannainstitute.org/food-safety-guide-download/ Accessed 30 January 2025.
Schoof, N., et al. (2021). Sheep in the vineyard: first insights into a new integrated crop–livestock system in Central Europe. *Sustainability*, 13(22), 12340.
Sehested, J., et al. (2004). Mixed grazing with sows and heifers effects on animal performance and pasture. In: Hermansen, J.E., et al. (eds.) *Ecological animal husbandry in the nordic countries*, DARCOF Report, Tjele, vol. 2, pp. 35–40.
Sestini, A. (1963). *Il Paesaggio*. Touring Club Italiano.
Smil, V. (2001). *Enriching the earth* (F. Haber, and C. Bosch, ed.). MIT Press.
Smith, V. H. and Glauber, J. W. (2019). Trade, policy, and food security. *Agricultural Economics*, (United 790 Kingdom), 51, 159–171. https://doi.org/10.1111/agec.12547.
Spencer, T. (2013). *Pastured poultry nutrition and forages*. National Sustainable Agriculture Information Service. https://www.sare.org/wp-content/uploads/Pastured-Poultry-Nutrition-and-Forages.pdf.
Sollenberger, L. E., et al. (2005). Reporting forage allowance in grazing experiments. *Crop Science*, 45(3), 896–900.
Suo, G. D., et al. (2019). Long-term effects of different surface mulching techniques on soil water and fruit yield in an apple orchard on the Loess Plateau of China. *Scientia Horticulturae*, 246, 643–651. https://doi.org/10.1016/j.scienta.2018.11.028.
Temel, S. and Pehluvan, M. (2015). Evaluating orchard and poplar leaves during autumn as an alternative fodder source for livestock feeding. *Ciencia e investigación agraria: revista latinoamericana de ciencias de la agricultura*, 42(1), 27–33.
Theofel, C. G., et al. (2020). Microorganisms move a short distance into an almond orchard from an adjacent upwind poultry operation. *Applied and Environmental Microbiology*, 86(15), e00573-20.

Thornton, P., et al. (2022). Impacts of heat stress on global cattle production during the 21st century: a modelling study. *The Lancet Planetary Health*, 6(3), e192–e201.

Tohiran, K. A., et al. (2019). Cattle-grazing in oil palm plantations sustainably controls understory vegetation. *Agriculture, Ecosystems & Environment*, 278, 54–60.

Tohiran, K. A., et al. (2023). Multi-species rotational grazing of small ruminants regenerates undergrowth vegetation while controlling weeds in the oil palm silvopastoral system. *Agricultural Systems*, 210, 103720.

Tramacere, L. G., et al. (2024). Effects of intercropping on the herbage production of a binary grass-legume mixture (Hedysarum coronarium L. and Lolium multiflorum Lam.) under artificial shade in Mediterranean rainfed conditions. *Agroforestry Systems*, 1–16. https://doi.org/10.1007/s10457-024-01012-8.

Tramacere, L. G., et al. (2025). Evaluating the impact of shade on Nitrogen fixation in sulla (Hedysarum coronarium L.): a 15N natural abundance study. *Italian Journal of Agronomy*, 100028. https://doi.org/10.1016/j.ijagro.2025.100028.

Vannucci, S. (2009). L'ulivo e l'olio, Storia dell'olivo, Diritti di sfruttamento economico: Bayer Crop Science S.r.l., Milano. http://www.colturaecultura.it Accessed 22 December 2017.

Vignozzi, N., et al. (2019). Soil ecosystem functions in a high-density olive orchard managed by different soil conservation practices. *Applied Soil Ecology*, 134, 64–76.

Wei, W., et al. (2024). Root spatial distribution and belowground competition in an apple/ryegrass agroforestry system. *Agricultural Systems*, 215, 103869.

Welker, W. V. and Glenn, D. M. (1989). Sod proximity influences the growth and yield of young peach trees. *Journal of the American Society for Horticultural Science*, 114(6), 856–859.

Westway, S. (2017). Silvopoultry: establishing a sward under the trees. Agroforestry innovation leaflet produced within the AGFORWARD project. https://www.agforward.eu/documents/leaflets/40_Silvopoultry_establishing_a_sward_under_the_trees.pdf. Accessed 7 October 2024.

Willett, L. B., et al. (1993). Mechanisms of movement of organochlorine pesticides from soils to cows via forages. *Journal of dairy Science*, 76(6), 1635–1644.

Wilson, L. M. and Hardesty, L. H. (2006). Chapter 11: targeted grazing with sheep and goats in orchard settings. In: Launchbaugh, K. et al. (eds.) *Targeted grazing: a natural approach to vegetation management and landscape enhancement*. American Sheep Industry Association, pp. 99–106.

Wolz, K. J. and DeLucia, E. H. (2018), Alley cropping: Global patterns of species composition and function, *Agriculture, Ecosystems and Environment*, 252, 61–8.

Yao, S., et al. (2009). Apple root growth, turnover, and distribution under different orchard groundcover management systems. *HortScience*, 44(1), 168–175.

Zamora, D. S., et al. (2009). Modeling cotton production response to shading in a pecan alleycropping system using cropgro. *Agroforestry Systems*, 76, 423–435. https://doi.org/10.1007/s10457-008-9166-x.

Zhou, Y., et al. (2018). Soil C:N:P stoichiometry responds to vegetation change from grassland to woodland. *Biogeochemistry*, 140(3), 341–357. https://doi.org/10.1007/s10533-018-0495-1.

Zuazo, V. D., et al. (2009). Soil conservation measures in rainfed olive orchards in south-eastern Spain: impacts of plant strips on soil water dynamics. *Pedosphere*, 19(4), 453–464.

Chapter 7

Types of silvopastoral systems: adding trees to pasture/rangelands

M. J. Beltran, National Institute of Agriculture Technology (INTA)/National University of San Antonio de Areco (UNSADA), Argentina; G. S. Cambareri, INTA Balcarce Research Station, Argentina; N. Banegas, Semi Arid Chaco National Institute of Animal Research (IJACS) - INTA, Argentina; P. L. Peri, INTA Rio Gallegos Research Station/National University of Patagonia Austral (UNPA)/National Council for Scientific and Technical Research (CONICET), Argentina; L. Colcombet, INTA Montecarlo Research Station, Argentina; G. J. Martínez Pastur, M. V. Lencinas, J. Benítez and J. Rodríguez-Souilla, Austral Center for Scientific Research (CADIC) – CONICET, Argentina; J. M. Cellini, Wood Research Laboratory (LIMAD) – National University of La Plata (UNLP), Argentina; and V. R. N. Cosentino, National Institute of Agriculture Technology (INTA)/CONICET, Argentina

1 Introduction

Silvopastoral Systems (SPS) have been identified as an efficient land management strategy to increase the performance of animal and tree production, livestock welfare (Esquivel, 2017), and improve carbon (C) sequestration in soil and in trees to offset livestock emissions (Figueiredo et al., 2017). These systems are a type of livestock agroforestry, which associates

http://dx.doi.org/10.19103/AS.2024.0141.12

trees and/or shrubs with pastures (Mosquera-Losada et al., 2006; Chará et al., 2020). The system comprises a woody component, which involves any type of tree, interacting with conventional components (herbaceous forage or pastures) under an integrated management system (Gándara et al., 2021). Trees provide timber, fruit, firewood, seeds, and shade generators (Luccerini et al., 2013). SPS are alternative management systems which both mitigate and adapt to climate change (Feliciano et al., 2018). SPS are a sustainable way of land management that has been both politically and economically promoted worldwide (CAP 2014–2020) as a tool to increase carbon sequestration in soils among other benefits that agroforestry provides (Mosquera-Losada et al., 2018).

In SPS, there are both ecological and economic interactions between the different components. Thus, SPS are those that have been designed to improve beneficial–ecological interactions, which may be shown as improvement in yield per unit area, resource use efficiency, and/or enhancement in environmental issues. Silvopastoral advantages can be enumerated as the provision of multiple products (e.g. food, wood, fodder, mulch, medicinal plants) or services (e.g. maintenance of soil fertility, erosion control, microclimate improvement, biodiversity enhancement, watershed protection, reduction of fire risk, carbon sequestration) by the trees (Peri et al., 2016a; Mosquera-Losada et al., 2005).

In some regions or areas, SPS are used according to the basic premise that these systems can be more biologically productive, profitable, and sustainable than forestry or animal production monocultures depending on the environment. The trees in SPS enhance nutrient uptake from the soil, which is then returning to the grass through the degradation of organic matter, improving both soil fertility and forage quality. The trees, furthermore, strengthen animal welfare due to their moderating effects on climate extremes (Dube et al., 2012; Peri, 2011).

2 Types of silvopastoral system

The combinations of trees and/or shrubs with pastures and animals are present in several forms, generating different SPS. Many of these systems are related to the customs, traditions, and natural landscapes of the regions. The silvopastoral systems' arrangements could be two or three strata SPS where grass, trees, and shrubs could be combined; live fences and barriers (Villanueva et al., 2009); windbreaks (Peri and Bloomberg, 2002); and native forests (Peri et al., 2019).

Different examples are mentioned below, some of which have already been implemented in Argentina, elsewise they could perfectly be applied worldwide considering, of course, the local pedoclimatic conditions.

2.1 Two strata system

This is composed of pastures or natural grassland combined with timber or fruit tree species. It is a system designed for the distribution of low-density trees per hectare proposed at the beginning of the plantation thus promoting the development of pasture and producing quality wood for sawmills or fruit. There are different SPS designs of this type, some of which are mentioned below. These designs aim to produce forage with the least negative effects of tree shading and quality wood destined for sawmills. Different designs include:

- **SPS with homogeneous tree distribution**: trees evenly distributed with a specific density which avoids negative effects on grass production due to shading (Figure 1 and 2). The distance between rows of trees can vary between 5 m and 7 m allowing sunlight to enter the system that generates a correct forage production (Gándara et al., 2021).
- **The strip system**: trees are planted in four or five rows at a distance of 4 m or 5 m and separated by alleys of between 20 m and 40 m approximately where the pasture grows without the shade of the trees (Figure 3 and 4).
- **Paired logs system**: trees are distributed in two rows at distances of 4 m or 5 m, separated by alleys between 8 m and 20 m (Figure 5 and 6, Esquivel, 2017)

Figure 1 and 2 SPS with homogeneous distribution in the field in the Delta zone, Buenos Aires, Argentina (Photo provided by Marcelo Beltrán.)

Figure 3 and 4 SPS distributed in strip design in Buenos Aires province, Argentina. (Photo provided by Marcelo Beltrán)

Figure 5 and 6 Trees in paired logs in Buenos Aires province, Argentina. (Photo provided by Luis Colcombet)

2.2 Triple strata system

Shrubs are added to the trees and pastures stratum. The upper arboreal stratum consists of a great diversity of multipurpose trees. The middle layer is mainly represented by shrub species of high protein value, planted in high density (about 5000 shrubs/ha) for animals' grazing (Alonso, 2011). Legume shrubs contribute organic matter to the soil and fix a large amount of nitrogen from the atmosphere. Finally, the last of the three strata is the pasture or grassland.

2.3 Native forest under SPS

In the same land unit, exotic tree species can be incorporated or the native forests can be managed as farming systems allowing the simultaneous production of wood and livestock. This system can generate synergies and trade-offs among ecosystem services, biodiversity, and/or ecosystem functions (Peri et al., 2021a). There is evidence that both overstocking and inappropriate silviculture practices, coupled with the fragility of the environment, and other socioeconomic aspects led to the degradation of native forests by reducing productivity, biodiversity, and ecosystem services (Peri et al., 2017; Barral et al., 2020; Domínguez-Núñez et al., 2020). SPS reduce the degradation of the native forests (Peri et al., 2021a).

2.4 Windbreaks

Where strong winds are a constraint to agricultural production, live windbreaks are often planted in agricultural fields to protect crops, livestock, and soils from wind hazards. The major factors that determine the efficiency of windbreaks are height, porosity (which implies the amount of wind that goes through the line of trees), orientation, length, and location in the landscape (Peri and Bloomberg, 2002). Windbreaks have a positive impact on soil quality as they reduce soil blasting and therefore wind erosion (Hudson, 2006).

3 Environmental services delivered by SPS

There are various environmental services that SPS offer, some of which are described below.

3.1 Carbon sequestration

SPS are particularly interesting as they combine the potential to increase biomass and soil organic carbon (SOC) storage while maintaining agricultural production (Cardinael et al., 2017). Agroforestry trees produce organic matter (OM), whose inputs to the soil might enhance SOC stocks (Jordan, 2004; Peichl et al., 2006) and could enhance SOC stocks. Restoring SOC levels is often seen as a wisely successful strategy to mitigate climate change while improving soil quality, potentially leading to higher crop yields (Lal, 2006). Several types of SPS possess different rates of above-ground and SOC sequestration (Corbeels et al., 2019; Feliciano et al., 2018). Leaf litter and pruning residues are left on the soil, whereas OM originating from root decomposition and root exudates can be incorporated much deeper into the soil as agroforestry trees may have a very deep rooting to minimize the competition with the pastures (Cardinael et al., 2015; Mulia and Dupraz, 2006).

3.2 Biodiversity conservation

SPS help to conserve the diversity of plants and animals and contribute to the survival of different species of native fauna (Schroth et al., 2004; Varah et al., 2013; Thevathasan and Gordon, 2004). Trees, shrubs, and other plant species provide food and shelter for birds and insects of all kinds (Zerbino et al., 2008; Cabrera et al., 2011; Barnes et al., 2014).

3.3 Water regulation and conservation of water sources

Trees increase the retention, infiltration, circulation, and storage capacity of water in the soil, protecting the soil against erosion, and conserving springs and rivers (Bergeron et al., 2011; Tully et al., 2012).

3.4 Improvement of soil productivity

Since most trees have a well-developed root system, they extract nutrients from the depths of the soil and deposit them on the surface, resulting in better nutrient recycling and greater soil productivity (Abrahams, 2002; Khan et al., 2017).

3.5 Combined environmental services

Hence, SPS provide options to mitigate climate change with the possibility of increasing pasture yields, and providing other positive environmental outcomes such as improvements in soil health and climate change adaptation (Coulibaly et al., 2017; Waldron et al., 2017). Soil health includes physical, chemical, and biological components of soil, and integrates ecosystem functioning and services provisioned by soils, including the capacity of soils to promote and sustain agricultural productivity (Kibblewhite et al., 2008). Thus, concerning the multitude of positive environmental and ecosystem services, agroforestry practices may directly contribute to the achievement of several United Nations Sustainable Development Goals (SDGs), labeled "no hunger", "renewable energy", "sustainable cities and communities", "responsible consumption", "climate action", "life on land", and often neglected, "partnerships for the goals" and may benefit other SDGs indirectly (Hubner et al., 2021).

4 SPS case studies: deciduous fruit trees and cattle

Some SPS examples in Argentina under different environmental conditions (weather and type of soil) are described below. Information about SPS focus on soil carbon sequestration, which is one of the main environmental services of the system, is shown in four different case studies (Table 1). This case study discusses AR1: BALCARCE.

Table 1 General information of the SPS case studies in Argentina.

Case study	AR1	AR2	AR3	AR4
Location	Balcarce, Buenos Aires, Argentina	Leales, Tucumán, Argentina	Tierra del Fuego, Argentina	Montecarlo, Misiones, Argentina
Coordinates	37°45'48.7"S 58°17'26.4"W	27°11' S 65°17' W	54° 0.1′ S 67° 41′ W	26° 38' 19.50" S 54° 30' 33,44" W
Type of production	Deciduous fruit trees (Pecan) and cattle (Dairy)	Native trees (*Neltuma alba*) on tropical grass (*Chloris gayana*) for cattle (beef)	Extensive cattle (beef) with different types of grassland and native forest management	Three strata SPS (pine trees, with two types of forage), for beef production
Mean rainfall	882 mm	880 mm	300 mm	1980 mm
Mean T (C°)	14.5°C	19.0°C	5.6°C	21.0°C
Type of soil	Molisoll	Molisoll	Inceptisol	Entisol

AR = Argentina.

Integration between deciduous fruit trees and cattle, functioning as SPS are not that frequent in Argentina, but their economic and environmental benefits should be taken into account (Kaeser et al., 2010, Coello Gomez, 2019). For instance, in the Southeast of Buenos Aires where the overall agricultural production matrix is composed of winter cereals and soybeans (MECON, 2023), the inclusion of deciduous trees-based SPS offer an opportunity to counterbalance the negative of such a matrix. On account of the previous statements, information has been gathered from a particular case study. This study is located in the Southeastern part of Buenos Aires province, at the fields belonging to an INTA Research Station (37°45'48.7"S 58°17'26.4"W). Soils of this region are dominated by the *Molisoll* family, with the soil in the case study being a Typic Argiudoll having 32.3 g C kg soil^{-1}, 2.5 g TKN kg soil^{-1}, pH = 6.8, and 351 g clay kg soil^{-1} at the 0–20 cm depth layer at the moment of planting. The local climate has cool summers (mean January temperature = 21.4°C) and mild winters (mean July temperature = 8°C) and year-evenly distributed precipitation (media of annual precipitations = 882 mm) (SIGA, 2023). Under this edapho-climatic context, a multi-cultivar pecan (*Carya illinoinensis*) orchard was planted in 2013, at a 123.45 trees ha^{-1}density (d), with a 9 × 9 m plantation frame, having alleys wide enough to be grazed by cattle of younger categories as the pecan tree trunks grow thick in diameter. The orchard planting was accompanied by pasture seeding, the latter composed of a mixture of white clover (*Trifolium repens*) and tall fescue (*Festuca arundinacea*). Grazing is eventual, with American Holstein (*Bos primigenius Taurus)* heifers and steers belonging to a nearby dairy farm (Fig. 7). The farm herd is 119 head of cattle, with a pregnancy index of 0.77 and an average animal weight of 600 kg. In

Figure 7 Ground view with heifers grazing (a) and soil sensors implementation (b) in the case study. (Photo provided by Gustavo Cambareri)

2021, the milk production of the farm was 22 l cow^{-1} day^{-1}, with a minimum of 18 and a maximum of 40 l cow^{-1} day^{-1}, respectively.

To characterize the study site during the years 2022 and 2023, dasymetric measures were taken such as the diameter at breast height (DBH), the basal area ($g = DBH^2 * 0.25 \pi * d$), and the canopy area (estimated through images captured by drone flights), to support the data of nut yield (Table 2). Regarding forage yield, the pasture is currently degraded, with an annual production estimated at less than 4 t MS yr^{-1}.

The indicators mentioned above depict the still juvenile stage of the orchard, inasmuch as DBH and cover are low. Nonetheless, despite this, the orchard was capable of bringing associated environmental benefits such as C sequestration. In order to assess that, C stocks in biomass and soil were estimated during fall 2022 and fall 2023 (Figure 9). This was first calculated by selecting 36 trees of the orchard, discarding trees in the border zone, and modeling tree biomass according to Smith and Wood (2006) model with a local correction, using DBH as input. Secondly, SOC was measured out of 12 composed soil samples (four taken in the pasture and eight taken near the trees) collected at the 0–30 cm

Table 2 Biomass production indicators in two different years (2022 and 2023).

Orchard indicators	2022	2023
Diameter at breast height (DBH)	9.14 cm	11.4 cm
Basal area ($g = DBH^2*0.25 \pi d$)	0.86 m^2	1.25 m^2
Total canopy área	618.8 m^2	786 m^2
Total canopy cover	11.9%	15.2%
Nut production (rank)	0.04–3.72 kg $tree^{-1}$	0.02–3.26 kg $tree^{-1}$

DBH = Diameter at breast height.

Figure 8 (a) Canopy covers evolution. Pecan orchard during April 2022. (b) Pecan orchard during April 2023. Orthomosaics are created through drone images. (Photo provided by Gustavo Cambareri)

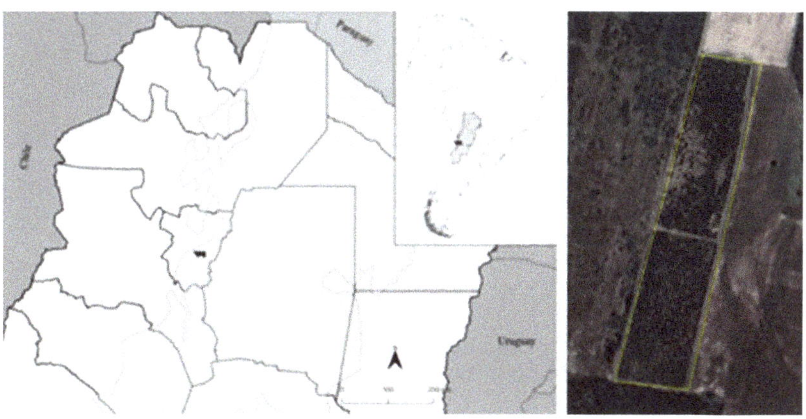

Figure 9 Location of the case study (a) and Silvopastoral system with *Chloris gayana cv Finecut* (grass) and *Prosopis alba* (b).Source: Nasca et al., 2020

depth layer and analysed in a LECO CR-12 device at an external Lab. Carbon stocks showed an increase in tree biomass, but a decrease in soil (Table 3) as a presumable consequence of pasture degradation and progressive loss of forage value. Regardless of this, an increase in total C sequestration was registered.

The results revealed that even in a marginal environment such as the one in the Southeast of Buenos Aires province, pecan associated with cattle in SPS was capable of providing environmental services such as C sequestration (Figure 10). This work is still under investigation and more data will be collected in the following years to support the initial hypothesis.

Table 3 Soil organic carbon stock and sequestration by the SPS.

	2022	2023
Biomass (t C ha^{-1})	1.67 ± 0.2	3.83 ± 0.42
Soil (t C ha^{-1})	98.75 ± 0.14	98.6 ± 1.58
Total C stock	100.43 ± 0.22	102.43 ± 0.32
C sequestration (t CO2eq ha^{-1})	7.35 ± 1.13	

C = carbon; CO_2eq = carbon dioxide equivalent.

Figure 10 Images of SPS with Neltumba alba y Chloris gayana cv Finecut. (Photos provided by Natalia Banegas)

5 SPS case studies: native trees on grass for beef cattle

The following example of SPS (LEALES (AR2)) (*Chlorisgayana*cv Finecut and *Prosopis alba*) located at the Animal Research Institute of the Semiarid Chaco (IIACS) dependent of the National Institute of Agricultural Technology (INTA), located in Leales at the southeast of Tucumán province, Argentina (27°11′ S, 65°17′ W). This area corresponds to a depressed saline plain and has an altitude of 335 m above sea level. The mean annual precipitation is 880 mm and rainfalls are concentrated between November and April. The mean annual temperature reaches around 19°C, ranging from 25°C in January to 13°C in July. The climate is subtropical subhumid with a dry season, matching up the latter with the relatively colder period of the year spanning from April to October. The soil type subsumed into the class Fluvacuentic Haplustoll following the US Soil Taxonomy System. The former study involves changes in SOC and nitrogen (N) stock after trees' implantation, and greenhouse gasses (GHG) emission from soil, fences, and urine spots.

In 2015, the national agencies that promote conservation (Ministry of Environment and Sustainable Development) and production (Ministry of Agroindustry), used general guidelines for Forest Management Incorporating Livestock in areas where sustainable management is allowed (Category II of Law 26.331). These guidelines were adopted with the aim of organizing productive activities in native forests while maintaining their capacity to provide ecosystem services (Peri et al., 2018). To this end, only low-impact interventions on the tree and shrub layers are allowed, and the SPS represent an alternative management in this context for livestock production. Also, tropical grass pastures (like *Chloris gayana* cv Finecut) are typically constrained in their capacity to store SOC due to limited plant-available N in soils under pure grass pastures and frequent overgrazing, which leads to low primary biomass production and SOC losses (Dalal and Carter, 2000; Banegas et al., 2019a). The introduction of legumes, in this case native legume trees, could increase SOC and total nitrogen by enhancing the microbial activity and N concentration in the topsoil (0–20 cm), particularly the most labile form that promotes improvement in grass growth and quality, attributed to N-rich legume leaf deposition, leaf recycled via animal feces and nodule-N turnover from biological N fixation (Radrizzani et al., 2011; Banegas et al., 2019a).

In this context, the aim of this trial has been to evaluate the stock and dynamics of SOC and N by the introduction of native trees (*Neltuma alba*) on tropical grass (*Chloris gayana*). In an area of 12 hectares with grass, in January 2017, the trees were planted, spaced 3 m apart within rows and with 8 m among rows. Intensive grazing is eventual with Braford cows at a stocking rate of 3 animals ha^{-1}. During the growing season, paddocks are grazed for 20 ± 5 days with a rest period of 35 ± 6 days. In the winter, deferred grass was grazed for 30 ± 15 days. Animals received carbohydrate and protein supplements during the winter and energy supplements during the summer. A completely randomized design was employed. Soil samples were collected at two depths: 0–10 cm and 10–30 cm in 2017 (before tree plantation) and 2023. Ten soil subsamples were collected per plot for each depth in both years and combined to obtain a composite sample for determinations. SOC and total nitrogen were determined in LECO CN 628.

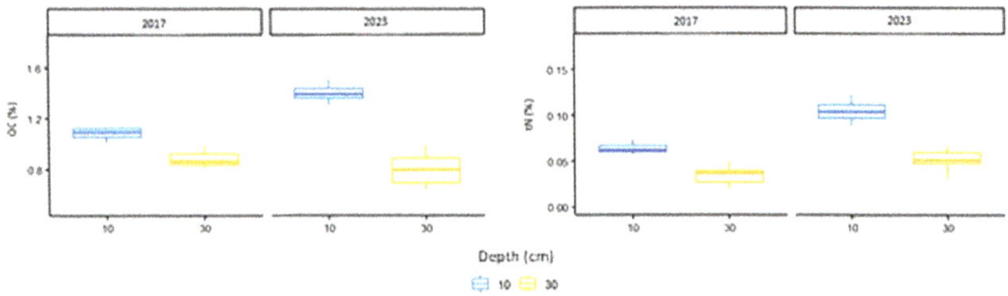

Figure 11 Soil organic carbon (OC) and total nitrogen (tN) at 0-10 and 10-30 cm in a silvopastoral system with *Neltuma alba* and *Chloris gayana* cv Finecut.

For 0–10 cm of soil in 2017, SOC and total nitrogen values were 1.08% ± 0.0 and 0.06% ± 0.01, respectively. While in 2023 an increase of these contents was measured being 1.4% ± 0.057 and 0.10% ± 0.01 for SOC and total nitrogen respectively (Fig. 11).

At 10–30 cm of soil depth, SOC values did not show differences among years, with values of 0.88% ± 0.07 and 0.80% ± 0.17 for 2017 and 2023, respectively. Nevertheless, total nitrogen registered an increment of 0.03%% ± 0.01 and 0.05% ± 0.01 for 2017 and 2023, respectively. At this depth, a see-saw effect was detected for the C:N ratio (data not shown), explained by higher amounts of nitrogen per unit of carbon at the intermediate layer. This phenomenon is probably tied to a decoupled dynamics between both variables. Since processes behind the accumulation of SOC and total nitrogen can vary independently of each other such as different mechanisms of losses and gains involved, a partial decoupling could be expected at a given time, and thus, C and N stocks might change in opposite directions during this period (Asner et al., 1997; Piñeiro et al., 2010; Banegas et al., 2019a)

At an equivalent mass of soil, SOC and total nitrogen stocks were 35.5 ± 2.56 and 1.80 ± 0.09 for 2017, and 39.32 ± 1.38 and 2.60 ± 0.25 (Fig. 12). Heretofore, the gain rate was 0.74 Mg SOC.ha^{-1} and 0.17 Mg total nitrogen ha^{-1}.

Tree implantation in a tropical pasture increased SOC and total nitrogen content in soils and promoted substantial capture of SOC in 0–10 cm attributed to a greater abundance and diversity of trees and grass roots. The introduction

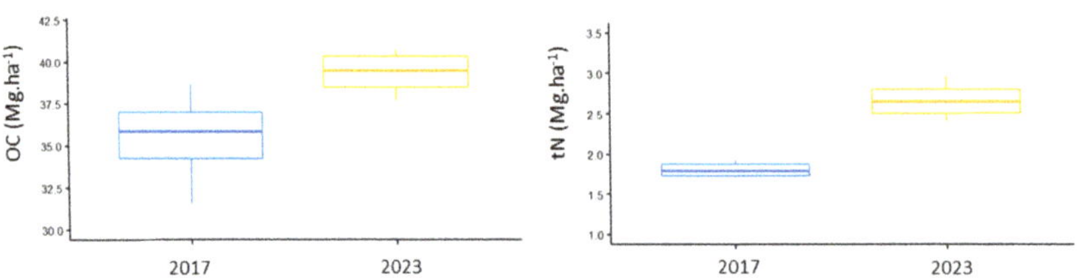

Figure 12 Soil OC and tN stock at the same soil mass in a SPS with *Neltuma alba* and *Chloris gayana* cv Finecut.

of trees also enhanced N concentration in 0–10 and 10–30 cm, attributed to N-rich *Neltumba* leaf deposition, leaf recycled via animal feces and nodule-N turnover from biological N fixation (Banegas et al., 2019b).

6 SPS case studies: extensive beef cattle with different types of grassland and native forest management

European colonization of southern Patagonia modified the native forests, altering the provision of ecosystem services, e.g. changing species to increase growth and palatability for livestock (Martínez Pastur et al., 2021; Peri et al., 2022). One of the most affected forest landscapes was the *Nothofagus antarctica* (ñire) forests in ecotone areas, owing to presented relevant features for grazing and timber for rural building aims (Peri et al., 2016b). SPS (TIERRA DEL FUEGO (AR3)) were proposed as the more sustainable alternative, combining management and conservation objectives, e.g. obtaining poles for fences and lumber, and open the forest canopy to stimulate the understory growth but maintaining enough shelter for cattle during winter (Peri et al., 2016a; Martínez Pastur et al., 2021, 2022). Besides, SPS were also pointed as the best management alternative to increase the in situ conservation of biodiversity (Martínez Pastur et al., 2020) and regulation of ecosystem services (e.g. carbon stocks maintenance) (Aravena Acuña et al., 2023).

One permanent long-term study site of the PEBANPA network (Peri et al., 2016b) was established at El Roble ranch, located in the central area of Tierra del Fuego (Argentina) at 54° 0.1′ SL and 67° 41′ WL. The ranch management was based on extensive cattle activities at full round year, with rotation of livestock between paddocks, seasonally or annually, depending on the vegetation availability, seasons, accessibility, and management objectives. This study site included natural and managed environments: (i) ñire forests without management as controls (CF) including young forests (FGP) and mature forests (MAT), (ii) SPS stand managed through thinning (LH = light harvesting, HH = heavy harvesting), whereas thinning below crown cover (TBC) and thinning outside the crown cover (TOC) were studied (Figure 13). We also included management practices that were employed in the past to convert the forests to grasslands, including (iii) clear-cuts (CC) and (iv) fires (FIRE). Finally, we also analyzed (v) dry grasslands (DG) dominated by *Festuca gracillima* and *Empetrum rubrum*, and (vi) wet grasslands (WG) dominated by *Juncus scheuzeroides*, *Carex curta*, *C. macloviana*, and *Caltha sagittata* (Martínez Pastur et al., 2021, 2022). The cattle for meat production was Hereford (*Bos Taurus Taurus*).

The effect of the thinning to promote SPS was measured in a 3 ha plot, where forest structure was monitored for a period of five growth seasons (2019–2023). The original stand presented 52.0 ±4.9 m^2 ha^{-1} of basal area with 1811

Figure 13 SPS (left) and conventional grassland (right) in Tierra del Fuego, Argentina. (Photos provided by Pablo Peri).

± 218 trees ha^{-1} of 10.1 ± 0.3 m height (Fig. 14). The total over-bark volume (TOBV) was reduced from 306 ± 42 m^3 ha^{-1} to 140.9 m^3 ha^{-1}, generating an increase of the mean diameter at breast height (DBH) from 15.6 cm to 27.1 cm. Thinning generates an homogenization of the forest structure, with different TOBV/DBH ratios, and promotes the increase of individual tree growth, which were not greatly affected at stand level (2.2 ± 0.6 m^3 ha^{-1} yr^{-1} in CF compared to 2.0 ± 0.3 m^3 ha^{-1} yr^{-1} in thinned stands). Before the interventions, the crown cover of the stands was 82.0 ± 0.9%, and was reduced to 54.9 ± 1.7% measured through hemispherical photographs (Martínez Pastur et al., 2011). Soil samples were collected at two depths: 0–10 cm and 10–30 cm. Ten soil subsamples were collected per each plot for each depth in each moment, and combined to obtain a composite sample for determination. SOC was determined in a LECO CR-12.

The understory biomass production was measured using cages (n = 20) with a 5 x 5 cm mesh and was collected every 45 days across the growing season (November to April) for three consecutive years (2020–2023). We found differences in the multiple analysis of variance (ANOVA) among the main effects without significant interactions (Table 4). The growing season influence over understory biomass, e.g. 2020–2021 was influenced by El Niño and the following years by La Niña. However, the last year was influenced by strong positive values of the Southern Hemisphere Annular Mode Index (Rodríguez-Souilla et al., 2023). The biomass quickly increased during late spring (0 to 90 days) and reached the maximum during late summer (day 135). This biomass was differently shown in the studied environments (Table 5). Thinned stands generated greater biomass, which was influenced by the tree cover, e.g. 13.3% higher with the influence of tree crowns (TBC) and 78.3% higher without the influence of tree crowns (TOC). The grasslands were greatly affected by La Niña seasons (2021–2022 and 2022–2023), highlighting the resilience of the forested landscapes. In our monitoring, the dry grasslands (DG) presented higher values

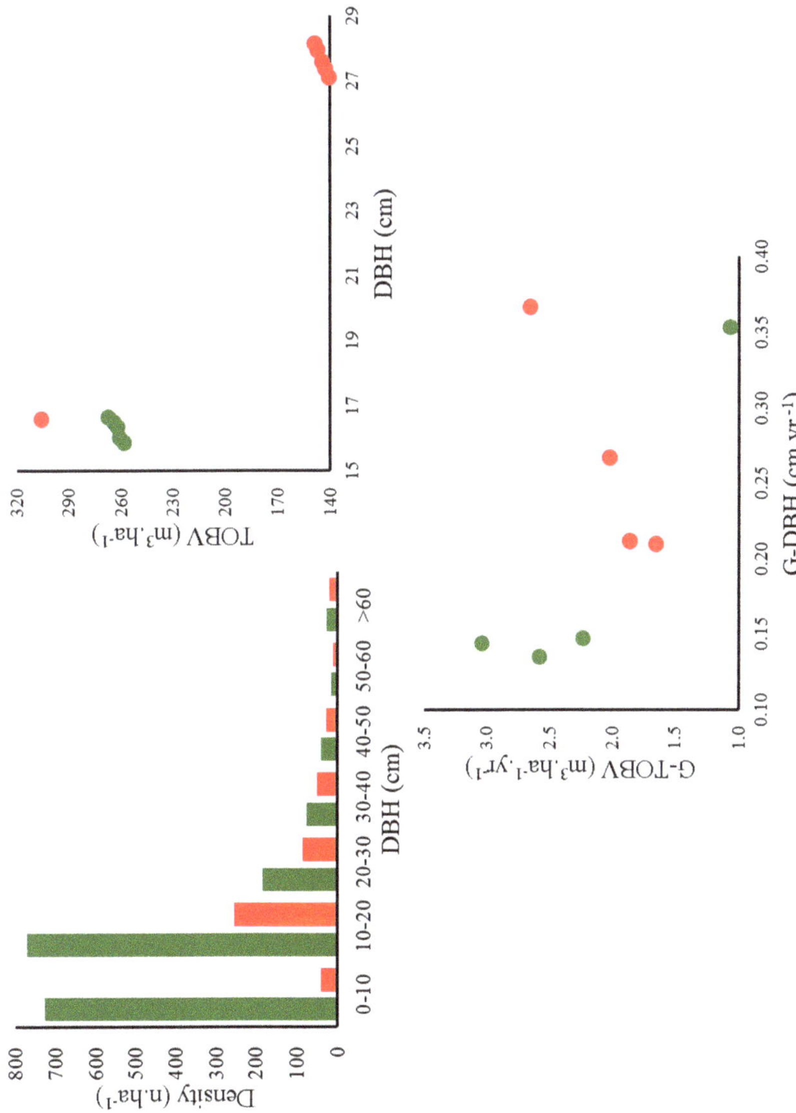

Figure 14 Comparison of forest structure between control forests (dark green) and thinning treatments (red) along different growing seasons (2019-2023). DBH = diameter at breast height (cm), TOBV = total over bark volume (m^3 ha^{-1}), G-DBH = growth of DBH (cm yr^{-1}), TOBV = growth of TOBV (m^3 ha^{-1} yr^{-1}).

Table 4 Multiple ANOVA for understory biomass production for different growing seasons (GS), environment types (ET), and days.

	Source	SS	n	MS	F	p
Principal effects	A: GS	13.18	2	6.59	7.36	0.001
	B: ET	84.74	4	21.19	23.66	<0.001
	C: Days	7.62	3	2.54	2.84	0.040
Interactions	A x B	11.92	8	1.49	1.66	0.110
	A x C	6.58	6	1.10	1.23	0.295
	B x C	4.71	12	0.39	0.44	0.946
	A x B x C	15.59	24	0.65	0.73	0.822
Residuals		161.18	180	0.90		
Total		305.53	239			

SS = sum of squares, MS = mean of squares, F = Fisher test, p = probability.

Table 5 Multiple ANOVA for understory biomass production (ton.ha -1) for different growing seasons (GS), environment types (ET), and days (November to April).

Treatment	Level	N	Mean	Difference	SE
GS	20_21	80	2.10	b	0.11
	21_22	80	2.16	b	0.11
	22_23	80	1.64	a	0.11
ET	CF	48	1.66	ab	0.14
	TBC	48	1.88	b	0.14
	TOC	48	2.96	c	0.14
	DG	48	2.16	b	0.14
	WG	48	1.17	a	0.14
Days	45	60	1.66	a	0.12
	90	60	2.02	ab	0.12
	135	60	2.12	b	0.12
	180	60	2.05	ab	0.12

CF = control forests, TBC = thinning under crown cover, TOC = thinning outside the Crown cover, DG = dry grasslands, WG = wet grasslands. SE = standard error. Different letters showed differences by Tukey test at p 0.05.

than the control forests (CF) and thinning areas with the influence of the tree canopies (TBC), but lower than areas between trees (TOC). Wet grasslands (WG) are considered the most productive environments for livestock, but here presented lower values on average.

The capacity of the forests to maintain long-term carbon sequestration, biodiversity, and other ecosystem services has been recognized by the scientific, governmental, and civil society sectors around the world in recent years (Aravena Acuña et al., 2023). The trade-offs of the management proposals must be considered in the decision-making. Forest carbon management (FCM) has been proposed to increase or maintain the amount of carbon sequestered in managed stands (e.g. decreasing forest harvest intensity to decrease carbon losses to the atmosphere) (Birdsey et al., 2000), and SPS are the best strategy for ñire forests. We determine carbon stocks in different unmanaged, managed, and transformed environments to determine the changes in the carbon stocks, including above-ground (AG_T) and below-ground tree components (BG_T), above-ground (AG_D) and below-ground deadwood (BG_D), plant understorey (UND), and the soil layer between 0 cm and 30 cm (SOIL).

Carbon stocks in the studied environments changed in magnitude and distribution among the considered components (Fig. 15). The carbon stocks were maximum in control forests (FGP and MAT), and decreased according to the impact level (LH > HH > CC – FIRE) which mainly affected the above-ground biomass (Aravena Acuña et al., 2023). Open lands presented lower carbon stocks compared to the forested environments, where wet grasslands (WG) presented major contents than dry grasslands (DG). The average contribution of the soil/tree components varied according to the treatments, where both equally contributed to the unmanaged forests, but the soil importance increased with the impact level. Carbon stocks in understory biomass were very small (0.1% to 1.2%), and presented a close relationship with crown cover and tree densities. Aravena Acuña et al. (2023) discussed the advantages of SPS considering the carbon stocks in the forest components and the increasing understory biomass. SPS offer the most adequate balance for multi-purpose objectives, which constitute an alternative with synergistic characteristics, since it improves the growth of remaining trees and enhances regeneration establishment, and at the same time improves forage production for animal production (Yanai et al., 2003). This management strategy is compatible with FCM (Bravo et al., 2017) by improving carbon sequestration in managed stands (Aravena Acuña et al., 2023).

7 SPS case studies: three strata SPS

The Northern Misiones province receives on average 1.980 mm +/– 450 mm precipitation evenly distributed over the months (Colcombet, 2022). Because of the low latitude (26° S), soil moisture is in high demand during the summertime (over 5–7 mm/day). The presence of winter frosts (–1°C to –4°C) from early June to late September can be a challenging obstacle for tropical and subtropical species as well. The soils are clayish laterites (Entisol) with Al contents that tend to become acid when organic matter is lost from nearly 4%–5,4% under the native forests to even under 2.8%–1.8% in 2 to 3 years

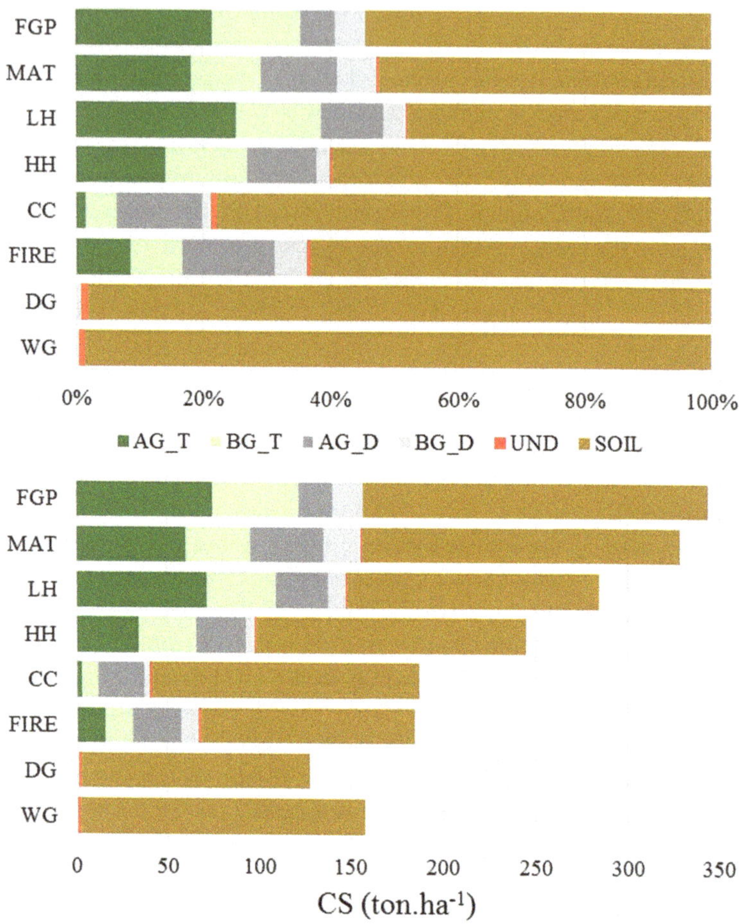

Figure 15 Carbon stock (CS, ton.ha $^{-1}$) classified by components (AG_T = above-ground trees, BG_T = below-ground trees, AG_D = above-ground deadwood, BG_D = below-ground deadwood, UND = understorey, SOIL = soil layer between 0-30 cm) and environments, including control forests (FGP = young forests, MAT = mature forests), impact levels (LH = light harvesting, HH = heavy harvesting, CC = clear-cuts, FIRE = stands with fire impacts), and grasslands (DG = dry, WG = wet).

when they are poorly managed after conversion to crops or pastures. Because of the loss of labile organic matter, acidity grows from a soil pH of 5–5.4 to even under 4,3, .solubilizing aluminum and fixing phosphorus (P). Therefore, the introduction of species that can contribute toproviding readily available P and better cation exchange capacity could contribute to agricultural and cattlegrowing sustainability. *Tithonia diversifolia* was selected because of its relatively high protein content (Pérez et al., 2009) and higher cation content (Navas Panadero, 2019).

The trial site (MONTECARLO (AR4)) is a commercial farm located in Montecarlo (26° 38′ 19.50″ S; 54° 30′33,44″ W), Misiones province, 540 m above sea level on shallow soils (useful Depth under 0.5m) in a hilly topography with a clay acid soil (pH 5). Although the three strata SPS can be more complicated to manage, the proposal consisted of integrating a conventional two-strata SPS, a third species that should elevate the relatively poor star grass protein content (8.1%–9.4%, Loto et al., 2022) of a site, with the objective of elevating the cattle performance and lower greenhouse gases net emissions. In 1983, the star grass (*Cynodon plectostachyus*) was planted on land that had been cleared from its original native forest. In 2015, the Hybrids Pine trees (*Pinus caribae hondurensis x Pinus elliottii*) were planted on the degraded grass paddock, in 6 m between rows x 2.5 m within the rows configuration. At the same time, a number of Tithonia was planted in the same rows among trees. Grazing by heifers or steers resumed two years later. During the late spring and early summer 2019–2020 and 2021–2022, the productivity of the three strata SPS integrated was evaluated. The rainfall of the first cycle was within the average (−1%) and the second cycle the second driest string summer in 95 years; record with only 54% rain falls compared to the average (Colcombet, 2022). The animal breed of the case study is Bradford (*Cebu Brahman x Hereford*).

During the first cycle, which consisted of a continuous grazing, *Thithonia* tended to overrun the star grass (Loto et al., 2022). Thus, grazing was adjusted during the second cycle to 1 week of use plus three weeks recovery period. Because of the severe drought during the second cycle, the SPS forage production decreased by 17% and the meat production 6.8%, compared to the 25.4 % and 9.3 % of the OS conditions, respectively (Table 6 and 7).

During both cycles with different characteristics (normal rainfall and continuous grazing for the first cycle and 46% under average rainfall and rotational grazing for the second cycle), in spite of the SPS showing slightly

Table 6 Forage average offer in kg of dry matter (DM) and forage quality of star Grass and Tithonia during the first and second trial cycles under open sky and SPS conditions.

	Cycle	Forage offer (kg DM ha^{-1})	PB (%)	NDF (%)	ADF (%)	DMD (%)
Stargrass Open Sky	1	5.980 ± 1.671	8.1	76.1	39.1	50.1
	2	4.460 ± 972	9.4	72.3	34.2	57.5
Star Grass SPS	1	3.331 ± 1.771	12.4	72.6	38.1	55.2
	2	2.930 ± 833	11.5	69.4	32.3	65.7
Tithonia SPS	1	1.130 ± 280	20.4	46.8	36.4	70.1
	2	754 ± 148	16.7	40.1	28.9	82.2

PB (%): protein content; NDF (%): neutral detergent fiber; acid detergent fiber (ADF); DMD (%): dry matter digestibility.

Table 7 Initial live weight, live final weight, daily live weight increase (DLWI) and meat production of the first and second cycles under open sky and SPS conditions.

	First cycle (2019-2020)			Second cycle (2021-2022)		
	SPS	OS	P value	SPS	OS	P value
Initial live weight (kg)	242.6 ± 24.9	234.4 ± 22.9	0.19	245 ± 17	248 ± 15	0.41
Final live weight (kg)	332.4 ± 26.5	321.7 ± 31.6	0.17	296 ± 21	295 ± 18	0.88
DWG (kg animal^{-1} day^{-1})	0,640 ± 0.09	0,620 ± 0.12	0.39	0.460 ± 0.1	0.420 ± 0.9	0.13
Live meat production (kg LW ha^{-1})	220	214		205	194	

OS: open sky grassland; SPS: silvopastoral system; DWG: daily weight gain; LW: living weight.

better performance, no statistically significant differences were found. On the other hand, the 548 live trees in 2019, which generated a 39% shading that year, were commercially thinned (48% pulp rounds –52% pruned logs for high uses as furniture) to 191 trees in 2020, yielding 60 m^3 ha^{-1} and a remaining stock of 49 m^3 ha^{-1} that provided 33% shading at the end of the second-period trial that is in only 7 years.

8 Conclusion

To summarize, despite the fact that the case studies were in different environments and concerning different species, SPS enhanced the soil organic carbon. Therefore, SPS are working as a mitigation climate change practice, supporting the SDG 13 "Climate action". In Latin America, there are several cases of agroforestry, where the highest means of soil carbon sequestration rate (6.54 t ha^{-1}) occurs in SPS and the lowest (−4.04 t ha^{-1}) in woodlots (Feliciano et al., 2018). In Colombian deforested Amazonia (tropical area), Rodriguez et al. (2021) also observed that SPS are more efficient to increase soil C storage (131 t ha^{-1}) in aggregated structures than the pastures (112 t ha^{-1}) without trees. For further information, there are several worldwide meta-analyses about the effect of SPS on SOC sequestration. Through these studies, different situations can be observed regarding carbon sequestration in the soil. There is a wide range of situations from a slight decrease to a significant increase in SOC (Mayer et al., 2022; Hubner et al., 2021; Corbeels et al., 2019; Rathore et al., 2022).

To conclude, there are manifold SPS that present a wide variety of environmental services. In each of the cases studied, the SPS increased the C input to the system. For instance, in Balcarce (AR 1) the results revealed that even in a marginal environment, such as the one in the Southeast of Buenos Aires province, a fruit tree associated with cattle in SPS improved the C sequestration in one year and is expected to increase even more in the future. In the case of Montecarlo (AR 4), an improvement in the quality of the forage could also be observed as well as an increment in the daily gain of weight of the animals. In the case of SOC sequestration, as it was previously shown in the cases studied, there is a range of results which are positive on the whole. Regarding Leales (AR 2), an upsurge in soil nitrogen is also observed, thus improving its chemical fertility and its C sequestration potential. The SOC sequestration depends on trees, grass species, type of soil, and weather. In the Tierra del Fuego (AR 3) trial, the importance of measuring the SOC content of natural areas was observed, in order to evaluate the impact of the forest and grasslands management. Henceforth, it is extremely important to continue the study of SPS to detect optimal combinations of trees and grass for each environment that enhance ecosystem benefits, in spite of the fact that the results show improvements in different aspects.

Although the data were obtained in Argentina, these results can be useful and adopted by farmers worldwide, always considering the edaphic and climatic variables of each area. Soil characteristics such as texture, pH, salinity, and chemical fertility, together with the humidity and temperature conditions of the environments might affect the production and interaction between trees and pastures. Therefore, in order to implement the appropriate SPS, the previously mentioned weather and soil variables should be taken into account. Thus, the correct combination of trees and pastures could be chosen according to each environment.

9 Acknowledgments

Part of the research realized in the case studies was provided by:

The project SENSE "Synergies in integrated systems: Improving resource use efficiency while mitigating GHG emissions through well-informed decisions about circularity" from the 2021 JOINT CALL Cofund ERA-Nets SusCrop (Grant N° 771134), FACCE ERA-GAS (Grant N° 696356), ICT-AGRI-FOOD (Grant N° 862665) y SusAn (Grant N° 696231).

The project 2023-PD-L02-I132 "Strategies to increase national SOC reserves" and the project 2023-PE-L03-I068 "Implemented silvopastoral systems and MBGI" from the National Institute of Agricultural Technology of Argentina.

10 Where to look for further information

Further information about the project can be found in https://www.argentina.gob.ar/inta

11 References

Abrahams, P.W. (2002), Soils: their implications to human health. *Sci. Total Environ.*, 291, 1–32. https://doi.org/10.1016/S0048-9697(01)01102-0.

Alonso J. (2011), Los sistemas silvopastoriles y su contribución al medio ambiente. *Rev. Cuba. Cienc. Agríc, Tomo,* 45(2), pp 107–115.

Aravena Acuña M.C., et al. (2023), Forest carbon management strategies influence storage compartmentalization in Nothofagus antarctica forest landscapes. *Can. J. For. Res.* https://doi.org/10.1139/cjfr-2023-0009.

Asner G.P., et al. (1997), The decoupling of terrestrial carbon and nitrogen cycles. *BioScience,* 47(4), 226–234.

Banegas N.R., et al. (2019a), Long term impact of grazing and tillage on soil quality of semi-arid Chaco (Argentina). *Spanish J. Soil Sci.*, 9(1), 24–41.

Banegas N.R., et al. (2019b), Leucaena leucocephala introduction into a tropical pasture in the Chaco region of Argentina. Effects on soil carbon and total nitrogen. *Trop. Grassl.-Forrajes Trop.*, 7(4), 295–302.

Barnes A., et al. (2014), Consequences of tropical land use for multitrophic biodiversity and ecosystem functioning. *Nat. Commun.*, 5, 5351. https://doi.org/10.1038/ncomms6351.

Barral M. P., et al. (2020), Widespread and major losses in multiple ecosystem services as a result of agricultural expansion in the Argentine Chaco. *J Appl Ecol.*, 57, 2485–2498. https://doi.org/10.1111/1365-2664.13740.

Bergeron M., et al. (2011), Reduced soil nutrient leaching following the establishment of tree-based intercropping systems in eastern Canada. *Agrofor. Syst.*, 83, 321–330.

Birdsey R., et al. (2000), Mitigation activities in the forest sector to reduce emissions and enhance sinks of greenhouse gasses. In The impact of climate change on American forests: a technical document supporting the 2000 USDA Forest Service RPA assessment; Joyce, L.A.; Birdsey, R.A., Eds.; US Department of Agriculture, Rocky Mountain Research Station. *For. Serv.,* 59, 112–128.

Bravo F., et al. (2017), Forest carbon sequestration: the impact of forest management. In: Bravo, F., Jandl, R., LeMay, V., (Eds.), *Managing Forest Ecosystems: The Challenge of Climate Change*. Cham: Springer, pp. 251–275.

Cabrera G., et al. (2011), Composición funcional de la macrofauna edáfica en cuatro usos de la tierra en las provincias de Artemisa y Mayabeque, Cuba. *Pastos y Forrajes,* 34, 331–346.

Cardinael R., et al. (2015), Impact of alley cropping agroforestry on stocks, forms and spatial distribution of soil organic carbon–a case study in a Mediterranean context. *Geoderma,* 259–260, 288–299. http://dx.doi.org/10.1016/j.geoderma.2015.06.015.

Cardinael R., et al. (2017), Increased soil organic carbon stocks under agroforestry: a survey of six different sites in France. *Agricul. Ecosyst. Environ.*, 236, 243–255. http://dx.doi.org/10.1016/j.agee.2016.12.011.

Chará J., et al. (2020), Sistemas silvopastoriles y su contribución al uso eficiente de los recursos y a los Objetivos de Desarrollo Sostenible: Evidencia desde América Latina. CIPAV, FAO, Editorial CIPAV, Cali, p. 60.

Coello Gomez J. (2019), Los Sistemas agroforestales modernos: ¿qué son y cuál es su fundamento?, Dossier técnico n°99. Departamento de Agricultura, Ganadería, Pesca y Alimentación de Cataluña (DARP). https://www.researchgate.net/publication/340538678.

Colcombet L. (2022), Statistical meteorological analysis of Montecarlo rainfall records 1927-2022 provided by researcher Fidelina Silva. https//repositorio.inta.gob.ar/handle

Corbeels M., et al. (2019), The 4 per 1000 goal and soil carbon storage under agroforestry and conservation agriculture systems in sub-Saharan Africa. *Tillage Res.*, 188, 16–26. https://doi.org/10.1016/j.still.2018.02.015.

Coulibaly J.Y., et al. (2017), Adoption of agroforestry and the impact on household food security among farmers in Malawi. *Agric. Syst.*, 155, 52–69.

Dalal R.C. and Carter J.O. (2000), Soil organic matter dynamics and carbon sequestration in Australian tropical soils. In: Lal, R., et al. (Eds.), *Global Climate Change and Tropical Ecosystems*. Boca Raton, FL: CRC Press, pp. 283–314

Domínguez-Núñez J.A., et al. (2020), Efficient use of soil in silvopastoral systems of native forests. In: Kumar, S., et al. (Eds.), *Resources Use Efficiency in Agriculture*. Singapore: Springer, pp. 481–518. https://doi.org/10.1007/978-981-15-6953-1_14.

Dube F., et al. (2012), Productivity and carbon storage in silvopastoral systems with Pinus ponderosa and Trifolium spp., plantations and pasture on an Andisol in Patagonia, Chile. *Agrofor. Syst.*, 86, 113–128. https://doi.org/10.1007/s10457-011–9471-7.

Esquivel J. and Colcombet L. (2017), Aporte de los sistemas silvopastoriles a la ganadería sustentable en el Nordeste de Argentina. Conferencia presentada en la Jornada de Producción Animal en Sistemas Agro-Silvo-Pastoriles. 40° Congreso Argentino de Producción Animal. Córdoba, Argentina, 6 al 9 de noviembre.

Feliciano D., et al. (2018), Which agroforestry options give the greatest soil and above ground carbon benefits in different world regions? *Agric. Ecosyst. Environ.*, 254, 117–129.

Figueiredo E.B., et al. (2017), Greenhouse gas balance and carbon footprint of beef cattle in three contrasting pasture-management systems in Brazil. *J. Clean. Product.*, 142, 420–431. http://dx.doi.org/10.1016/j.jclepro.2016.03.132.

Gándara L., et al. (2021), ¿Can silvopastoral systems improve carbon balances in Argentina?. XI Congreso Internacional de Sistemas Silvopastoriles y I Congreso de la Red Global de Sistemas Silvopastoriles. México, 3 a 5 de noviembre de.

Hubner R., et al. (2021), Soil carbon sequestration by agroforestry systems in China: a meta-analysis. *Agricul. Ecosyst. Environ.*, 315, 107437.

Hudson N. (2006), *Conservación del suelo*. Barcelona: Editorial Reverté SA.

Jordan C.F. (2004), Organic farming and agroforestry: alley cropping for mulch production for organic farms of southeastern United States. *Agrofor. Syst.*, 61–62, 79–90.

Kaeser A., et al. (2010), Agroforesterie moderne en Suïsse. Vergers novateurs:productivité et rentabilité. Rapport ART 725. Ed. Station de recerche Agroscope Reckenholz-Tänikon, Tänikon.

Khan S., et al. (2017), Impact of different tillage practices on soil physical properties, nitrate leaching and yield attributes of maize (Zea mays L.). *J. Soil Sci. Plant Nutr.*, 17, 240–252. https://doi.org/10.4067/s0718- 95162017005000019.

Kibblewhite M.G., et al. (2008), Soil health in agricultural systems. *Philos. Trans. R. Soc. Lond., B, Biol. Sci.*, 363, 685–701. https://doi.org/10.1098/rstb.2007.2178.

Lal R. (2006), Enhancing crop yields in the developing countries through restoration of the soil organic carbon pool in agricultural lands. *Land Degrad. Dev.*, 17, 197–209.

Loto M., et al. (2022), Utilización del Botón de Oro (Tithonia diversifolia) en Sistemas Silvopastoriles Triestratos durante la recría de bovinos en la Provincia de Misiones, Argentina. II Cuban Seminar of tihonia diversifolia, Varadero, Cuba.

Luccerini S.A., et al. (2013), Sistemas Silvopastoriles: una alternativa productiva para nuestro país. Apuntes Agroeconómicos, Facultad de Agronomía de la Universidad de Buenos Aires N° 8, Buenos Aires.

Martínez Pastur G., et al. (2011), Canopy structure analysis for estimating forest regeneration dynamics and growth in Nothofagus pumilio forests. *Ann. For. Sci.*, 68, 587–594. https://doi.org/10.1007/s13595-011-0059-1.

Martínez Pastur G., et al. (2020), Conservation values of understory vascular plants in even- and uneven-aged Nothofagus antarctica forests. *Biodiv. Conserv.*, 29(13), 3783–3805. https://doi.org/10.1007/s10531-020-02049-8.

Martínez Pastur G., et al. (2021), Changes in forest structure values along the natural cycle and different management strategies in Nothofagus antarctica forests. *For. Ecol. Manage.*, 486, e118973. https://doi.org/10.1016/j.foreco.2021.118973.

Martínez Pastur G., et al. (2022), Changes in forest structure modify understory and livestock occurrence along the natural cycle and different management strategies in Nothofagus antarctica forests. *Agrofor. Syst.*, 96, 1039–1052. https://doi.org/10.1007/s10457-022-00760-9.

Mayer S., et al. (2022), Soil organic carbon sequestration in temperate agroforestry systems – a meta-analysis. *Agricul. Ecosyst. Environ.*, 323 https://doi.org/10.1016/j.agee.2021.107689.

Ministerio de Economía (MECON). (2023), Datos de Agricultura y Ganadería. https://datos.magyp.gob.ar/dataset.

Mosquera-Losada M.R., et al. (2006), Pasture, tree and soil evolution in silvopastoral systems of Atlantic Europe. *For. Ecol. Manag.*, 232(1–3), 135–145.

Mosquera-Losada M.R., et al. (2005), The herbaceous component in temperate silvopastoral systems. In *Silvopastoralism and Sustainable Land Management*. Wallingford: CABI, Digital Library, pp. 93–100.

Mosquera-Losada M.R., et al. (2018), Deciduous plantations established on former agricultural land in northwest of Spain as silvopastoralism: tree growth; pasture production and vascular plant biodiversity. *Catena*, 169, 1–10.

Mulia R. and Dupraz C. (2006), Unusual fine root distributions of two deciduous tree species in southern France: what consequences for modelling of tree root dynamics? *Plant Soil*, 281, 71–85.

Nasca J., et al. (2020), Evaluación productiva y de emisiones de gases de efecto invernadero en dos sistemas de cría bovina contrastantes del Chaco Seco argentino. *Rev. Agron. Noroeste Argent.*, 40(1), 13–22.

Navas Panadero A. and Montaña V. (2019), Comportamiento de Tithonia diversifolia bajo condiciones de bosque húmedo tropical. *Rev. Investig. Vet. Perú*, 30(2), 721–732. https://dx.doi.org/10.15381/rivep.v30i2.15066.

Peichl M., et al. (2006), Carbon sequestration potentials in temperate tree-based intercropping systems, southern Ontario, Canada. *Agrofor. Syst.*, 66, 243–257.

Pérez A., et al. (2009), Tithonia diversifolia (Hemsl.) A. Gray. *Pastos y Forrajes*, 32(1), 1. http://scielo.sld.cu/scielo.php?script=sci_arttext&pid=S0864-03942009000100001&lng=es&tlng=es.

Peri P.L. (2011), Carbon storage in cold temperate ecosystems in Southern Patagonia, Argentina. In: Atazadeh, I. (Ed.), *Biomass and Remote Sensing of Biomass*. Croacia: InTech Publisher, pp. 213–226.

Peri P.L. and Bloomberg M. (2002), Windbreaks in southern Patagonia, Argentina: a review of research on growth models, wind speed reduction, and effects on crops. *Agroforest. Syst.*, 56, 129–144.

Peri P.L., et al. (2016a), Silvopastoral systems under native forest in Patagonia, Argentina. In: Peri, P.L., et al. (Eds.), *Silvopastoral Systems in Southern America*. Series Advances in Agroforestry 11. Bern: Springer, pp. 117–168.

Peri P.L., et al. (2016b), Biodiversity and ecological long-term plots in Southern Patagonia to support sustainable land management: the case of PEBANPA network. *J. Nat. Conserv.*, 34, 51–64. https://doi.org/10.1016/j.jnc.2016.09.003.

Peri P.L., et al. (2017), Carbon sequestration in temperate silvopastoral systems, Argentina. In: Montagnini, F., (Ed.), *Integrating Landscapes: Agroforestry for Biodiversity Conservation and Food Sovereignty*. Advances in Agroforestry 12, pp. 453–478. Cham: Springer International Publishing.

Peri P.L., et al. (2018), Modelling soil carbon content in South Patagonia and evaluating changes according to climate, vegetation, desertification and grazing. Sustainability 10 (2), 438, https//doi.org/10.3390/su10020438

Peri PL., et al. (2019), Implementación y producción en SSP de Sudamérica como alternativa productiva: Beneficios, limitaciones y desafíos. X Congreso Internacional de Sistemas Silvopastoriles, Paraguay, 24 al 26 de septiembre.

Peri P.L., et al. (2021), Ecosystem services in Patagonia: a multi-criteria approach for an integrated assessment. In: *Natural and Social Sciences of Patagonia*. Cham: Springer. https://doi.org/10.1007/978-3-030-69166-0.

Peri P.L., et al. (2022), Management strategies for silvopastoral system in native forests. *Ecol. Austral.*, 32, 749–766. https://doi.org/10.25260/EA.22.32.2.1.1872.

Piñeiro G., et al. (2010), Pathways of grazing effects on soil organic carbon and nitrogen. *Rangeland Ecol Manag.*, 63(1), 109–119.

Radrizzani A., et al. (2011), Soil organic carbon and total nitrogen under Leucaena leucocephala pastures in Queensland. *Pasture Science*, 2011, 62, 337–345.

Rathore S.R., et al. (2022), Integrated agroforestry systems improve soil carbon storage, water productivity, and economic returns in the marginal land of the semi-arid region. *Saudi J. Biol. Sci.*, 29. https://doi.org/10.1016/j.sjbs.2022.103427.

Rodriguez L., et al. (2021), Agroforestry systems impact soil macroaggregation and enhance carbon storage in Colombian deforested Amazonia. *Geoderma*, 384. https://doi.org/10.1016/j.geoderma.2020.114810.

Rodríguez-Souilla J., et al. (2023), Variable retention harvesting and climate variations influence over natural regeneration dynamics in Nothofagus pumilio forests of Southern Patagonia. *For. Ecol. Manage.*, 544, e121221. https://doi.org/10.1016/j.foreco.2023.121221.

Schroth G., et al. (2004), *Agroforestry and Biodiversity Conservation in Tropical Landscapes*. Washington, DC: Island Press.

SIGA. (2023), Sistema Integrado de Gestíon de datos Agrometeorológicos. https://siga.inta.gob.ar/#/data.

Smith M.W. and Wood B.W. (2006), Pecan tree biomass estimates. *HortScience*, 41(5), 1286–1291.

Thevathasan N.V. and Gordon A.M. (2004), Ecology of tree intercropping systems in the North temperate region: experiences from southern Ontario, Canada. *Agrofor. Syst.*, 61–62, 257–268. https://doi.org/10.1023/B:AGFO.0000029003.00933.6d.

Tully K.L., et al. (2012), More trees less loss: nitrogen leaching losses decrease with increasing biomass in coffee agroforests. *Agric. Ecosyst. Environ.*, 161, 137–144.

Varah A., et al. (2013), Enhanced biodiversity and pollination in UK agroforestry systems. *J. Sci. Food Agric.*, 93, 2073–2075.

Villanueva C., et al. (2009), Sistemas silvopastoriles: una herramienta para la adaptación al cambio climático de las fincas ganaderas en América Central. Ed CATIE, Turrialba (Costa Rica). https://repositorio.catie.ac.cr/handle/11554/10149.

Waldron A., et al. (2017), Agroforestry can enhance food security while meeting other sustainable development goals. *Trop. Conserv. Sci.*, 10, 1–6.

Yanai R.D., et al. (2003), Soil carbon dynamics after forest harvest: an ecosystem paradigm reconsidered. *Ecosystems*, 6(3), 197–212. https://doi.org/10.1007/s10021-002-0206-5.

Zerbino S., et al. (2008), Evaluación de la macrofauna del suelo en sistemas de producción en siembra directa y con pastoreo. *Agrociencia*, 12, 44–55.

Part 3

Advances in silvoarable and other systems

Chapter 8

Types of silvoarable system: multifunctional windbreaks and riparian buffers

Oskar Englund, Mid Sweden University, Sweden; and Pål Börjesson, Mid Sweden University/ Lund University, Sweden

1 Introduction: multifunctional biomass production systems

The complexity and interconnected nature of environmental, climate, and socio-economic problems require new strategies based on holistic perspectives. Biomass production, to provide bioenergy and bioproducts, is such an example. One primary aim of the increased production and utilization of bioenergy is to replace fossil fuels and mitigate climate change. But the increased production of biomass can also lead to other, both positive and negative, environmental effects, due to land-use changes in forestry and agriculture. It is essential to avoid negative environmental effects from increased biomass production using unsustainable systems. This can be done by implementing rules and guidelines through legislation, such as the EU Renewable Energy Directive (RED) (EC,

http://dx.doi.org/10.19103/AS.2024.0141.13

2023), and voluntary certification systems, such as the Forest Stewardship Council (FSC, 2024).

However, complementary to the implementation of rules and guidelines to reduce the risk of negative environmental effects, the promotion of biomass production systems capable of delivering positive, additional environmental benefits is an equally important task. This approach focuses on so-called multifunctional biomass production systems which, through well-chosen siting, design, management, and system integration, can support ecosystem services and mitigate existing environmental impacts (Berndes et al., 2008; Englund et al., 2020). These multifunctional systems can be divided roughly into two categories. The first consists of more site-specific and specialized areas of cultivation, such as the two systems that are assessed in this chapter:

- Energy forestry plantations as dedicated wind breaks (for reducing wind erosion); and
- Riparian buffer zones (for reducing nutrient leaching).

The second category consists of more widespread areas of perennial energy crop cultivation in the agricultural landscape, leading to more general ecosystem services, e.g. increased soil carbon accumulation and enhanced biodiversity. Regardless of the category, multifunctional production systems usually provide multiple environmental co-benefits[1] when introduced in intensively managed arable landscapes dominated by annual crops.

While the concept of multifunctional biomass production systems might appear as a recent invention, the underlying idea also influenced land-use strategies in the past. When, e.g., lignocellulosic biomass feedstock production systems, such as perennial grasses and short-rotation woody crops, were introduced in agriculture during the 1980s and 1990s as a response to the oil crisis of the 1970s, they were suggested as potential opportunities for remediating various environmental problems. Besides reducing nutrient leaching and soil erosion using riparian buffers and windbreaks (as presented in this chapter), multifunctional biomass production systems can provide, e.g., cadmium removal from contaminated arable land by phytoextraction, treatment of nutrient-polluted drainage water in vegetation filters, and increased biodiversity in the agricultural landscape (Börjesson, 1999a).

The example of large-scale introduction of multifunctional biomass cultivation systems in Sweden during the 1990s also showed their economic value, including both reductions in costs to farmers (direct costs) and to

1 We distinguish between primary benefit and co-benefits. The former is the main incentive for establishing a multifunctional system and the latter are other environmental benefits that are provided by the system 'for free.' Sometimes, co-benefits can be substantial – even exceeding the needs for impact mitigation (Englund et al., 2021; Englund et al., 2022).

society as a whole (external costs) (Börjesson, 1999b). Research regarding perennial biomass production systems and additional environmental effects, or ecosystem services, was undertaken during the 1980s and 1990s in several different countries and regions, such as in North America (see e.g. Pimentel and Krummel, 1987; Ranney and Mann, 1994; Christian et al., 1994; McLaughin and Walsh, 1998; Kort et al., 1998) and in Europe (see e.g. Rijtema and de Vries, 1994; Makeschin, 1994; Sage and Robertson, 1994; Obernberger et al., 1997). This area of research has continued and there is now a growing body of literature that investigates the opportunities associated with multifunctional biomass production systems based on perennial crops, also referred to as beneficial land-use change by strategic perennialization (Englund et al., 2020).

One key challenge and critical issue in this kind of research is to understand and quantify all the various ecosystem services that can be provided by the implementation of multifunctional biomass production systems in intensively managed arable landscapes. Previous studies have often focused on a specific ecosystem service. However, the aggregated knowledge from the now vast research base has made it possible to perform more comprehensive assessments that cover multiple environmental benefits. This was showcased in a recent study of how a strategic expansion of new biomass plantations could reduce environmental impacts from EU agriculture, covering reduced nitrogen emissions to water, soil loss by erosion, recurring floods, and losses of soil organic carbon (SOC), in addition to the provision of renewable energy (see Fig. 1) (Englund et al., 2020). The potential environmental benefits will, however, vary in size due to local and regional conditions and the historical and current land use. One critical factor is, e.g., the dominance of annual crops in the landscape, where a higher share of annual crops normally leads to increased environmental benefits from strategic perennialization (Englund et al., 2020). Since the benefits vary depending on location, the spatial perspective is critical to further advance research. We will return to this several times in the chapter.

2 Multifunctional bioenergy buffers and windbreaks

One type of multifunctional biomass production system that has attracted particular recent focus is bioenergy buffers, which can be defined as linear landscape elements cultivated with perennial herbaceous or woody biomass crops placed along arable field margins and watercourses. Ferrarini et al. (2017a) adopted an impact assessment methodology to quantify the effects of various ecosystem services delivered by bioenergy buffers. The results show several environmental benefits leading to the conclusion that payments for ecosystem services obtained from bioenergy buffers can ultimately improve the economics of sustainable bioenergy and help to achieve the environmental goals of EU policies on water, soil, and biodiversity protection. Another study

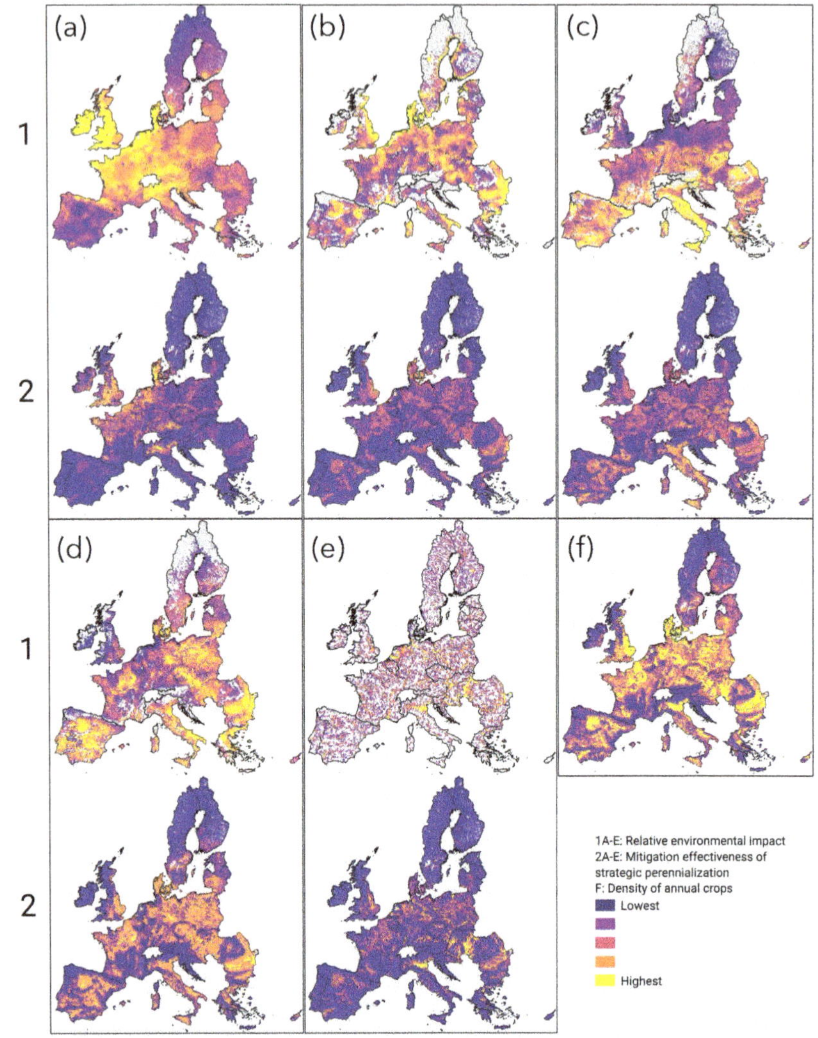

Figure 1 Relative degree of nitrogen emissions to water (a), soil loss by wind erosion (b), soil loss by water erosion (c), accumulated losses of soil organic carbon (d), and recurring floods (e). Map 2 for each impact a–e indicates the degree to which the impact can be mitigated using a strategic introduction of perennial crops in the agricultural landscape (strategic perennialization). Map f shows the density of annual crops in each landscape (i.e. crop area relative to landscape area). All maps are constituted by the same >81 000 sub-catchments as used in the case study presented in this chapter. See Englund et al. (2020) for more information.

by Christen and Dalgaard (2013) confirms the effectiveness of multifunctional buffers in temperate European agriculture.

A case study in southern Sweden, in the county of Scania, estimated the ecosystem services of climate regulation, energy provisioning, and water purification, from riparian buffer zones cultivated with willow for bioenergy (Styles et al., 2016). The results from this case study showed that implementation of willow buffer plantations in the intensive arable landscape of Scania could be one important strategy to reduce nutrient loading to water, while at the same time producing renewable energy and mitigating climate change, also when considering potential effects of displaced food production and indirect land-use change.

Studies of bioenergy buffers have also been performed in North America, e.g. in Iowa, where the environmental benefits of riparian buffers within an agricultural watershed were studied (Tufekcioglu et al., 2003). Another study by Fortier et al. (2010) quantified the ecosystem services provided by hybrid poplars in multiclonal agricultural riparian buffer strips in southern Quebec, Canada. The economics of bioenergy buffers have also been analysed in a case study in the USA Midwest corn belt, where growing willow as bioenergy buffers was examined (Ssegane et al., 2016). While this case study showed that it was difficult to provide positive revenues for farmers, net costs per unit of nitrogen removed were comparable to other conservation practices.

The many studies that have now quantified ecosystem services provided by different multifunctional production systems at different locations have been valuable for better understanding under what conditions multifunctional production systems can provide the desired environmental benefits and to what extent co-benefits can be expected. Along with substantial improvements in computational capacity and the availability of high-resolution geospatial data, this has paved the way for new kinds of studies, that can explore the potential of multifunctional production systems at a larger scale using geographically explicit methods (Englund et al., 2020, Englund et al. 2021, Englund et al., 2022).

The overall research questions analyzed and discussed in this chapter are how multifunctional windbreaks and riparian buffers in the form of perennial energy crop cultivation can lead to different environmental benefits within the EU agricultural production sector, the magnitude of these benefits, and the amount of biomass feedstock that they can deliver (see Fig. 1). We present a recent case study (Englund et al., 2021) that shows how these research questions can be answered using large-scale, geographically defined studies and provides new insights into how riparian buffers and windbreaks can provide multiple environmental benefits while providing biomass for the emerging bioeconomy. We will also discuss how large-scale implementation of riparian buffers could impact agricultural production at different scales, the climate benefits, and policy implications. Finally, while it is now possible to make pan-European studies of multifunctional production systems with quantitative results, there are notable limitations and thereby interesting opportunities for further research. The chapter therefore concludes with a section discussing research trends and needs in which we also discuss the use of perennial production systems for strengthening ecological connectivity – an urgent and global priority for supporting biodiversity.

3 Modelling widespread deployment of multifunctional riparian buffers and windbreaks: a European case study

In this case study, we modelled scenarios for the introduction of windbreaks and riparian buffers across Europe (EU27+UK), using different biomass crops and management systems, and quantified the corresponding (i) areas, (ii) amounts of biomass produced and the corresponding energy content, (iii) primary environmental benefit, and (iv) environmental co-benefits. This section provides sufficient methodological background to properly interpret the results. For full methodological details, see Englund et al. (2021).

3.1 Primary benefits of multifunctional riparian buffers and windbreaks

The primary benefit of riparian buffers was assumed to be reduced nitrogen (N) emissions to water. Five different riparian buffer systems were defined based on three different widths: 5 m ('Narrow'), 21 m ('Wide'), and 50 m ('Double-wide'). The first two systems were modelled with either SRC or with grass. The third system was modelled only with SRC.

The potential for riparian buffers to mitigate nitrogen emission to water in each landscape was calculated by comparing the buffer strip efficiency (BSE; Ferrarini et al., 2017b) of the different buffer designs with the current degree of the impact (Fig. 1).

The primary aim of windbreaks was assumed to be a reduction of soil loss by wind erosion. Two windbreak systems were defined:

- 'SRC windbreaks' referring to the establishment of short rotation coppice willow or poplar, with a rotation period of 3–4 years and a height of 5 m.
- 'SRF windbreaks' referring to the establishment of short rotation forestry poplar, with a rotation period of 15 years and a height of 20 m.

In both the SRC and the SRF design, a windbreak plantation width of 50 m is assumed, for practical and economic reasons, since larger cropping sites

Published by Burleigh Dodds Science Publishing Limited, 2025.

reduce management costs. The distance between windbreaks is assumed to be 20H, i.e. 20 times the windbreak height. In both designs, half of the windbreak is harvested at a time, allowing for constant windbreak functionality.

For windbreaks, it was observed that soil erosion is classified as low (rarely very low) in a vast majority of agricultural landscapes across Europe (Fig. 1). Windbreaks were therefore assumed to be able to reduce soil loss by wind erosion to a low level, i.e. less than 5 t ha^{-1} $year^{-1}$, but not further. An assumption was made that the two windbreak designs, if fully implemented on all agricultural fields in a landscape that are under annual crop production, could mitigate soil loss from a high to a low erosion level.

3.2 Policy scenarios

To illustrate different outcomes of the widespread implementation of riparian buffers and windbreaks, three conceptual policy scenarios were designed: 'Biomass', 'Low-impact', and 'Food-first', as summarized below. What they entail in practice is described later along with the results.

The Biomass scenario: There are incentives for implementing multifunctional biomass production systems in all landscapes where it is likely that the introduction of the assessed system can effectively reduce the primary environmental impact (reduced nitrogen leaching). Farmers are assumed to select the option expected to result in the highest mitigation, while also maximizing biomass output from the multifunctional biomass production system, i.e. using extensive management options with the highest yielding crop.

The Low-impact scenario: Incentives are stronger to mitigate the primary impact (reduced nitrogen leaching), but only down to a defined 'low' level at the landscape scale. Local and regional circumstances will determine whether farmers prefer to favour biomass production or to minimise effects on current agricultural production. Different crops or management systems may therefore be implemented in different landscapes.

The Food-first scenario: Incentives are identical to the Low-impact policy scenario, with the addition that impacts on food production are disincentivized. Farmers will therefore strive to reduce the primary impact down to a 'low' level, using as small an area as possible for multifunctional buffers or windbreaks.

3.3 Areas of buffers and windbreaks in the different scenarios

For each buffer option in the three deployment scenarios, the buffer area in each individual landscape was estimated as the product of buffer width (times two, assuming that it is established on both sides of the watercourse) and the total length of rivers in each landscape (European Environment Agency, 2012). This resulted in the maximum buffer area for the three buffer designs, which was used in the Biomass scenario. In the Low-impact scenario, the necessary buffer area needed to achieve a low impact in each landscape was calculated by assuming that the BSE, at the landscape scale, is proportional to the share of the maximum buffer area. For example, if a BSE of 30% is necessary to reduce N emissions down to a low impact level, 50% of the maximum area for narrow buffers (having a BSE of 60%) is needed and 30% of the maximum area for wide or double-wide buffers (having a BSE of 100%). In the Food-first scenario, in which the smallest buffer area possible is desired, buffer areas were calculated as in the Low-impact scenario, with the addition that narrow buffers are prioritized and wide buffers are only implemented in landscapes where narrow buffers are insufficient to mitigate nitrogen emissions down to a low level (Fig. 2).

For windbreaks, the necessary implementation level was first calculated for each landscape by comparing the mitigation needs with the mitigation potential. The total windbreak area needed to reduce wind erosion to a low level was calculated as the product of (i) implementation level, (ii) the share of land under annual crops needed for establishing windbreaks at 100% implementation, and (iii) the total area under annual crops. The total share of agricultural land under annual crop production needed for establishing windbreaks, at 100% implementation, was calculated as the quotient of windbreak width and windbreak distance, i.e. one-third for SRC windbreaks and one-ninth for SRF windbreaks.

3.4 Biomass production in buffers and windbreaks

Biomass production from riparian buffers was estimated for each landscape and for each buffer option in the three deployment scenarios, as the product of buffer areas and the yield of the highest yielding SRC and grass option, respectively (Dees et al., 2017). The energy output was calculated as the product of biomass production and energy content of the harvested biomass; 18.7 MJ kg^{-1} DM (dry matter), for both SRC (Nordborg et al., 2018a) and grass (Baxter et al., 2014).

Biomass production from windbreaks was estimated as for riparian buffers, but considering only SRC willow and SRC poplar. The yield for SRF poplar was assumed to be identical to SRC poplar (Aylott et al., 2008; Dimitriou and Mola-Yudego, 2017). The same energy content, 18.7 MJ kg^{-1} DM, was used for both SRC willow and SRC/SRF poplar (Nordborg et al., 2018a; Nordborg et al., 2018b).

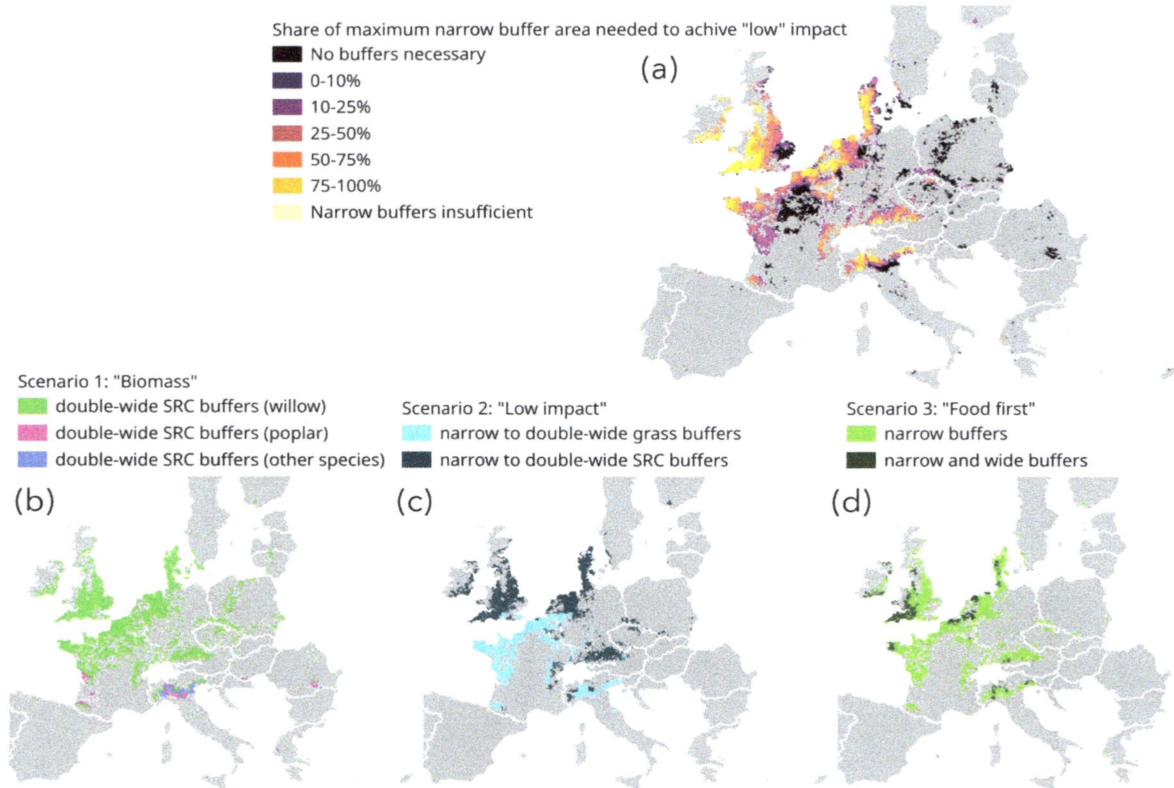

Figure 2 Widespread deployment of multifunctional riparian buffers across Europe. (a) Share of maximum narrow buffer area needed to achieve a low level of nitrogen emissions to water, and (b–d) the type of riparian buffers established in the three policy scenarios. Figure originally published in Englund et al. (2021).

3.5 Avoided greenhouse gas emissions

The biomass production from buffers and windbreaks was translated to GHG emission savings by assuming that the biomass is used for biofuel, electricity, or heat, assuming default GHG emissions savings values from the EU Renewable Energy Directive (European Commission, 2023). These default values do not include any net-carbon emissions (or savings) from potential land-use change.

3.6 Environmental co-benefits

We initially modelled the introduction of riparian buffers and windbreaks, respectively, in individual landscapes across Europe, in three different scenarios, based on their potential to mitigate a primary environmental impact. Here we estimate which additional environmental benefits can be expected based on these land-use changes.

Buffers established on agricultural land can be expected to increase SOC, mitigate water erosion, and retain sediments. These benefits were assessed and quantified. Buffers can also be expected to mitigate flooding and may possibly also reduce wind erosion. These benefits were also assessed but not quantified.

Windbreaks can be expected to increase SOC and reduce N emissions to water. These benefits were assessed and quantified. Flood mitigation was assessed but not quantified.

For methodological details, see the original article (Englund et al., 2021).

4 Environmental benefits of multifunctional riparian buffers

The degree to which 'narrow' riparian buffers can suffice to mitigate nitrogen emissions to water down to a 'low' level in each individual sub-catchment is shown in Fig. 2. The impact mitigation potential for narrow buffers varies notably. In about 25% of all landscapes, the impact is already at a low level (black), while in about 12% of all landscapes, narrow buffers are insufficient for reducing the impact to a low level (yellow). The mitigation potential primarily depends on the degree of the impact and the existence of watercourses in the landscape.

In all scenarios, multifunctional riparian buffers are predominantly established in north-western Europe (Fig. 2b–d). In most locations, SRC was identified as the highest yielding buffer option, with willow as the highest yielding SRC species. In some areas, however (e.g. in France, Belgium, and Italy), grass was identified as the highest yielding buffer option.

Since riparian buffers are established to a varying degree and in different ways in the different policy scenarios, the buffer area, biomass production, and environmental benefits from a widespread deployment vary considerably, as described below.

4.1 Riparian buffers in the Biomass policy scenario

In this scenario, the farmers are assumed to favour the 'double-wide' option, since this would maximise biomass production from the system. In most parts of Europe, SRC willow is the preferred option since it is higher yielding than poplar or other species. A total of 7574 landscapes, covering about 58 million hectares (Mha), were identified by the model as suitable for riparian buffers. In these landscapes, the model established a total of 1.4 Mha of 'double-wide' SRC buffers, covering 4.6% of the area under annual crops in the landscapes where riparian buffers were established, and 1.3% of the total area under annual crops in Europe. These buffers are estimated to reduce nitrogen emissions to water by about 900 thousand tonnes (kt) while delivering over 16 million tonnes (Mt) of biomass, annually.

Relative to the business-as-usual scenario with continued land-use, the buffers in the Biomass policy scenario are estimated to increase soil carbon by almost 33 Mt by 2050, corresponding to 4 Mt of CO_2 emissions savings, annually.

The amount of soil loss by water erosion that can be avoided by establishing riparian buffers in this scenario is 3.3 Mt year^{-1}, corresponding to 6.3% of the soil loss by water erosion on cropland in the landscapes where buffers are established, and 1.2% of total soil loss by water erosion on cropland in Europe. In the median landscape, riparian buffers contribute with 26% of the necessary reductions in soil loss through water erosion to achieve a low-impact level at the landscape scale. The buffers also retain a total of 49 Mt of soil that would be eroded on cropland within their respective sub-catchments, corresponding to 19% of all soil loss by water on cropland in Europe.

4.2 Riparian buffers in the Low-impact policy scenario

Since multifunctional riparian buffers, in this scenario, are only eligible for compensation if N emissions to water are greater than the threshold for what can be considered a 'low'-impact riparian buffers are established in fewer (5707) sub-catchments than in the Biomass scenario, covering a smaller area (about 43 Mha).

The total buffer area differs since farmers can freely decide which buffer option to implement. It ranges from about 69 kha (if only 'narrow' buffers are implemented) to 431 kha (if only 'double-wide' buffers are implemented). This corresponds to 0.3–2.1% of the area under annual crops in the landscapes where riparian buffers could be established and 0.1–0.4% of the total area under annual crops in Europe. The resulting annual biomass production ranges between 0.8 Mt 4.9 Mt of SRC biomass or 0.8–2.1 Mt of grass biomass.

Regardless of buffer options, the buffers in this scenario can reduce N emissions to water by approximately 370 kt N year^{-1}.

Relative to a business-as-usual scenario with continued land use, the buffers in the Low-impact scenario are estimated to increase soil carbon by 1.1–6.3 Mt by 2050, corresponding to 0.1–0.8 Mt CO_2 emissions savings, annually.

The total avoided soil loss by water erosion ranges from 192 kt (only narrow buffers) to 1150 kt (only double-wide buffers), corresponding to 0.5–3.1% of all soil loss by water erosion on cropland where buffers are established, and 0.1–0.4% of total soil loss by water erosion on cropland in Europe. In the median landscape, the avoided water erosion amounts to 2–11% of what is necessary to achieve a low-impact level at the landscape scale. Additionally, sediment retention amounts to 13–16 Mt year^{-1}. Direct and indirect avoided soil loss by water erosion within the buffer landscapes thus amounts to 38–50%. At the European level, buffers retain 5–6% of all soil loss by water erosion on cropland.

4.3 Riparian buffers in the Food-first policy scenario

Multifunctional riparian buffers are here established in the same sub-catchments as in the Low-impact policy scenario but on a smaller area. A total of 49 kha of 'narrow' buffers and 52 kha of 'wide' buffers are established, corresponding to 0.5% of the area under annual crops in the landscapes where riparian buffers could be established and 0.1% of the total area under annual crops in Europe. As in the Low-impact policy scenario, these buffers result in 371 kt of avoided nitrogen emissions to water per year. Since buffers in this scenario are only implemented in the most efficient way (in terms of impact mitigation relative to the area used for buffers), the output of biomass is lower, about 1.2 Mt year^{-1}.

Relative to a business-as-usual scenario with continued land use, the buffers in the Food-first policy scenario are estimated to increase soil carbon by 2.2 Mt by 2050, corresponding to 0.3 Mt of CO_2 emissions savings, annually.

Avoided water erosion amounts to 291 kt year^{-1}, from narrow (126 kt) and wide (165 kt) buffers combined, corresponding to 0.8% of total soil loss by water erosion on cropland where buffers are established, and 0.1% of total soil loss by water erosion on cropland in Europe. In the median landscape, avoided water erosion amounts to 2.2% of what is necessary to achieve a low-impact level at the landscape scale. Additional sediment retention amounts to 16 Mt year^{-1}, totaling 42.6% direct and indirect avoided soil loss by water erosion within the buffer landscapes. At the European level, buffers retain 5.6% of all soil loss by water erosion on cropland.

4.4 Discussion of the environmental benefits from riparian buffers

The avoided nitrogen emissions to water in the Biomass policy scenario are equivalent to about 11% of the gross surplus of nitrogen on agricultural land in EU27 + UK, and 33% of total nitrogen emissions to water in Europe. In the other scenarios, corresponding avoided nitrogen emissions to water are equivalent to 4% of total gross nitrogen surplus and 13% of total nitrogen emissions to water.

The potential of riparian buffers to enhance SOC differs between locations depending on how much SOC can be enhanced by introducing perennial crops, which largely depends on the soil type and historical land use. At the landscape scale, the total SOC increase also depends on the total area used for buffers. Given the large variation of both factors, there is also a large variation in the degree to which riparian buffers can contribute to enhancing SOC as a co-benefit, both at the landscape scale and within individual landscapes.

The annual CO_2 emissions savings presented here are average values over 30 years. In reality, SOC increases are greater during the first 10 years after establishment, meaning higher short-term CO_2 emissions savings from increased SOC than what is indicated here.

While these SOC increases can be considered substantial in absolute terms, riparian buffers are unlikely to play an important role in restoring accumulated losses of SOC on agricultural soils across Europe. SOC losses are widespread and substantial, and riparian buffers can only enhance SOC in the locations where they are established, meaning that most landscapes in Europe are unaffected, as is most of the land within the landscapes where buffers are established. To effectively restore SOC on a large scale, more extensive changes in agricultural practices are necessary, e.g., widespread introduction of perennial ley crops in crop rotations (Englund et al., 2022).

Although there is a notable variation between different landscapes, countries, and regions (Fig. 2), riparian buffers are generally considered as also having limited potential for reducing soil loss by water erosion at the European scale. However, the potential for avoiding streambank erosion, which was not modelled in this study, could be substantial (IPCC, 2019). The potential for retaining eroded soil in buffers and thus avoiding sedimentation in watercourses appears substantial, not only within landscapes where buffers are established but also at the European level. It should, however, be noted that sediment retention in buffers does not mitigate negative effects of water erosion on eroded cropland, such as reduced soil fertility.

5 Environmental benefits of multifunctional windbreaks

The degree to which windbreaks are implemented in each landscape in the different policy scenarios is shown in Fig. 3. Contrary to the case of riparian buffers, windbreaks are here implemented in the same sub-catchments (a total of 1168 sub-catchments, covering 10 Mha) in all policy scenarios. In the Biomass policy scenario (Fig. 3a), farmers need not limit implementation based on the degree of wind erosion, meaning that the implementation level is always 100%. In the Low-impact and Food-first policy scenarios (Fig. 3b), implementation is determined by what is needed to reduce wind erosion to a 'low impact' but not beyond. In most landscapes, this means implementing windbreaks to a lesser extent than in the Biomass policy scenario although in some landscapes to a considerably greater extent.

In the Biomass and Low-impact policy scenarios (Fig. 3c), the highest yielding options are always used: generally, willow in the northern parts of Europe and poplar in the southern parts. In the Food-first policy scenario, the selection is also affected by the goal of minimising total windbreak area, meaning that SRF windbreaks, with a greater distance between the windbreak strips (or 400 m compared with 100 m for SRC) and thereby a lower land occupation, are favoured. As for riparian buffers, the size of environmental benefits achieved, and amount of biomass produced, varies considerably between different sub-catchments.

5.1 Windbreaks in the Biomass policy scenario

In the Biomass policy scenario, the total windbreak area ranges between about 1.7–2.3 Mha, depending on whether SRC or SRF windbreaks are used. This corresponds to a small share (1.6–2.1%) of the total area under annual crops in Europe, but a large share (30–33%) of the area under annual crops in the landscapes where windbreaks could be established. These windbreaks are estimated to reduce soil loss by wind erosion by 5–13 Mt, annually, corresponding to 9–23% of total soil loss by wind erosion in Europe. The annual biomass production from these windbreaks sums up to 3–24 Mt.

The reduction in soil loss by water erosion varies from 1.9 Mt year^{-1} with only SRF windbreaks to 3.2 Mt year^{-1} with only SRC windbreaks. This corresponds to a very small share (1%) of all soil loss by water erosion in Europe, but a substantial share (10–33%) of all water erosion in the landscapes where windbreaks could be established. The average effect of windbreaks in reducing water erosion to a low impact level at the landscape scale is almost 100%.

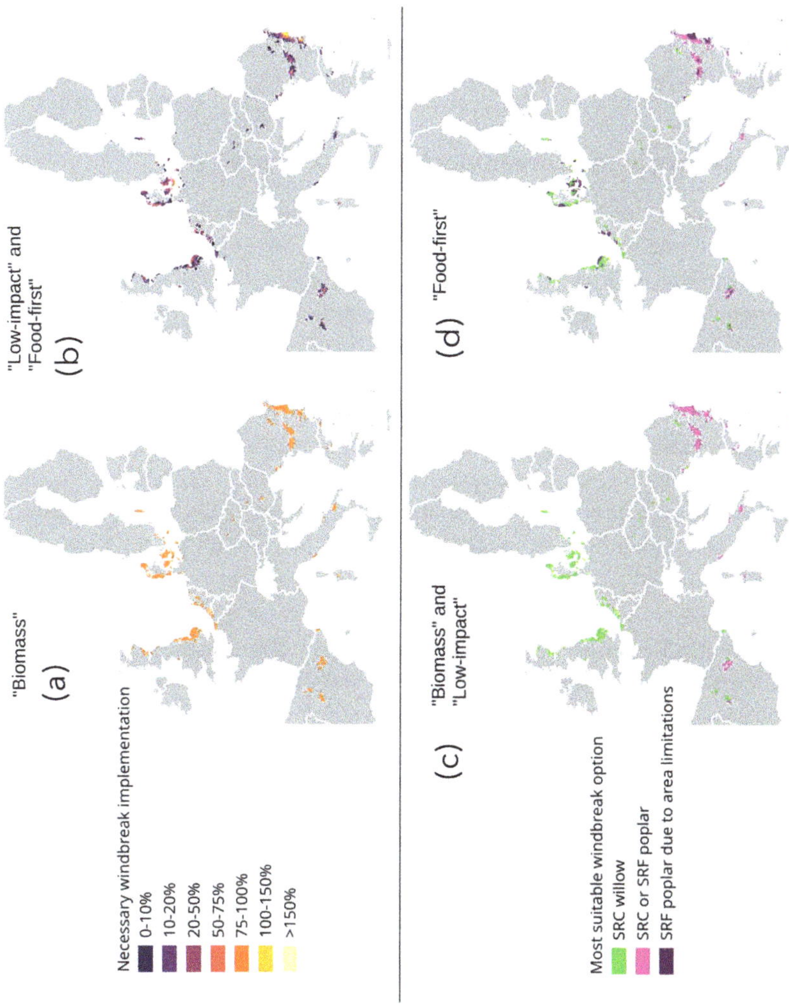

Figure 3 Widespread deployment of multifunctional windbreaks across Europe. Implementation levels in the different scenarios and implemented windbreaks options. Figure originally published in Englund et al. (2021).

In many landscapes, the impact of windbreaks on reducing nitrogen emissions to water is substantial. This is especially true in the Biomass policy scenario, in which windbreaks suffice to reduce the impact to a low level in most landscapes. In total, nitrogen emissions to water are here estimated to be reduced by a total of 20–22 kt, annually.

The impact of windbreaks on SOC at the landscape scale depends on the size of the windbreak area. In the Biomass policy scenario, the increase in SOC is naturally the highest, at 38–45 Mt C by 2050. This equates to an annual reduction in CO_2 emissions of 4.6–5.5 Mt.

5.2 Windbreaks in the Low-impact policy scenario

In the Low-impact policy scenario, the total windbreak area is notably smaller than in the Biomass policy scenario: 185–555 kha, corresponding to 0.2–0.5% of the current total area under annual crops in Europe, and 2.7–8.2% of the area under annual crops in the landscapes where windbreaks could be established. Regardless of how the windbreak options (SRC/SRF) are combined, these windbreaks are estimated to reduce soil loss by wind erosion by about 13 Mt, annually, or 23% of total soil loss by wind erosion in Europe. The annual biomass production from these windbreaks sums up to about 2–6 Mt DM.

The reduction in soil loss by water erosion varies from 0.2 Mt year^{-1} with only SRF windbreaks to 0.6 Mt year^{-1} with only SRC windbreaks. This represents an even smaller share of total water erosion, both in Europe (0.1–0.2%) and in the landscapes with windbreaks (2–7%). The average effect of windbreaks in reducing erosion to a low level at the landscape scale is between 4% and 12%.

The reduction in N emissions to water due to windbreaks is estimated to be 1.6–4.8 kt year^{-1}, but with large variations between different landscapes due to the variation in windbreak implementation. In some areas, where wind erosion is adverse and the windbreak implementation is high, N emissions are reduced to a low impact level and even beyond. In other areas, the effect is minimal.

The total increases in SOC where windbreaks are established are estimated at 4–11 Mt C, corresponding to annual CO_2 emissions savings of 0.5–1.3 Mt.

5.3 Windbreaks in the Food-first policy scenario

In the Food-first scenario, SRF windbreaks are favoured due to their greater land-use efficiency. The total windbreak area is here 312 kha, of which 190 kha are SRC windbreaks and 212 kha are SRF windbreaks. This corresponds to 0.3% of the current area under annual crops in Europe, and 4.6% of the area under annual crops in the landscapes where windbreaks could be established. Wind erosion mitigation is identical to the Low-impact policy scenario and the annual biomass production is about 3 Mt DM.

The reduction in soil loss by water erosion due to windbreaks is estimated at 0.3 Mt year^{-1}, which, naturally, corresponds to the smallest share of total water erosion, both at the European scale (about 0.1%) and in the areas with windbreaks (5%). The average effect of windbreaks in reducing water erosion to a low level at the landscape scale is approximately 10%.

The reduction in N emissions to water due to windbreaks is about 2.9 kt, but with even larger variations between different landscapes than in the Low-impact policy scenario.

The total increases in SOC are estimated at 6 Mt C, corresponding to annual CO_2 emissions savings of 0.7 Mt.

5.4 Discussion of the environmental benefits from windbreaks

In the Biomass policy scenario, there might not be a need for further actions to reduce water erosion in the landscapes where windbreaks are implemented. In the other scenarios, where windbreaks are generally implemented to a lesser extent, the contribution to water erosion mitigation at the landscape scale is more limited, although not insignificant. To estimate sediment retention, it is necessary to know the orientation of windbreaks relative to slope, as this strongly influences the sediment trapping efficiency. While this is technically possible, sediment retention in windbreaks has not been assessed here.

Windbreaks can be as effective as riparian buffers in reducing N emissions to water in many landscapes where they are implemented. However, on a European scale, their effect is small compared to that of riparian buffers (up to 22 kt N versus about 900 kt N). This difference is mainly because areas prone to N emissions to water are not necessarily subject to wind erosion. Despite this, it demonstrates how a single measure can address multiple environmental issues and thereby improve overall land-use efficiency by reducing the need for cropland dedicated to impact mitigation. This underscores the importance of considering multiple environmental goals and adopting a landscape perspective on land-use management.

Windbreaks can be effective in enhancing SOC in areas where they are used, especially in the Biomass scenario where a substantial portion of cropland is repurposed for windbreaks, leading to notable SOC restoration. In the other scenarios where the windbreak area is variable and generally smaller, windbreaks still play an important role in SOC restoration in landscapes with severe wind erosion and high windbreak implementation. The potential for SOC restoration through windbreaks could be further increased by shifting their location in the landscape, as SOC restoration decreases over time. This would, however, increase management costs.

At the European scale, the positive impact on SOC from the implementation of windbreaks is relatively minor since most landscapes are not subject to adverse wind erosion and windbreak implementation. Consequently, the contribution of windbreaks to achieving climate neutrality is limited. In the Biomass scenario, the annual emissions savings relative to the total GHG emissions from all sectors in Europe in 2018 is only about 0.14%.

6 Implications of widespread implementation of multifunctional riparian buffers and windbreaks: balancing food production with other ecosystem services

Riparian buffers and windbreaks can offer several potential co-benefits in addition to what has been quantified here, but accurately measuring these benefits requires considering specific characteristics of the landscape. For example, both riparian buffers and windbreaks can substantially mitigate the impact of recurring floods, by mediating water flows within landscapes and increasing percolation (Fig. 4). However, to quantify such effects, hydrological modelling based on the scenario maps we have created would be necessary, or at least the use of suitable indicators, such as detailed data on topography and soil structure. To date, no studies that provide suitable methods for the latter have been found.

Similarly, riparian buffers have the potential to reduce wind erosion by acting as natural barriers, depending on how they are oriented relative to prevailing winds. While quantifying these benefits in precise biophysical terms was beyond the scope of this study, we did estimate the likelihood that riparian buffers and windbreaks could mitigate floods and that riparian buffers could mitigate wind erosion. Our analysis suggests that flood mitigation is 'highly likely' to occur in 14–20% of landscapes with riparian buffers, depending on the policy scenario, and in about 17% of landscapes with windbreaks, in all scenarios (Fig. 4). The potential for riparian buffers to substantially mitigate wind erosion is lower, with a 'high' likelihood in only 2–4% of the landscapes.

Furthermore, the introduction of buffer strips of perennial crops on agricultural land can enhance biodiversity and support the conservation of sensitive species, such as semi-aquatic amphibians. However, the effectiveness depends on local conditions, and riparian buffers may have adverse effects if they replace existing natural riparian zones. Additionally, while these perennial plantations can modify local moisture regimes, microclimates, and vegetation structures – potentially serving as fire barriers – they can also accumulate dry biomass, creating fire risks. This highlights the need for careful management and species selection, especially in fire-prone areas. The water demands of fast-growing grasses and SRC species also need to be considered, particularly in dry regions where water availability is a concern.

Published by Burleigh Dodds Science Publishing Limited, 2025.

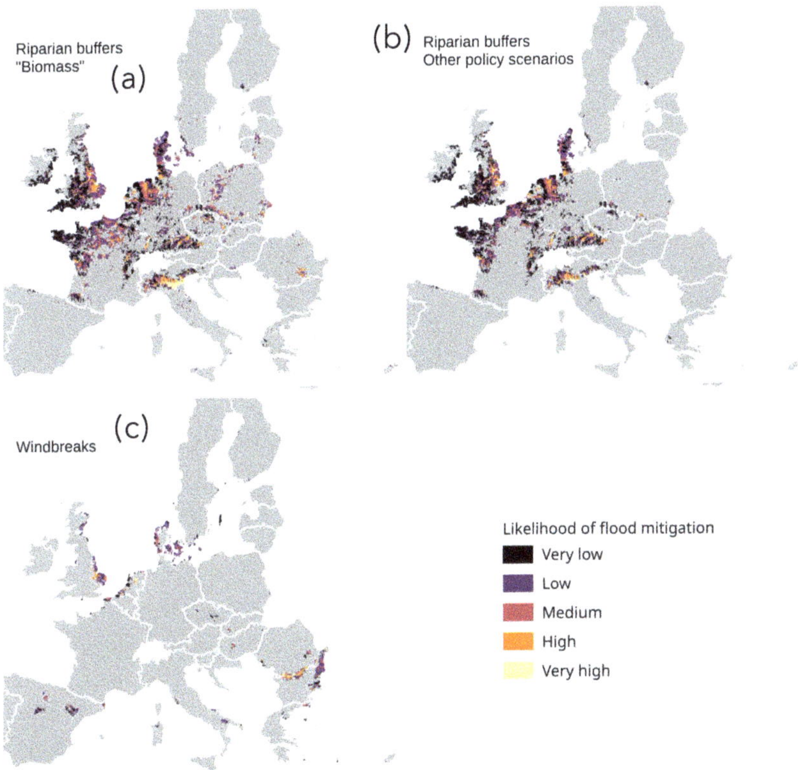

Figure 4 Flood mitigation by implementation of riparian buffers and windbreaks. Likelihood of mitigated flooding events due to implementation of riparian buffers (a = Biomass policy scenario, b = other policy scenarios) and windbreaks (c = all scenarios). Figure originally published in Englund et al. (2021).

Thus, to fully understand the benefits and potential trade-offs of strategic perennialization, detailed analyses covering a wide range of environmental aspects at a finer spatial scale are necessary, as well as using high-resolution data. This includes examining potential negative impacts of both direct and indirect land-use change, as discussed below.

As described above, widespread implementation of riparian buffers and windbreak can, if managed and localised appropriately, reduce nitrogen emissions to water and soil loss by wind erosion, respectively, to low levels in Europe, using only a small amount of agricultural land. Both production systems can also provide substantial environmental co-benefits and biomass for the emerging bioeconomy.

6.1 Impacts on agricultural food production

Using agricultural land for non-food biomass production (such as riparian buffers and windbreaks) can impact food production negatively, but measuring such impacts is complex. For instance, riparian buffers can be fully on farmland or partly on pre-existing, unmanaged buffer zones, which reduces the need to convert farmland. In situations where cropland is turned into buffers or windbreaks, production of annual crops may decrease due to a reduced area of cropland. However, remaining farmland may be subjected to less soil erosion or better flood protection, resulting in higher yields, which could reduce net farmland displacement.

For windbreaks, a substantial portion of the land used for annual crops could be converted into windbreaks (one-third for SRC windbreaks and one-ninth for SRF windbreaks at full implementation), depending on the scenario. This loss can be partially balanced by increased yields in areas protected from wind erosion. Moreover, the various additional benefits of riparian buffers and windbreaks, such as increases in SOC, may enhance yield levels both in the short and long term.

The effects on food production also largely depend on what the biomass is ultimately used for. If the biomass from some types of energy crops can be processed into both energy carriers and products that were previously produced from annual crops, the demand for annual crops (and thus the area needed for annual crop production) is reduced. This can substantially reduce the cropland displacement effect. A recent article (Englund et al., 2022) showed that widespread deployment of grass in crop rotations, despite causing a substantial reduction in area for annual crop production at the European level, could result in net cropland *savings*. This is possible when the grass biomass is used as feedstock for production of protein concentrate that replaces soymeal in combination with biogas for energy purposes.

6.2 Climate benefits

When SRC and SRF are utilised as dedicated riparian buffers and windbreaks, about 300 PJ and 450 PJ biomass, respectively, are produced in the Biomass policy scenario. Combined, these cultivations represent a total of 3.4% of the area under annual crops in the EU. When this total amount of biomass is utilised for energy purposes replacing fossil fuels, the GHG emissions savings could be substantial. For example, if the biomass is converted to either Fischer-Tropsch diesel (displacing fossil diesel) or electricity and heat (displacing coal and oil), the GHG emissions savings are estimated to be in the range of 29–44 Mt CO_2 equivalents year^{-1}. In addition, around 9–10 Mt CO_2 equivalents year^{-1} would be stored in soils until 2050. In total, in the Biomass scenario, riparian buffers and windbreaks enable GHG emissions savings of about 38–54 Mt CO_2 equivalents year^{-1}, corresponding to roughly 1–1.5% of total GHG emissions (all sectors) in EU in 2018 (Eurostat, 2020).

The greenhouse gas (GHG) emissions savings from an increased production and utilization of biomass can, however, be expected to change over time. This is because both product substitution patterns and GHG intensity of substituted products will change as consumption and production systems develop towards lower GHG intensity. Still, fossil fuels are expected to be used for a considerable time and can thereby potentially be displaced by biomass-based alternatives on the margin. Another development may lead to increased GHG emissions savings from biomass-based energy in the longer term when large-scale biomass conversion is combined with carbon capture and storage (bio-CCS) (Béres et al., 2024). This enables removal and long-term storage of atmospheric biogenic carbon dioxide in geological reservoirs, leading to net negative GHG emissions. Such biomass applications lead to high mitigation effects also when fossil fuels have been phased out.

6.3 Policy implications

Despite the early and growing knowledge about multifunctional biomass production systems and how these systems could reduce various negative environmental impacts from today's conventional agricultural practices, large-scale implementation of such production systems has not yet occurred. A study by Dale et al. (2016) describes an approach to landscape design that focuses on integrating bioenergy production with other components of environmental, social, and economic systems. One conclusion of this study is that barriers to implementation of landscape design include high costs, the need to consider diverse land-management objectives from a wide array of stakeholders, up-front planning requirements, and the complexity and level of effort needed for successful stakeholder involvement. Furthermore, there has been a lack of dedicated incentives in agricultural policies promoting multifunctional biomass production systems (Englund et al., 2020). Since multifunctional biomass production systems relate to several policy areas – not only agriculture but also energy, climate, and environment – incentives need to be harmonised to effectively promote wider implementation of multifunctional systems.

A conclusion from the study presented in this chapter is that multifunctional riparian buffers and windbreaks can provide substantial environmental benefits with limited negative effects on current agricultural production in the EU. On local and regional levels, however, these establishments can represent a substantial amount of arable land which can influence the existing local economy and individual farmers. At the same time, the local concentration of the proposed biomass production systems can generate a critical mass of land that ensures an essential market development for the resulting biomass (Mola-Yudego et al., 2014). The lack of markets and associated dedicated market-based incentives for promoting lignocellulosic biomass for energy purposes is another important barrier to wider implementation of multifunctional systems (Englund et al., 2020). Realising the deployment scenarios presented in this case study thus requires substantial policy efforts to generate sufficient incentives, both production based and market based. One potential opening regarding production-based incentives is the inclusion of Eco-schemes in the EU Common Agricultural Policy (CAP). Eco-schemes will offer farmers the possibility to obtain direct payments to adapt practices beneficial for climate, water, soil, air, and biodiversity (EC, 2021). This is a practical example of how to combine climate, energy, environmental, and agriculture policy goals.

Depending on the prioritization of specific environmental concerns and targets in various countries and regions, modelling different deployment scenarios of strategic perennialization, as illustrated in this case study, can act as important support in strategic decision making. If, e.g., local biomass production for improved energy security and high climate change mitigation are both prioritized in an EU member state that is highly dependent on fossil fuels, then the Biomass policy scenario in the presented case study can denote the upper level of the potential biomass supply. On the other hand, if food production is prioritized in a member state with a limited area of agriculture land per capita, then the Food-first policy scenario can indicate the deployment potential. Based on this, spatial modelling combined with scenario analysis can give valuable input regarding the development of target levels in individual countries and regions and the design of appropriate incentives.

If implementing bioenergy buffers leads to losses in farmed cropland that cannot be fully compensated by yield improvements, as discussed above, the risk of indirect land-use changes must be considered. Thus, indirect land-use changes causing biodiversity impacts and ecosystem carbon losses should be addressed in appropriate ways in developed policies and incentives (Sunfleth et al., 2020; Khanna et al., 2017). The same concerns exist when lower-yielding alternative farming practices are introduced, aiming at similar environmental benefits as multifunctional systems, as exemplified by studies associating organic food with additional emissions or foregone carbon sequestration, due to the need for more arable land to compensate for lower yields (Smith et al., 2019: Searchinger et al., 2018). Alternative energy systems to bioenergy and biofuels used for transportation, such as electric vehicles powered with carbon-intensive electricity, represent another example of potential carbon leakage in the energy sector (Andersson and Börjesson, 2021). Thus, decision-making in the context of societal sustainability transitions needs to reflect a holistic view on both the supply-side and the demand-side mitigation and should be based on data and insights from multiple scientific disciplines and complementary methodologies covering a multitude of sustainability indicators.

7 Future research trends

To understand the effects of large-scale implementation of multifunctional perennial production systems, large-scale, geographically explicit, studies are necessary. The case study presented here is, to our knowledge, unparalleled in terms of scale. There is therefore a good potential for conducting similar studies in other world regions. A limiting factor is, however, the availability of necessary data. For example, the detailed geospatial data on relevant environmental impacts and yield simulations for relevant species that were necessary for conducting this study may currently not be available (at least not with the same quality and resolution) for all world regions.

Decades of experimental and theoretical research have created a solid foundation for researching multifunctional production systems using new methods with spatial analysis and modelling. This development has enabled much needed detail of where different kinds of environmental impacts could be mitigated using different kinds of multifunctional production systems. However, in the case study presented here, and in the similar articles by the authors and others, results are aggregated to the landscape level (Englund et al., 2017) and provide little information on what is happening inside the landscape. We therefore need new ways to map scenarios for strategic perennialization within individual landscapes. This involves mapping, with a high level of resolution, precision and accuracy, of (i) environmental impacts, (ii) optimal localisation of multifunctional production systems that can mitigate adverse impacts by supporting targeted ecosystem services, and (iii) the outcome in terms of environmental impact of realising such land-use scenarios. This is also necessary in order to use modelling as a basis for developing new policy instruments that can effectively target specific environmental impacts by promoting multifunctional production systems.

Fortunately, such studies are becoming increasingly viable with the general advancement of spatial analysis tools, increased computation capacity both at the workstation and cluster level, and, most importantly, more and more geospatial data of higher quality and better resolution. While the current development within artificial intelligence (AI) indeed raises concerns of ethical and even existential nature, the massive advancements in machine learning provide new opportunities for creating more advanced models. Given that land use is characterised by multiple, often conflicting, objectives, an interesting approach for future studies could be (AI-aided) multi-objective (or pareto) optimisation (Todman et al. 2019).

While much experimental data has been generated over the years, there are still some gaps that need to be filled. For example, there is a need for better empirical data to support quantifying flood mitigation, regardless of whether this is done using hydrological modelling or more simple indicators. Most importantly, there is a need for long-term experimental field trials in which different kinds and designs of perennial production systems are established to mitigate different kinds of environmental impacts. Models, such as presented in this case study, can be used to identify suitable locations and setups for such field trials.

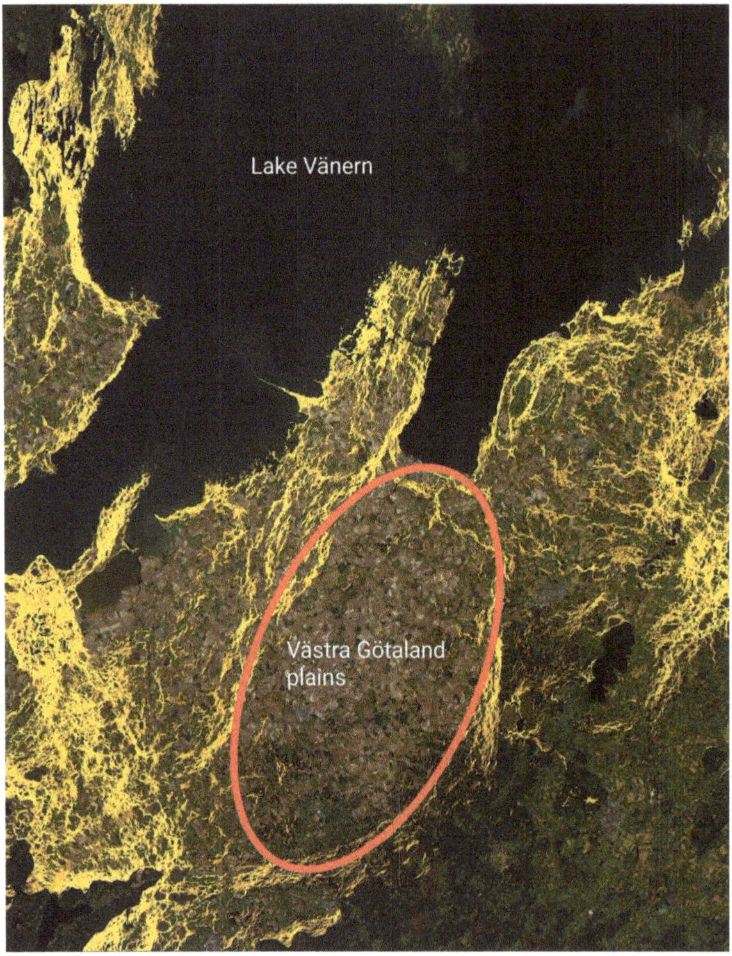

Figure 5 Simulated ecological corridors in Västra Götaland county (orange/yellow), southern Sweden. Västra Götaland plains is one of the most important agricultural areas in Sweden and the analysis indicates notably weak ecological connectivity. Source: Adapted from Franzén (2020).

Published by Burleigh Dodds Science Publishing Limited, 2025.

This study used conceptual policy scenarios that provided theoretical incentives for farmers to implement riparian buffers and windbreaks. It therefore provides quantitative information on what happens if these incentives are effective. However, further studies are needed to understand, e.g., how farmers actually respond to different kinds of incentives, to be able to design incentives that are effective. It is also necessary to better understand the economics of multifunctional biomass production. The additional cost of managing land to achieve desired environmental benefits must be compensated for to enable viable business models (Dale et al. 2016; Englund et al., 2020b).

Finally, there is a need to understand how multifunctional production systems can be designed to support biodiversity. This opportunity has been suggested by many authors, but usually based on theory rather than empiricism, with a few exceptions (Baum et al., 2012; Vanbeveren and Ceulemans, 2019). A promising approach could be to explore how perennial production systems could be established to strengthen ecological connectivity, i.e. the extent to which species can spread between habitats – an urgent and global priority for supporting biodiversity (IUCN, 2017; UNEP, 2015; Lemieux, et al., 2022). Agriculture is a main driver of biodiversity loss, partly due to insufficient connectivity in intensively managed landscapes (Kehoe et al., 2017; Outhwaite et al., 2022). This is illustrated in Fig. 5, showing an intensively managed agriculture region in southern Sweden with a notable lack of ecological corridors. Solutions for strengthening ecological connectivity across agricultural landscapes are thus urgently needed to provide better conditions for biodiversity. If we could better understand the mechanisms of ecological connectivity and how to design effective ecological corridors in agricultural landscapes, land-use strategies can be developed to minimise fragmentation, strengthen connectivity, and support biodiversity while delivering biomass to the bioeconomy, using multifunctional biomass production systems.

8 Conclusions

Multifunctional riparian buffers and windbreaks can effectively reduce nitrogen emissions to water and soil loss by wind erosion, with substantial environmental co-benefits. Strategically implementing these systems on 1–2 % of the land currently under annual crops in Europe can mitigate agricultural environmental impacts while providing necessary biomass resources for the bioeconomy.

While the amount of land necessary for buffers and windbreaks may appear small, effects at local and regional levels can be considerable. Concentrated local implementation can substantially influence the agricultural landscape but may also enable sufficient markets for the biomass. Cropland displacement can be minimised partly by establishing buffers and windbreaks on non-farmland, where appropriate, and partly by increased yields due to improved soil fertility and reduced erosion and flooding. The overall impact on food production also depends on how the biomass is utilised. If biomass replaces products previously derived from annual crops, the demand for annual crops and associated land use elsewhere decreases.

The climate benefit would, however, be maximised by instead using the biomass for bioenergy. While the climate benefit of bioenergy can decrease over time, as consumption and production systems develop towards lower GHG intensity, long-term climate benefits may increase if bioenergy is combined with carbon capture and storage (bio-CCS or BECCS), enabling net negative emissions.

Widespread adoption requires harmonised incentives across agricultural, energy, climate, and environmental policy areas. Further research should explore farmers' responses to incentives and the economic viability of multifunctional biomass production, including compensation mechanisms for delivering environmental benefits.

More detailed, geographically specific studies at finer scales within landscapes are necessary to understand large-scale implementation impacts. Additionally, long-term experimental field trials are needed to accurately quantify multiple environmental benefits simultaneously, including but not limited to flood mitigation. Finally, research on designing multifunctional systems that support biodiversity, particularly through enhancing ecological connectivity, is essential.

9 Where to look for further information

- Englund, O., Börjesson, P., Mola-Yudego, B., Berndes, G., Dimitriou, I., Cederberg, C., Scarlat, N., (2021). Strategic deployment of riparian buffers and windbreaks in Europe can co-deliver biomass and environmental benefits. *Communications Earth & Environment*, 2:176, https://doi.org/10.1038/s43247-021-00247-y.
- Englund, O., Mola-Yudego, B., Börjesson, P., Cederberg, C., Dimitriou, I., Scarlat, N., Berndes, G., (2022). Large-scale deployment of grass in crop rotations as a multifunctional climate mitigation strategy. *GCB Bioenergy*, 15(2), 166-184. https://doi.org/10.1111/gcbb.13015.
- Englund, O., Dale, V.H., Kline, K.L., Dimitriou, I., Mola-Yudego, B., Murphy, F., English, B., McGrath, J., Busch, G., Negri, M.C., Brown, M., Goss, K., Jackson, S., Parish, E.S., Cacho, J., Zumpf, C., Quinn, J., Mishra, S.K., (2020). Multifunctional perennial production systems for bioenergy: performance and progress. *WIREs Energy & Environment*, 9(5), e375. https://doi.org/10.1002/wene.375.

- Englund O, Börjesson P, Berndes G, Scarlat N, Dallemand J-F, Grizzetti B, Dimitriou I, Mola-Yudego B, Fahl F, (2020). Beneficial land use change: strategic expansion of new biomass plantations can reduce environmental impacts from EU agriculture. *Global Environmental Change*, 60, 101990. https://doi.org/10.1016/j.gloenvcha.2019.101990.

10 References

Andersson Ö. and Börjesson P. (2021). The greenhouse gas emissions of an electrified vehicle combined with renewable fuels: life cycle assessment and policy implications. *Applied Energy*, 289, 116621.

Aylott M. J., et al. (2008). Yield and spatial supply of bioenergy poplar and willow short-rotation coppice in the UK. *New Phytologist*, 178, 358–370.

Baum S., et al. (2012). Short Rotation Coppice (SRC) Plantations provide additional habitats for vascular plant species in agricultural mosaic landscapes. *Bioenergy Research,* 5, 573–583.

Baxter X. C., et al. (2014). Miscanthus combustion properties and variations with Miscanthus agronomy. *Fuel*, 117, 851–869.

Béres R., et al. (2024). Assessing the feasibility of CO_2 removal strategies in achieving climate-neutral power systems: insights from biomass, CO_2 capture, and direct air capture in Europe. *Advances in Applied Energy*, 14, 100166.

Berndes G., et al. (2008). Multifunctional bioenergy production systems – an introduction with presentation of specific applications in India and Sweden. *Biofuels, Bioproducts and Biorefining*, 2, 16–25.

Börjesson P. (1999a). Environmental effects of energy crop cultivation in Sweden – Part I: identification and quantification. *Biomass and Bioenergy,* 16, 137–154.

Börjesson P. (1999b). Environmental effects of energy crop cultivation in Sweden – Part II: economic valuation. *Biomass and Bioenergy,* 16, 155–170.

Christen B. and Dalgaard T. (2013). Buffers for biomass production in temperate European agriculture: a review and synthesis on function, ecosystem services and implementation. *Biomass and Bioenergy*, 55, 53–67.

Christian D., et al. (1994). Perspectives on biomass energy tree plantations and changes in habitat for biological organisms. *Biomass and Bioenergy* 6, 31–39.

Dale V., et al. (2016). Incorporating bioenergy into sustainable landscape designs. *Renewable and Sustainable Energy Reviews*, 56, 1158–1171.

Dees M., et al. (2017). *A spatial data base on sustainable biomass cost-supply of lignocellulosic biomass in Europe – methods & data sources*. S2BIOM Project Report. https://s2biom.wenr.wur.nl/doc/S2Biom_D1_6_v1_2_FINAL_20_04_2017_CP.pdf.

Dimitriou I. and Mola-Yudego B. (2017). Poplar and willow plantations on agricultural land in Sweden: area, yield, groundwater quality and soil organic carbon. *Forest Ecology and Management*, 383, 99–107.

EC, European Commission. (2021). *List of potential agricultural practices that Eco-schemes could support*. European Commissions. https://agriculture.ec.europa.eu/document/download/2445d833-7a35-4974-8ecb-ee7d5a0332e6_en?filename=factsheet-agri-practices-under-ecoscheme_en.pdf.

EC, European Commission. (2023). *Directive (EU) 2023/2413 of the European Parliament and the Council of 18 October 2023 amending Directive (EU) 2018/2002, Regulation (EU) 2018/1999 and Directive 98/70/EC as regards the promotion of energy from renewable sources, and repealing Council Directive (EU) 2015/652*. https://eur-lex.europa.eu/eli/dir/2023/2413/oj/eng

Englund O., et al. (2017). How to analyse ecosystem services in landscapes – a systematic review. *Ecological Indicators*, 73, 492–504.

Englund O., et al. (2020a). Beneficial land use change: Strategic expansion of new biomass plantations can reduce environmental impacts from EU agriculture. *Global Environmental Change*, 60, 101990.

Englund O., et al. (2021). Strategic deployment of riparian buffers and windbreaks in Europe can co-deliver biomass and environmental benefits. *Communications Earth & Environment*. https://doi.org/10.1038/543247-021-00247-y.

Englund O., et al. (2020b). Multifunctional perennial production systems for bioenergy: Performance and progress. *WIREs Energy & Environment*, 9(5), e375. https://doi.org/10.1002/wene.375.

Englund, O., et al. (2022). Large-scale deployment of grass in crop rotations as a multifunctional climate mitigation strategy. *GCB Bioenergy*, 15(2), 166–184. https://doi.org/10.1111/gcbb.13015

European Environment Agency. (2012). European catchments and Rivers network system (Ecrins), v. 1.1. https://www.eea.europa.eu/data-and-maps/data/european-catchments-and-rivers-network.

Eurostat. (2020). Greenhouse gas emission statistics – emission inventory. https://ec.europa.eu/eurostat/statistics-explained/index.php?title=Greenhouse_gas_emission_statistics_-_emission_inventories.

Ferrarini A., et al. (2017a). Multiple ecosystem services provision and biomass logistics management in bioenergy buffers: A state-of-the-art review. *Renewable and Sustainable Energy Reviews*, 73, 277–290.

Ferrarini A., et al. (2017b). Impacts of willow and miscanthus bioenergy buffers on biogeochemical N removal processes along the soil–groundwater continuum. *GCB Bioenergy*, 9, 246–261.

Fortier, J., et al. (2010). Biomass and volume yield after 6 years in multiclonal hybrid poplar riparian buffer strips. *Biomass and Bioenergy*, 34(7), 1028–1040. https://doi.org/10.1016/j.biombioe.2010.02.011

Franzén W. (2020). *Mapping connectivity in the Swedish agricultural landscape*. Östersund: Mid Sweden University. http://urn.kb.se/resolve?urn=urn:nbn:se:miun:diva-39253.

FSC. (2024). Forest Stewardship Council. https://fsc.org/en.

International Union for Conservation of Nature (IUCN). (2017). *Connectivity conservation*. Gland. https://www.iucn.org/theme/protected-areas/wcpa/what-we-do/connectivity-conservation.

IPCC. (2019). *Climate change and land: an IPCC special report on climate change, desertification, land degradation, sustainable land management, food security, and greenhouse gas fluxes in terrestrial ecosystems*. [P.R. Shukla, J. Skea, E. Calvo Buendia, V. Masson-Delmotte, H.- O. Pörtner, D. C. Roberts, P. Zhai, R. Slade, S. Connors, R. van Diemen, M. Ferrat, E. Haughey, S. Luz, S. Neogi, M. Pathak, J. Petzold, J. Portugal Pereira, P. Vyas, E. Huntley, K. Kissick, M. Belkacemi, J. Malley, (eds.)]. https://doi.org/10.1017/9781009157988.001

IUCN. (2017). *Connectivity conservation*. Gland: International Union for Conservation of Nature. https://www.iucn.org/theme/protected-areas/wcpa/what-we-do/connectivity-conservation.

Lemieux C., et al. (2022). Transformational changes for achieving the Post-2020 Global Biodiversity Framework ecological connectivity goals. *FACETS*, 7, 1008–1027.

Kehoe L., et al. (2017). Biodiversity at risk under future cropland expansion and intensification. *Nature Ecology & Evolution*, 1, 1129–1135.

Khanna M., et al. (2017). The social inefficiency of regulating indirect land use change due to biofuels. *Nature Communications*, 8, 15513.

Kort J., et al. (1998). A review of soil erosion potential associated with biomass crops. *Biomass and Bioenergy*, 14, 351–360.

Makeschin F. (1994). Effects of energy forestry on soils. *Biomass and Bioenergy*, 6, 63–80.

McLaughin SB. and Walsh ME. (1998). Evaluating environmental consequences of producing herbaceous crops for bioenergy. *Biomass and Bioenergy*, 14, 317–324.

Mola-Yudego B., et al. (2014). A conceptual framework for the introduction of energy crops. *Renewable Energy*, 72, 29–38.

Nordborg M., et al. (2018a). Energy analysis of poplar production for bioenergy in Sweden. *Biomass and Bioenergy*, 112, 110–120.

Nordborg M., et al. (2018b). Energy analysis of willow production for bioenergy in Sweden. *Renewable and Sustainable Energy Reviews*, 93, 473–482.

Obernberger I., et al. (1997). Concentrations of inorganic elements in biomass fuels and recovery in the different ash fractions. *Biomass and Bioenergy*, 12, 211–224.

Outhwaite C., et al. (2022). Agriculture and climate change are reshaping insect baliodiversity worldwide. *Nature*, 605, 97–102.

Pimentel D. and Krummel J. (1987). Biomass energy and soil erosion: assessment and resource costs. *Biomass and Bioenergy*, 14, 15–38.

Ranney JW. and Mann LK. (1994). Environmental considerations in energy crop production. *Biomass and Bioenergy*, 6, 211–228.

Rijtema P. and de Vries W. (1994). Differences in precipitation excess and nitrogen leaching from agricultural lands and forest plantations. *Biomass and Bioenergy*, 6, 103–115.

Sage R. and Robertson P. (1994). Wildlife and game potential of short-rotation coppice in the UK. *Biomass and Bioenergy*, 6, 41–48.

Searchinger T., et al. (2018). Assessing the efficiency of changes in land use for mitigating climate change. *Nature*, 564, 249–253.

Smith L.G., et al. (2019). The greenhouse gas impacts of converting food production in England and Wales to organic methods. *Nature Communications*, 10, 4641.

Ssegane H., et al. (2016). The economics of growing shrub willow as a bioenergy buffer on agricultural fields: a case study in the Midwest Corn Belt. *Biofuels, Bioproducts and Biorefining*, 10, 776–789.

Styles D., et al. (2016). Climate regulation, energy provisioning and water purification: quantifying ecosystem service delivery of bioenergy willow grown on riparian buffer zones using life cycle assessment. *Ambio*, 45, 872–884.

Sunfleth B., et al. (2020). Recent developments in low iLUC policies and certification in the EU biobased economy. *Sustainability*, 12, 8147.

Todman, L.C., et al. (2019). Multi-objective optimization as a tool to identify possibilities for future agricultural landscapes. *Science of The Total Environment,* 687, 535–545. https://doi.org/10.1016/j.scitotenv.2019.06.070.

Tufekcioglu, A., et al. (2003). Biomass, carbon and nitrogen dynamics of multi-species riparian buffers within an agricultural watershed in Iowa, USA. *Agroforestry Systems* 57, 187–203.

UNEP. (2015). *UNEP launches global connectivity conservation project.* Nairobi. http://www.unep-wcmc.org/news/unep-launches-global-connectivity-conservation-project.

Vanbeveren S. and Ceulemans R. (2019). Biodiversity in short-rotation coppice. *Renewable and Sustainable Energy Reviews*, 111, 34–43.

Published by Burleigh Dodds Science Publishing Limited, 2025.

Chapter 9

Modelling and optimising associated biodiversity in alley cropping systems

Mário Santos, CITAB (Centre for the Research and Technology of Agro-Environmental and Biological Sciences) - University of Trás-os-Montes e Alto Douro, Portugal/Federal Institute of Education, Science and Technology of Para, Brazil; Berta Gonçalves and André Fonseca, CITAB (Centre for the Research and Technology of Agro-Environmental and Biological Sciences) - University of Trás-os-Montes e Alto Douro, Portugal; Reinaldo Lucas Cajaiba, Federal Institute of Education, Science and Technology of Para, Brazil; Henrique Trindade, Fátima Gonçalves and Cristina Carlos, CITAB (Centre for the Research and Technology of Agro-Environmental and Biological Sciences) - University of Trás-os-Montes e Alto Douro, Portugal; Darinka Gonzalez,CITAB (Centre for the Research and Technology of Agro-Environmental and Biological Sciences) - University of Trás-os-Montes e Alto Douro, Portugal/University of Vigo, Spain; Mário Gonzalez Pereira and Lívia Madureira, CITAB (Centre for the Research and Technology of Agro-Environmental and Biological Sciences) - University of Trás-os-Montes e Alto Douro, Portugal; Mariane Cristina Inocente and Fátima Piña-Rodrigues, Federal University of São Carlos (UFSCar), Brazil; João Santos, CITAB (Centre for the Research and Technology of Agro-Environmental and Biological Sciences) - University of Trás-os-Montes e Alto Douro, Portugal; and María Rosa Mosquera-Losada, University of Santiago de Compostela, Spain

http://dx.doi.org/10.19103/AS.2024.0141.15

1 Introduction

Agricultural landscapes have replaced natural ecosystems in 40% of the world's land area and are associated with 70% of global freshwater withdrawals (Bonsch et al., 2015). Modern, intensive, large-scale agriculture is also perceived as a critical threat to our common future, due to the disruption of socio-ecological systems, biodiversity loss and degradation of ecosystem services (Vanbergen et al., 2020). The global increase in the area occupied by industrial/large-scale agriculture has been jeopardising the integrity of ecosystems and the way they function. Due to its huge dependence on fossil fuels, agrochemicals and machinery, modern agriculture is a significant contributor to climate change, soil erosion, environmental pollution and other problems (DeLonge et al., 2016).

Modern intensive agriculture can be contrasted with alternative, more sustainable farming systems. Although definitions vary, the common foundation for what are commonly seen as 'sustainable' alternatives to modern, intensive systems is the concept of 'agroecology'. This has been defined as an integrated approach applying ecological (as well as social) principles to the design and management of food and agricultural systems (https://www.fao.org/agroecology/home/en/). Agriculture, in this context, is viewed as an adapted form of ecosystem in which an understanding of fundamental biogeochemical flows/cycles and functional interactions between organisms within this type of ecosystem can be used to optimise its ability to deliver a range of ecosystem services, including food production. Examples of 'sustainable' practices promoting agroecosystem function include optimising the physical, chemical and biological health of soils to promote crop growth and health (reducing the need for external inputs such as fertilisers and pesticides), promoting biodiversity (e.g. to encourage beneficial species such as pollinators or predators of crop pests), and ensuring diversity in agricultural landscapes (to optimise functional interactions between species) (Nicholls et al., 2016; Wezel et al., 2020).

Because they seek to work with and enhance natural processes, these alternative, more 'sustainable' practices are usually recognised as more beneficial to the environment not only from the point of view of crop production but also from the point of view of optimising delivery of ecosystem services (Schröder et al., 2019). Sustainable agriculture is considered pivotal for ensuring food for an increasing world population and for enhancing

Published by Burleigh Dodds Science Publishing Limited, 2025.

other ecosystem services such as the maintenance of soil health, carbon sequestration, management of water flow and quality as well as biodiversity conservation, among others (Santos et al., 2023). These services ultimately result in additional resilience and adaptation of agroecosystems and local communities to climate and environmental extremes, contributing to human well-being and health (Horrigan et al., 2002). Sustainable practices based on agroecological principles are seen as being able to outperform industrial agriculture by enhancing ecosystem services (including food production) while avoiding damage to the environment (Struik & Kuyper, 2017). Agroecology is seen as transformative as it can potentially integrate aspects of economics, ecology, health and social cohesion, thus meeting a range of the United Nations Sustainable Development Goals (SDGs) (Santos et al., 2022a).

Agroforestry is an agroecological approach that associates trees with livestock or crops in the same space. It is considered a particularly promising technique to promote sustainability by reconciling different needs such as agricultural production and ecological and societal sustainability. There are a number of different types of agroforestry system, including silvoarable systems, silvopastoral systems, forest farming, riparian buffer strips and windbreaks/hedgerows (Mosquera-Losada et al., 2023).

For many years, agroforestry was considered a traditional and outdated system by many decision-makers and underrated and undersupported by agricultural policies. Instead, governments and others actively encouraged removal of trees to create open fields to better enable large-scale intensive agriculture reliant, in part, on large machines such as combine harvestors. Agroforestry practices were widely abandoned and areas converted to open fields, pastures or forests (Den Herder et al., 2017). However, agroforestry is now increasingly seen as a key sustainable practice with the potential to reconcile agriculture production with maximising ecosystem services and reducing environmental impacts, including climate change (Santos et al., 2022b; Menezes et al., 2024). The regulating and supporting services provided by agroforestry may be particularly significant when facing potentially more extreme and unpredictable future climatic conditions. Trees, e.g. have the capacity to improve carbon sequestration, improve functional biodiversity, promote water regulation and soil fertility, optimise microclimate control and the adsorption of agrochemicals which then do not pollute neighbouring landscape features such as watercourses (Moreno et al., 2018).

Alley cropping agroforestry systems consist of alternating crop alleys, which constitute the arable part of the system, with rows of trees, typically accompanied by an understory vegetation strip. These systems are compatible with modern agriculture practices such as mechanisation and robotics (Garrett et al., 2021). The conservation and management of biodiversity in agroecosystems, and particularly in alley cropping, is currently an important topic (Nair et al., 2021). Even if alley cropping has the potential to contribute to biodiversity conservation in agricultural landscapes through structural complexity, habitat and landscape heterogeneity, a limited understanding of the interrelated effects of micro-habitats on diversity patterns reduces our ability to project its impacts on biodiversity (Santos et al., 2022a).

2 Using ecological models to predict biodiversity outcomes from alley croping

Ecological models are frequently employed to analyse space and/or time dynamics as well as the overall behaviour of complex systems (Bakiş et al., 2021). Spatial, dynamic and spatio-temporal ecological models might incorporate variables that are otherwise difficult to integrate. They can be used to simulate trade-offs between provisioning, regulating and supporting services as well as informing policies to promote biodiversity in alley cropping systems (Arosa et al., 2017). Nevertheless, models are dependent on a range of assumptions, supporting data quality and knowledge of how parameters change with environmental stochasticity (Rumohr et al., 2023).

To predict biodiversity outcomes and benefits from different agroforestry programmes involves linking different input variables (e.g. the species chosen by the farmer, such as different cultivated/bred species and plant/animal varieties) with associated biodiversity outputs (e.g. the species colonising an agroecosystem, including agriculturally beneficial, neutral and harmful organisms), in what can be termed 'planned biodiversity'. Planned biodiversity (to achieve specific outcomes) is one of the main levers to promote the success of agroforestry programmes because choosing the right species combination for each location will optimise provision of target ecosystem services, including crop production and yiels (Barrios et al., 2018). In this context, the term 'associated biodiversity' refers to species associated with ecosystem services (as well as ecosystem disservices). The term 'reactive biodiversity' relates to the diversity of species' responses to environmental stimuli. Both are interconnected and contribute to the resilience of ecosystems in the face of disturbances (Ulanowicz, 2018).

To better understand the interaction between planned biodiversity and associated biodiversity within alley cropping systems, a cell-based spatial modelling approach has been developed (Wilensky, 1999). This approach makes it possible to simulate a range of agroecosystems to explore the way different combinations deliver planned biodiversity targets to optimise associated biodiversity (and, therefore, the delivery of ecosystem services). Specific simulation objectives include:

1 establishing the main cause-effect relationships between crop/tree combinations and the associated biodiversity;

2 predicting biodiversity enhancement under different management paradigms and environmental conditions; and
3 discussing the added-value of alley cropping systems for biodiversity conservation and ecosystem services.

The following sections describe setting up and running a simulation.

3 Simulation in practice: materials and methods

3.1 Virtual field description

A virtual field was parameterised to represent different types of agroecosystems, including monocropping, intercropping, agroforestry and forestry systems, defined as follows:

1 monocropping (Fig. 1A) represents an agricultural field associated with a single use;
2 intercropping (Fig. 1B) mimics an agricultural field with two alternating in-line crops;
3 simple intensive alley cropping (Fig. 1C) depicts a field with rows of trees and/or shrubs, creating alleys within which agricultural or horticultural crops are produced;
4 simple biodiverse intensive alley cropping (Fig. 1D) depicts a field with rows of native trees and/or shrubs to create conservation/ecosystem services and alleys within which agricultural or horticultural crops are produced;

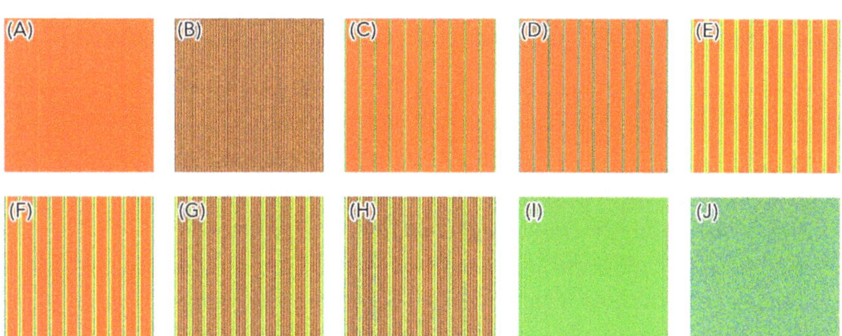

Figure 1 Interface of the model used to simulate the gradient of agroecosystems' planned biodiversity (Planned Biodiversity): (A) monocropping; (B) intercropping; (C) simple intensive alley cropping; (D) simple biodiverse intensive alley cropping; (E) simple extensive alley cropping; (F) biodiverse extensive alley cropping; (G) extensive alley cropping with intercropping; (H) biodiverse extensive alley cropping with intercropping; (I) simple forest; (J) conservation forest. Further explanations in Virtual Landscape Description.

5 simple extensive alley cropping (Fig. 1E) depicts a field with rows of trees and/or shrubs with understory vegetation strips, to create conservation/ ecosystem services and alleys within which agricultural or horticultural crops are produced;
6 biodiverse extensive alley cropping (Fig. 1F) depicts a field with rows of native trees and/or shrubs with understory vegetation strips, to create enhanced conservation/ ecosystem services and alleys within which agricultural or horticultural crops are produced;
7 extensive alley cropping with intercropping (Fig. 1G) depicts a field with rows of trees and/or shrubs with understory vegetation strips, to create enhanced conservation/ ecosystem services and alleys within which intercrops are produced;
8 biodiverse extensive alley cropping with intercropping (Fig. 1H) depicts a field with rows of two species of trees and/or shrubs (one of which is native) with understory vegetation strips, to create enhanced conservation/ ecosystem services and alleys within which intercrops are produced;
9 simple forest (Fig. 1I) simulates a single species tree production forest;
10 conservation forest (Fig. 1J) simulates a forest co-dominated by two tree species.

The conceptual description of the model follows a simplified version of the specifications of the standard Overview, Design Concepts and Details (ODD) protocol (Grimm et al., 2020). The software NetLogo 6.3 (Wilensky, 1999) was chosen to create virtual landscapes and to compute the biodiversity outcomes.

3.2 Purpose

The model investigates how associated biodiversity emerges from the interplay between planned biodiversity and management, separated within tree, crop or understory vegetation-dominated patches and integrating the influences of different production systems (ecological vs conventional). By partitioning patch biodiversity according to phytophagous richness, structural richness, shadow richness and edge/border richness, we intend to understand associated biodiversity patterns linked with different agricultural options and compare the outcomes with types of alley cropping systems (Garret et al., 2021).

3.3 Entities, state variables and scales

The model includes 10 000 patches (unit cells) that make up the 4 ha field (please see Fig. 1 and Table 1 for details). Each patch was characterised by its land use (tree, crop or understory vegetation) and the associated biodiversity

Table 1 Description of the variables of each conceptual entity used in the model

Entity	Variable	Brief description
Patches	*Number*	• 10 000 patches arranged in a square grid (100 × 100 patches), which corresponds to a total of 4 ha.
	Size	• Individual patch of 4 square metres (2 × 2 metres).
	Land-use Richness	• Each patch corresponds to a single dominant land use, crop, tree or understory vegetation, identified by different colours. • Associated biodiversity (SuR) was partitioned between phytophagous species richness (PhR), structural species richness (StR), shadow species richness (ShR) and edge/border richness (BoR). PhR, StR and ShR were defined considering the maximum values set related with each patch use (see please Methods – Diversity predictions). In the case of BoR, the calculations are made considering 8 as the maximum number of edges by patch (eight possible neighbours with different land uses).
	N_Rx	• Richness was normalised using equation 1.
	Field	• *Field* sets the agroecosystem, namely (A) monocropping; (B) intercropping; (C) simple intensive alley cropping; (D) simple biodiverse intensive alley cropping; (E) simple extensive alley cropping; (F) biodiverse extensive alley cropping; (G) extensive alley cropping with intercropping; (H) biodiverse extensive alley cropping with intercropping; (I) simple forest; and (J) conservation forest.
	Management	• *Management* defines the intensity of the production system applied to the field, specifically (X) ecological farming, (XX) conventional pest control of crops, and (XXX) conventional pest control of the overall field (including the tree/understory vegetation patches).

was gauged accordingly to the sum of the components considered (phytophagous richness, structural richness, shadow richness, and edge/border richness).

3.4 Simulating different agroecosystems

Each patch in the virtual agroecosystem represents a single use, either crop patches, tree patches or understory vegetation. Two types of crops, two types of trees, and an understory vegetation strip were considered, within multiple combinations, to illustrate different agroecosystems (planned biodiversity) (Fig. 1). To simulate pest control methods, three management/level conditions were defined, namely (Main et al., 2020; Gunstone et al., 2021; Stein-Bachinger et al., 2021; Serrão et al., 2022; Beaumelle et al., 2023):

1 ecological farming, with no implication for the associated biodiversity;
2 conventional pest control of crops, which implies a reduction in the associated biodiversity of crop patches; and
3 conventional pest control of the overall field, that leads to a reduction in the associated biodiversity of crop patches, tree patches, and understory vegetation patches.

3.5 Predictions of associated biodiversity

Associated biodiversity predictions per patch were partitioned according to (see e.g. Loreau et al., 2021; Wright et al., 2021):

- phytophagous species richness (PhR);
- structural species richness (StR);
- shadow species richness (ShR);
- edge/border species richness (BoR); and
- overall richness (SuR, sums the contributions of the previous estimates).

Species richness was parametrised from databases and previous publications in this scope (explained below).

PhR was defined for each type of crop (for each crop patch, from a random distribution considering a maximum value for each type: a more biodiverse crop and less biodiverse crop), tree (for each tree patch, from a random distribution considering a maximum value for each type: a more biodiverse tree and less biodiverse tree), and understory vegetation patches (for each understory vegetation patch, from a random distribution considering a maximum value) respectively, assuming phytophagous diversity increasing, in general, from crop to tree or understory vegetation, using relative values up to a limit supported on the Database of Insects and their Food Plants (https://dbif.brc.ac.uk/homepage.aspx).

In the case of StR, a similar approach was considered, supported by publications relating vegetation structure and richness of non-phytophagous species (Vasconcellos-Neto et al., 2017; Ebeling et al., 2018; Koricheva et al., 2018; Prather et al., 2020; Corcos et al., 2021; Kuchenbecker et al., 2022; Eo et al., 2023; Silva et al., 2023).

ShR was linked with the age/size of the tree patches, and their projection to the bordering patches, assuming an increasing relation between shadow and soil organism's richness (Coyle et al., 2017; Ashraf et al., 2018; Robinson et al., 2018; Pardon et al., 2019; Sáenz-Romo et al., 2019; Ampoorter et al., 2020; Le Provost et al., 2021).

On the other hand, BoR, associated with species using more than one habitat/species, was gauged only when the number of dissimilar bordering patches was equal to or greater than 1 (Pardon et al., 2019; Kranz et al., 2019; Gallé et al., 2020; Van Schalkwyk et al., 2020; Gayer et al., 2021; Cuperus et al., 2023). In this case, border richness is limited by the number of neighbouring patches different from the evaluated one.

All metrics of diversity (PhR, StR, ShR, BoR and SuR) were normalised (Rnorm,*i*, Eq. 1) using Rmax, the maximum value of Richness possible. The normalised values (NPhR, NStR, NShR, NBoR and NSuR), ranging from 0 to 1, allow direct comparisons between the different metrics and the different combinations (Gotelli & Colwell, 2001). The model was run 3000 times (combining 100 independent replicates of each Diversity and Management option) and the mean value was set aside by simulation (details in Supplementary Appendix A):

$$\mathrm{Rnorm},i = \mathrm{R}i/\mathrm{Rmax} \quad (1)$$

3.6 Comparing predicions of associated biodiversity

A generalised linear model (GLzM) was used to discriminate the magnitude of the normalised differences in the average richness between the agroecosystems and management options simulated (and their interaction), complemented by Partial Eta Square of the terms of the analysis of variance (ANOVA) of the GLzM, a standardised estimate of the effect size of factors (Nakagawa & Cuthill, 2007). Pairwise differences were determined using Tukey multiple comparisons of means with a 95% family-wise confidence level (Rafter et al., 2002). Statistical analyses were implemented using R 4.0.3 (R Core Team, 2020), through the effect size (v 0.4.0; Ben-Shachar et al., 2020), ggplot2 (Wickham, 2016), nmle (Pinheiro and Bates, 2023), and the ezr packages (Kanda, 2013).

4 Results: trends in associated biodiversity along a gradient of agroecosystem practices

Partial biodiversity metrics' predictions responded differently to the simulated agroecosystems planned biodiversity (Figs 2, 3, 4 and 5) (Casagrande et al., 2017). NPhR (Fig. 2) was associated with an increasing trend along the gradient considered from monocropping (A) until conservation forest (J). It is worth highlighting the value of NPhR for intercropping (B), whose median exceeds (or is equivalent to, depending on the management system) the following simple intensive alley cropping (C) and simple biodiverse intensive alley cropping (D), and the higher dispersion observed in forest systems ((I) simple forest and (J) conservation forest).

On the other hand, the normalised structural richness (NStR) (Fig. 3) is predicted to increase stepwise (with reductions associated with the intensity of the management system) in the following order:

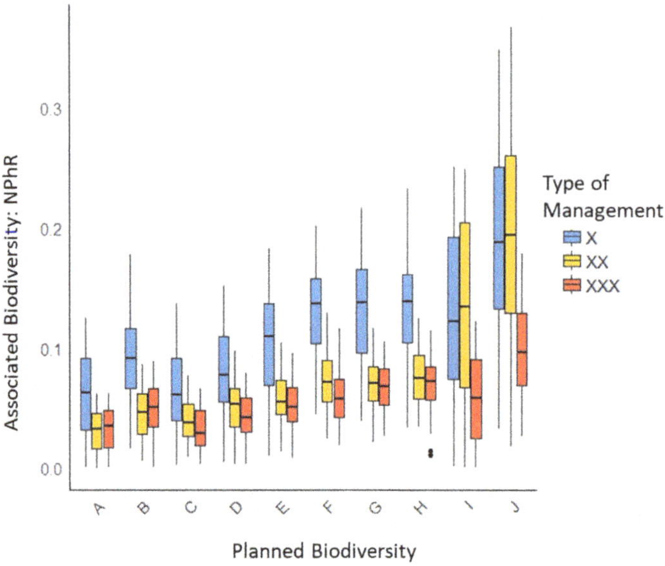

Figure 2 Box and Whisker plots expressing the differences in the normalised phytophagous richness (Associated Biodiversity: NPhR) associated with the different simulated agroecosystems/planned biodiversity (Planned Biodiversity) and management intensity/levels (Type of Management): (A) monocropping; (B) intercropping; (C) simple intensive alley cropping; (D) simple biodiverse intensive alley cropping; (E) simple extensive alley cropping; (F) biodiverse extensive alley cropping; (G) extensive alley cropping with intercropping; (H) biodiverse extensive alley cropping with intercropping; (I) simple forest; (J) conservation forest; (X) ecological farming; (XX) conventional pest control of crops; (XXX) conventional pest control of the overall field (including the tree/understory vegetation patches). Box-plot details: lower and upper box boundaries 25th and 75th percentiles, respectively, line inside box median, lower and upper lines limit 10th and 90th percentiles, respectively, filled circles data falling outside 10th and 90th percentiles.

- starting with crop-only systems ((A) monocropping and (B) intercropping);
- followed by the alley cropping systems ((C) simple intensive alley cropping);
- then, (D) simple biodiverse intensive alley cropping;
- (E) simple extensive alley cropping;
- (F) biodiverse extensive alley cropping;
- (G) extensive alley cropping with intercropping;
- (H) biodiverse extensive alley cropping with intercropping); and
- (highly dispersed again) forest systems ((I) simple forest and (J) conservation forest).

A similar response is predicted for the normalised shadow richness (NShR), even if with nil results for the crop-only systems (Fig. 4) (with a stepwise decrease associated with the management intensity).

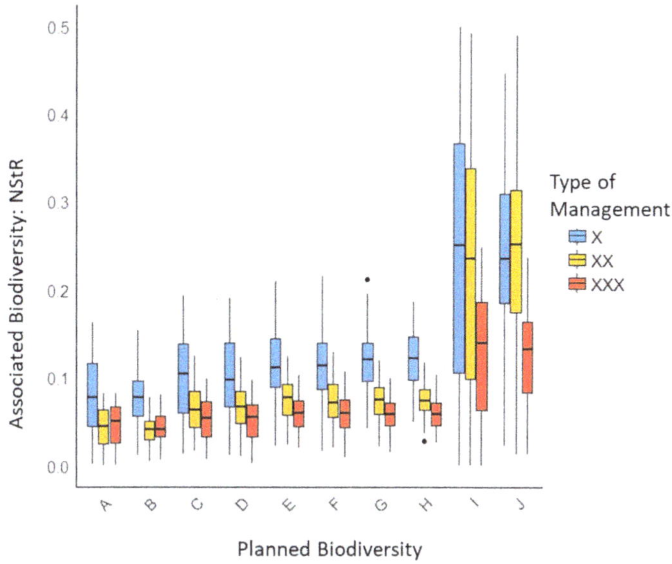

Figure 3 Box and Whisker plots expressing the differences in the normalised structural richness (Associated Biodiversity: NStR) associated with the different simulated agroecosystems/planned biodiversity (Planned Biodiversity) and management intensity/levels (Type of Management): (A) monocropping; (B) intercropping; (C) simple intensive alley cropping; (D) simple biodiverse intensive alley cropping; (E) simple extensive alley cropping; (F) biodiverse extensive alley cropping; (G) extensive alley cropping with intercropping; (H) biodiverse extensive alley cropping with intercropping; (I) simple forest; (J) conservation forest; (X) ecological farming; (XX) conventional pest control of crops; (XXX) conventional pest control of the overall field (including the tree/understory vegetation patches). Box-plot details: lower and upper box boundaries 25th and 75th percentiles, respectively, line inside box median, lower and upper lines limit 10th and 90th percentiles, respectively, filled circles data falling outside 10th and 90th percentiles.

Finally, the normalised edge/border richness (NBoR) (Fig. 5) is nil for the simple systems ((A) monocropping and (I) simple forest) intermediate for the other and higher for (B) intercropping, (G) extensive alley cropping with intercropping and (H) biodiverse extensive alley cropping with intercropping (with stepwise reductions associated with the management intensity/levels).

The overall associated biodiversity (NSuR) results are depicted in Fig. 6. Even if the management intensity/levels tend to reduce the NSuR obtained per agroecosystem, higher values were simulated for (B) intercropping, (G) extensive alley cropping with intercropping, (H) biodiverse extensive alley cropping with intercropping and (J) conservation forest.

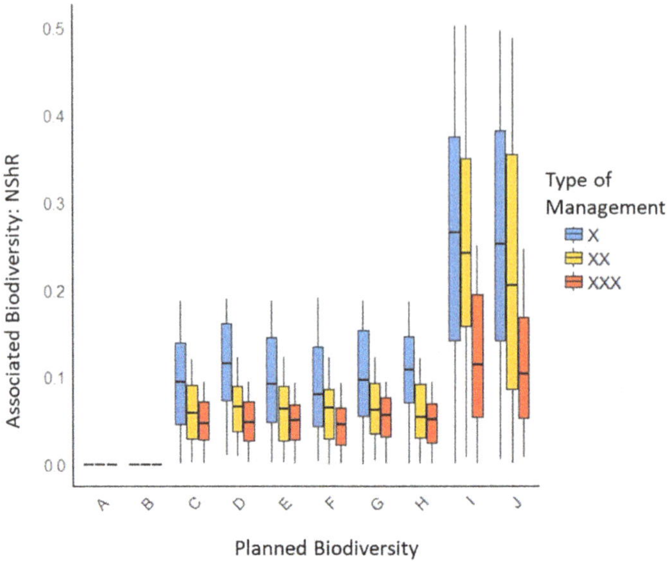

Figure 4 Box and Whisker plots expressing the differences in the normalised shadow richness (Associated Biodiversity: NShR) associated with the different simulated agroecosystems/planned biodiversity (Planned Biodiversity) and management intensity/levels (Type of Management): (A) monocropping; (B) intercropping; (C) simple intensive alley cropping; (D) simple biodiverse intensive alley cropping; (E) simple extensive alley cropping; (F) biodiverse extensive alley cropping; (G) extensive alley cropping with intercropping; (H) biodiverse extensive alley cropping with intercropping; (I) simple forest; (J) conservation forest; (X) ecological farming; (XX) conventional pest control of crops; (XXX) conventional pest control of the overall field (including the tree/understory vegetation patches). Box-plot details: lower and upper box boundaries 25th and 75th percentiles, respectively, line inside box median, lower and upper lines limit 10th and 90th percentiles, respectively, filled circles data falling outside 10th and 90th percentiles.

5 Results: comparing predictions of associated biodiversity

The global influence of the factors agroecosystems/planned biodiversity (Planned Biodiversity) and type of management (Type of Management) on the associated biodiversity (Associated biodiversity: NSuR) and their interaction are depicted in Table 2. Both factors and their interaction were associated with statistically significant effects in the NSuR.

The effect size for the factors of the ANOVA estimated by Eta square, a standardised estimate, is presented in Table 3.

As the simulated decrease in NSuR is proportional to the Management level defined (Type of Management X, XX or XXX), pairwise differences for NSuR (Associated Biodiversity: NSuR) considering only the Planned Biodiversity factor (including all Management levels) were assessed in Fig. 7 and Table 4.

Published by Burleigh Dodds Science Publishing Limited, 2025.

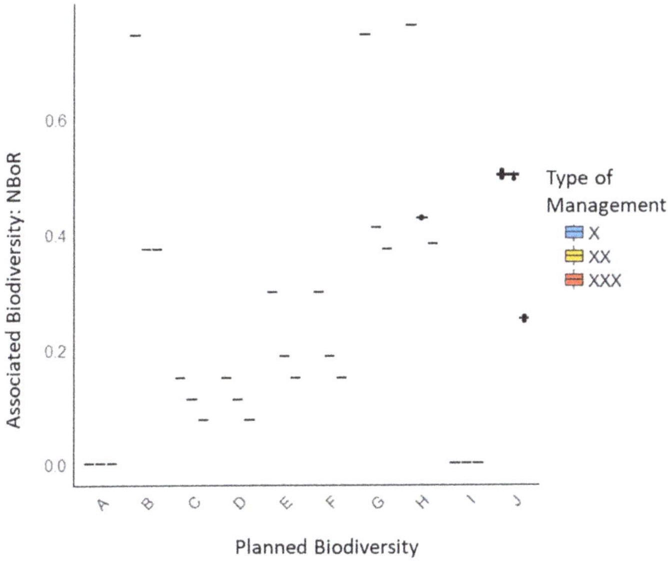

Figure 5 Box and Whisker plots expressing the differences in the normalised edge/border richness (Associated Biodiversity: NBoR) associated with the different simulated agroecosystems/planned biodiversity (Planned Biodiversity) and management intensity/levels (Type of Management): (A) monocropping; (B) intercropping; (C) simple intensive alley cropping; (D) simple biodiverse intensive alley cropping; (E) simple extensive alley cropping; (F) biodiverse extensive alley cropping; (G) extensive alley cropping with intercropping; (H) biodiverse extensive alley cropping with intercropping; (I) simple forest; (J) conservation forest; (X) ecological farming; (XX) conventional pest control of crops; (XXX) conventional pest control of the overall field (including the tree/understory vegetation patches). Box-plot details: lower and upper box boundaries 25th and 75th percentiles, respectively, line inside box median, lower and upper lines limit 10th and 90th percentiles, respectively, filled circles data falling outside 10th and 90th percentiles.

The pairwise differences for NSuR (Associated Biodiversity: NSuR), considering only the Management (Type of Management) factor (including all Planned Biodiversity agroecosystems) were assessed in Table 5.

6 Assessing implications for different management strategies to optimise associated biodiversity

The associated biodiversity (NSuR) results suggest that intercropping, extensive alley cropping with intercropping, biodiverse extensive alley cropping with intercropping and conservation forest are optimal choices for biodiversity conservation in the case of applying ecological management methods, as discussed by Santos et al. (2022b). In the case of the other management options, such as conventional pest control of crops or conventional pest control

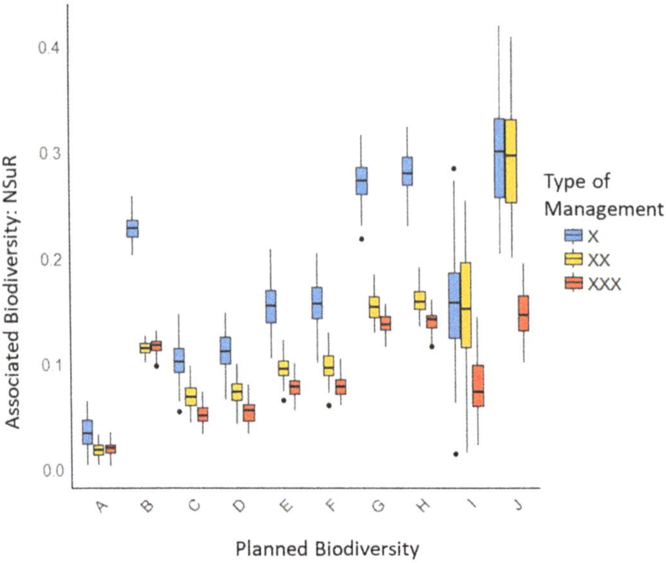

Figure 6 Box and Whisker plots expressing the differences in the normalised richness (Associated Biodiversity: NSuR) associated with the different simulated agroecosystems/planned biodiversity (Planned Biodiversity) and management intensity/levels (Type of Management): (A) monocropping; (B) intercropping; (C) simple intensive alley cropping; (D) simple biodiverse intensive alley cropping; (E) simple extensive alley cropping; (F) biodiverse extensive alley cropping; (G) extensive alley cropping with intercropping; (H) biodiverse extensive alley cropping with intercropping; (I) simple forest; (J) conservation forest; (X) ecological farming; (XX) conventional pest control of crops; (XXX) conventional pest control of the overall field (including the tree/understory vegetation patches). Box-plot details: lower and upper box boundaries 25th and 75th percentiles, respectively, line inside box median, lower and upper lines limit 10th and 90th percentiles, respectively, filled circles data falling outside 10th and 90th percentiles.

Table 2 Analysis of Variance of the Generalised Linear Model relating the factors associated with the agroecosystem/planned biodiversity (Planned Biodiversity) and type of management (Type of Management) with the normalised richness (Associated Biodiversity: NSuR)

Factors	*Df*	Sum Sq	Mean Sq	*F* value	Pr(>*F*|)
Planned Biodiversity	9	11.543	1.2825	2486.9	<2e−16
Type of Management	2	4.113	2.0563	3987.4	<2e−16
Planned Biodiversity × Type of Management	18	1.805	0.1003	194.4	<2e−16
Residuals	2970	1.532	0.0005		

Degrees of freedom (Df), Sum of squares (Sum sq), Mean squares (Mean Sq), *F* value (*F* value) and probability of *F* (Pr(>|*F*)). Variables explanation in Table 1.

Table 3 The effect size of factors of the Analysis of Variance (ANOVA) of the Generalised Linear Model relating the agroecosystem/planned biodiversity (Planned Biodiversity) and type of management (Type of Management) with the normalised richness (Associated Biodiversity: NSuR) (Table 2)

Factors	Eta2 (partial)	95% CI
Planned Biodiversity	0.88	[0.83, 1.00]
Type of Management	0.73	[0.72, 1.00]
Planned Biodiversity × Type of Management	0.54	[0.52, 1.00]

Partial Eta squared (Eta2 (partial)) and 95% confidence interval of the standardised estimate (95% CI). Variables explanation in Table 1.

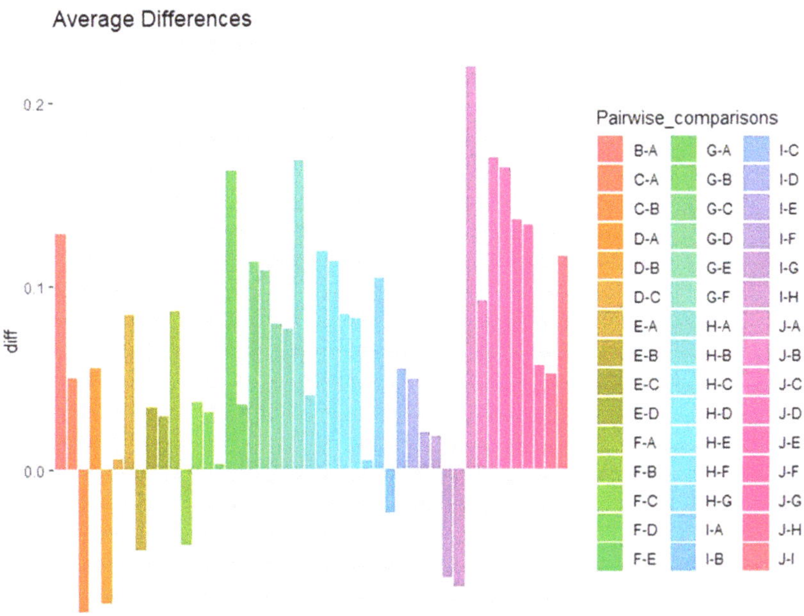

Figure 7 Bars expressing the average pairwise differences (diff) in the normalised richness (Associated Biodiversity: NSuR) associated with the different simulated agroecosystems/planned biodiversity (Planned Biodiversity) (considering all management intensity/levels): (A) monocropping; (B) intercropping; (C) simple intensive alley cropping; (D) simple biodiverse intensive alley cropping; (E) simple extensive alley cropping; (F) biodiverse extensive alley cropping; (G) extensive alley cropping with intercropping; (H) biodiverse extensive alley cropping with intercropping; (I) simple forest; (J) conservation forest. Details are provided in Table 4.

of overall field, only extensive alley cropping with intercropping, biodiverse extensive alley cropping with intercropping and conservation forest maintain a high level of associated biodiversity (even if reduced when compared with

Table 4 Pairwise average differences for the normalised biodiversity predictions (Associated Biodiversity: NSuR) between the different combinations simulated of planned biodiversity/agroecosystems (Planned Biodiversity) and associated significance accordingly with the Tukey multiple comparisons of means with a 95% family-wise confidence level (***$P < 0.001$).

AD (R–C)	A	B	C	D	E	F	G	H	I
A	–	–	–	–	–	–	–	–	–
B	0.128 (***)	–	–	–	–	–	–	–	–
C	0.049 (***)	–0.078 (***)	–	–					
D	0.055 (***)	–0.073 (***)	0.005 (n.s.)	–					
E	0.083 (***)	–0.044 (***)	0.034 (***)	0.028 (***)					
F	0.086 (***)	–0.041 (***)	0.036 (***)	0.031 (***)	0.002 (n.s.)				
G	0.163 (***)	0.035 (***)	0.113 (***)	0.108 (***)	0.079 (***)	0.076 (***)			
H	0.168 (***)	0.040 (***)	0.118 (***)	0.113 (***)	0.084 (***)	0.082 (***)	0.005 (n.s.)		
I	0.104 (***)	–0.024 (***)	0.054 (***)	0.048 (***)	0.020 (***)	0.017 (***)	–0.059 (***)	–0.064 (***)	
J	0.219 (***)	0.091 (***)	0.169 (***)	0.164 (***)	0.135 (***)	0.133(***)	0.056 (***)	0.051 (***)	0.115 (***)

AD, MSuR average differences; R, rows; C, columns. (A) monocropping; (B) intercropping; (C) simple intensive alley cropping; (D) simple biodiverse intensive alley cropping; (E) simple extensive alley cropping; (F) biodiverse extensive alley cropping; (G) extensive alley cropping with intercropping; (H) biodiverse extensive alley cropping with intercropping; (I) simple forest; (J) conservation forest.

Table 5 Pairwise average differences for the normalised biodiversity predictions (Associated Biodiversity: NSuR) between the management levels considered (Type of Management) and associated significance accordingly with the Tukey multiple comparisons of means with a 95% family-wise confidence level (****P* < 0.001)

AD (R-C)	X	XX
XX	−0.057 (***)	–
XXX	−0.089 (***)	−0.032 (***)

AD, NSuR average differences; R, rows; C, columns. (X) ecological (organic) farming; (XX) chemical pest control of crops; (XXX) chemical pest control of the overall field (including the tree/understory vegetation patches).

ecological management) (for additional information, please see Beaumelle et al., 2023).

The statistical tests corroborate the visual results, namely by depicting the importance of both factors (planned biodiversity and management) and their interaction for the associated biodiversity outcome. Additionally, the pairwise differences integrating all management levels are also particularly informative, namely considering the diversity in farmer's options and variations in management within an agricultural landscape (Delonge et al., 2016). The pairwise comparisons also call attention to the relevance of alley cropping with intercropping systems and the maintainnce of natural forests to enhance biodiversity in agricultural landscapes, as pointed out by several publications (Delonge et al., 2016; Casagrande et al., 2017; Gallé et al., 2020; Chen et al., 2022).

Even considering the pre-parameterisation of the patches' partial biodiversities, the results obtained for each planned biodiversity/agroecosystem combination, of the associated diversity components, were not fully anticipated.

Considering normalised phytophagous richness (NPhR), the performance of intercropping-associated biodiversity, higher than the simple intensive alley cropping and simple biodiverse intensive alley cropping, independent of management level, was unexpected (e.g. Kaiser et al., 2017; Jactel et al., 2021). Similarly, the higher values of NPhR for biodiverse extensive alley cropping, extensive alley cropping with intercropping and biodiverse extensive alley cropping with intercropping, when compared with simple forest (at least considering the ecological management) was also an interesting outcome, in agreement with the Database of Insects and their Food Plants (https://dbif.brc.ac.uk/homepage.aspx).

On the other hand, the NStR and the NShR results were anticipated (e.g. Le Provost et al., 2021; Eo et al., 2023; Silva et al., 2023), namely the crops' lower values (monocropping, intercropping), the intermediate results in the case of alley cropping (simple intensive alley cropping, simple biodiverse intensive alley cropping, simple extensive alley cropping, biodiverse extensive alley cropping, extensive alley cropping with intercropping, biodiverse extensive alley cropping with intercropping), and the higher structural and shadow richness related with the forests (simple forest, conservation forest).

Concerning the edge/border richness (NBoR), the most interesting result was the very high and almost equivalent values obtained (at least for the ecological management) for the intercropping, extensive alley cropping with intercropping, and biodiverse extensive alley cropping with intercropping, congruent with Gayer et al. (2021) outcomes.

7 The nexus between alley cropping, biodiversity and sustainable agriculture

The world faces the important challenge of feeding an increasing population but from an agriculture that is sustainable (Struik & Kuyper, 2017). In fact, sustainable agriculture could be a win–win solution: producing more food while reducing negative impacts of convential agriculture on the environment, including biodiversity loss and climate change (Mosquera-Losada et al., 2023). Alley cropping systems (and in general agroforestry) are a key approach integrating the three dimensions of sustainable development (economic, social and environmental) and jointly addressing food security and climate challenges.

Agroforestry has been recognised as a sustainable land use practice and system, mixing modern and traditional practices, the promotion of genetic resources of ancient crops or livestock breeds vulnerable to extinction and wildlife conservation (Nair et al., 2021). Moreover, agroforestry systems include high-value fruit and timber trees, soil protection and contribute to healthy diets and the well-being of consumers (Mosquera-Losada et al., 2023).

Our results support the hypothesis that alley cropping (and intercropping) has positive impacts on biodiversity, within several crop/tree species' (and understory vegetation) combinations (Gayer et al., 2021). Still, the simple model did not consider mechanisms operating at larger scales, and management issues were approached in a very basic way (Gotelli et al., 2001; Gunstone et al., 2021; Le Provost et al., 2021). The results obtained agree with the literature showing that management plays an essential role in shaping agroecosystem biodiversity. The shift to ecological (or at least organic) farming combined with alley cropping systems is linked with enhancement of planned biodiversity, but associated biodiversity also increases as a result of more complex microenvironments (e.g. Kovács-Hostyánszki et al., 2017).

Positive effects of alley cropping (and in general agroforestry) were noted for diverse taxonomic groups and systems (Pardon et al., 2019), improving valuable ecosystem services such as habitat for pollinators and natural

Published by Burleigh Dodds Science Publishing Limited, 2025.

enemies of crop pests (Hostyánszki et al., 2020). The crop(s) and tree(s) species chosen have a significant effect on associated biodiversity, namely when combinations include traditional crops and native tree species (and understory vegetation) (Santos et al., 2022b). Associated biodiversity is also coupled with species thresholds to environmental conditions impacting physiology, demography, dispersal, interspecific interactions and adaptation (Moreno et al., 2018).

The model presented here was relatively simple when compared with most agroforestry modelling studies, since we relied on parameterising dimensions of species richness from databases and previous publications. We also took into consideration that environmental measures are implemented at the farm level but recognise that many of the associated biodiversity species depend on upscaling to the agricultural landscape scale (Gallé et al., 2020).

The current study might be regarded as preliminary and with several limitations. Most notably, the model's assumptions and outputs need to be tested to demonstrate its usefulness and applicability (Santos et al., 2013). Performance comparisons might be achieved by parameterising the model to realistic species combinations and comparing the results with field studies (Santos et al., 2018). Several authors state that the fundamental step for framework acceptance is pre-defining the context of application and objectives (e.g. Grimm et al., 2020).

8 Conclusion

Agricultural intensification, as well as the abandonment of traditional agriculture practices, can result in reduced biodiversity and ecosystem services that are difficult to reverse because of the long-time lags characterising the dynamics of socio-ecological systems (Santos et al., 2022b). Societal and ecological mechanisms interact, generating changes in biological communities and degradation of ecosystem functions that are misunderstood and, at a human generation time-scale, mostly irreversible.

Our modelling approach complements previous work relating to agroforestry, management and biodiversity conservation. The simple demonstration, supported by partial contributions for overall diversity, reinforces the idea that agroforestry (in our case study, alley cropping) seems critical to halt biodiversity loss in agricultural landscapes. Despite the limitations inherent to a preliminary modelling exercise, the methodology developed provides a starting point to anticipate the changes in associated biodiversity induced by farmer choices and management options, guiding pertinent future strategies to integrate crop production with nature.

9 Acknowledgements

This work is supported by National Funds by FCT – Portuguese Foundation for Science and Technology, under the projects UIDB/04033/2020 (https://doi.org/10.54499/UIDB/04033/2020) and LA/P/0126/2020 (https://doi.org/10.54499/LA/P/0126/2020), associated with CITAB - Centre for the Research and Technology of Agro-Environmental and Biological Sciences and Inov4Agro - Institute for innovation, capacity building and sustainability of agri-food production. We would also like to acknowledge the Projects Innovative concepts and technologies for ECOlogically sustainable NUTRIent management in agriculture aiming to prevent, mitigate and eliminate pollution in soils, water and air (ECONUTRI) under the grant Horizon/101081858 (https://cordis.europa.eu/project/id/101081858), Integrated SERvices supporting a sustainable AGROecological transition (AGROSERV) under the grant Horizon/ 101058020 (https://cordis.europa.eu/project/id/101058020) and the INTERACT project – 'Integrative Research in Environment, Agro-Chains and Technology' [project number campbe-01-0145-FEDER-000017], in its line of research entitled BEST, co-financed by the European Regional Development Fund (ERDF) through NORTE 2020 (North Regional Operational Program 2014/2020).

10 Where to look for further information

For additional information, please refer to the Research Topic titled "Can the Trees Save the Crops? Predicting the Services Provided by Traditional and Novel Agroforests in Changing Mediterranean Landscapes". It can be accessed at: https://www.frontiersin.org/research-topics/12712/can-the-trees-save-the-crops-predicting-the-services-provided-by-traditional-and-novel-agroforests-in-changing-mediterranean-landscapes

11 Appendix: Modelling parameters

Part 1. Netlogo code of the model used to predict the associated biodiversity of agroecosystems

Global Variables:

```
Globals [Mean_diversity Median_diversity Modes_diversity dph25 dph65
  dph45 dph35 dph75 dst25 dst65 dst45 dst35 dst75 dsh ]
```

Patch variables:

```
Patches-own [phyto_richness phytotoabundance species_list structure_
  richness shadow_richness border_richness sum_richness man_richness
  distance_shadow]
```

Defining the model initial conditions:

```
to setup
clear-all
set dph25 random-float 1
set dph65 random-float 2
set dph45 random-float 4
set dph35 random-float 2
set dph75 random-float 4
set dst25 random-float 1
set dst65 random-float 3
set dst45 random-float 2
set dst35 random-float 1
set dst75 random-float 3
set dsh random-float 5
Agroforest
Management_Richness_estimation
PhytoRichness_estimation
StructureRichness_estimation
ShadowRichness_estimation
EdgeRichness_estimation
SumRichness_estimation
Diversity_estimations
reset-ticks
end
```

Defining the type of planned biodiversity:

```
to Agroforest
If Diversity = 1 [ask patches [set pcolor 25 ]]
If Diversity = 2 [ask patches [set pcolor 25 if pxcor mod 2 = 0 [set pcolor 35
  ]]]
If Diversity = 3 [ask patches [set pcolor 25 if pxcor mod 10 = 0 [set pcolor 65]]]
If Diversity = 4 [ask patches [set pcolor 25 if pxcor mod 10 = 0 [set pcolor 75]]]
If Diversity = 5 [ask patches [set pcolor 25 if pxcor mod 10 = 0 [set pcolor 65]
  if pxcor mod 10 = 9 or pxcor mod 10 = 1 [set pcolor 45]]]
If Diversity = 6 [ask patches [set pcolor 25 if pxcor mod 10 = 0 [set pcolor 75]
  if pxcor mod 10 = 9 or pxcor mod 10 = 1 [set pcolor 45]]]
If Diversity = 7 [ask patches [set pcolor 25 if pxcor mod 2 = 0 [set pcolor 35]
  if pxcor mod 10 = 0 [set pcolor 65] if pxcor mod 10 = 9 or pxcor mod 10
  = 1 [set pcolor 45]]]
If Diversity = 8 [ask patches [set pcolor 25 if pxcor mod 2 = 0 [set pcolor 35] if
  pxcor mod 10 = 0 [set pcolor 65 + 10 * random 2 ] if pxcor mod 10 = 9 or
  pxcor mod 10 = 1 [set pcolor 45]]]
If Diversity = 9 [ask patches [set pcolor 65 ]]
```

Published by Burleigh Dodds Science Publishing Limited, 2025.

```
If Diversity = 10 [ask patches [set pcolor 65 + 10 * random 2 ]]
end
Calculating the management effects on biodiversity:
to Management_Richness_estimation
ask patches [if Management = 1 [set man_richness 1] ifelse Management = 2
   and pcolor != 65 and pcolor != 75 [set man_richness 2][set man_richness
   1] if Management = 3 [set man_richness 2]]
end
Calculating the phytophagous associated biodiversity:
to PhytoRichness_estimation
ask patches [if pcolor = 25 [set phyto_richness (random-float dph25 )/ 4 /
   man_richness set species_list (list "a" "b" "c")]
if pcolor = 65 [ set phyto_richness (random-float dph65) / 4 / man_richness]
if pcolor = 45 [set phyto_richness (random-float dph45) / 4 / man_richness]
if pcolor = 35 [set phyto_richness (random-float dph35) / 4 / man_richness
   set species_list (list "d" "e" "f" "g" "h" "i")]
if pcolor = 75 [set phyto_richness (random-float dph75) / 4 / man_richness] ]
end
Calculating the species associated with the structural complexity of plants
   (associated biodiversity):
to StructureRichness_estimation
ask patches [if pcolor = 25 [set structure_richness (random-float dst25) / 3 /
   man_richness]
if pcolor = 65 [ set structure_richness (random-float dst65) / 3 / man_richness
   ]
if pcolor = 45 [set structure_richness (random-float dst45) / 3 / man_richness]
if pcolor = 35 [set structure_richness (random-float dst35)/ 3 / man_richness ]
if pcolor = 75 [set structure_richness (random-float dst75) / 3 / man_richness]]
end
Calculating the soil species associated with the shadow (associated
   biodiversity):
to ShadowRichness_estimation
ask patches [ (ifelse remainder abs (pxcor) 10 >= 5 [set distance_shadow (10
   - remainder abs (pxcor) 10)]
[set distance_shadow remainder abs (pxcor) 10] ) Ifelse (Diversity > 2 and
   diversity < 9)
[ ifelse pxcor mod 10 = 0 [set shadow_richness random-float dsh / 5 / man_
   richness][set shadow_richness random-float (dsh / (distance_shadow + 1
   )) / 5 / man_richness] ][set shadow_richness 0] if diversity = 9 or diversity =
   10 [set shadow_richness (random-float dsh) / 5 / man_richness ]]
end
```

Published by Burleigh Dodds Science Publishing Limited, 2025.

Calculating the generalist species taking advantage of edge systems (associated biodiversity):

```
to EdgeRichness_estimation
ask patches [set border_richness (count neighbors with [pcolor != [pcolor] of myself]) / 8 / man_richness] ;repeat world-width [ask patches [ask neighbors with [pcolor = [pcolor] of myself] [if number > [number] of myself [set number [number] of myself]]]]
end
```

Calculating the associated biodiversity:

```
to SumRichness_estimation
ask patches [ set sum_richness (phyto_richness + structure_richness + shadow_richness + border_richness) / 4 ]
end
```

Calculating world statistics:

```
to Diversity_estimations
set Mean_diversity mean [sum_richness] of patches
set Median_diversity median [sum_richness] of patches
set Modes_diversity modes [sum_richness] of patches
end
```

Visualising Phytophagous richness associated biodiversity:

```
to See_Phytorichness
ask patches [ set pcolor phyto_richness * 10 ]
end
```

Visualising Structural richness associated biodiversity:

```
to See_StructureRichness
ask patches [ set pcolor structure_richness * 10]
end
```

Visualising Shadow richness associated biodiversity:

```
to See_ShadowRichness
ask patches [ set pcolor shadow_richness * 10]
end
```

Visualising Edge/border richness associated biodiversity:

```
to See_BorderRichness
ask patches [ set pcolor border_richness * 10]
end
```

Visualising the associated biodiversity:

```
to See_SumRichness
ask patches [set pcolor sum_richness * 10]
end
```

Part 2 Example of a simulation

a) Interface of the Netlogo model used to predict the associated biodiversity of agroecosystems

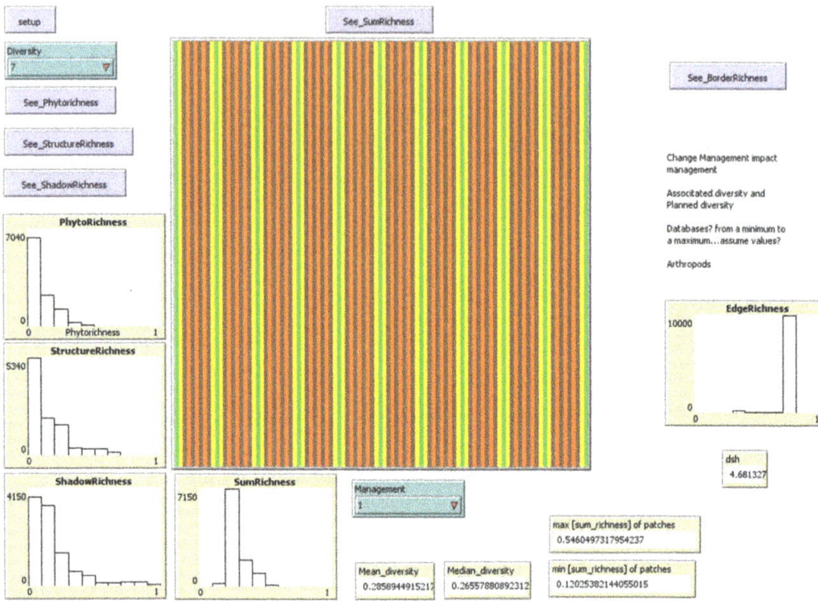

b) Phytophagous richness graphical results

c) Structural richness graphical results

d) Shadow richness graphical results

e) Border/edge richness graphical results

f) Associated Richness graphical results

12 References

Ampoorter, E., et al. (2020). Tree diversity is key for promoting the diversity and abundance of forest-associated taxa in Europe. *Oikos*, 129(2), 133–146.

Arosa, M. L., et al. (2017). Long-term sustainability of cork oak agro-forests in the Iberian Peninsula: a model-based approach aimed at supporting the best management options for the montado conservation. *Ecological Modelling*, 343, 68–79.

Ashraf, M., et al. (2018). Alley-cropping system can boost arthropod biodiversity and ecosystem functions in oil palm plantations. *Agriculture, Ecosystems & Environment*, 260, 19–26.

Bakış, A. L. P., et al. (2021). Is biodiversity linked with farm management options in vineyard landscapes? A case study combining ecological indicators within a hybrid modelling framework. *Ecological Indicators*, 121, 107012.

Barrios, E., et al. (2018). Contribution of trees to the conservation of biodiversity and ecosystem services in agricultural landscapes. *International Journal of Biodiversity Science, Ecosystem Services & Management*, 14(1), 1–16.

Beaumelle, L., et al. (2023). Pesticide effects on soil fauna communities–a meta-analysis. *Journal of Applied Ecology*, 60(7), 1239–1253.

Ben-Shachar, M., et al. (2020). Effectsize: estimation of effect size indices and standardized parameters. *Journal of Open Source Software*, 5(56), 2815.

Bonsch, M., et al. (2015). Environmental flow provision: implications for agricultural water and land-use at the global scale. *Global Environmental Change*, 30, 113–132.

Casagrande, M., et al. (2017). Enhancing planned and associated biodiversity in French farming systems. *Agronomy for Sustainable Development*, 37, 1–16.

Chen, Y., et al. (2022). Winegrowers' decision-making: a pan-European perspective on pesticide use and inter-row management. *Journal of Rural Studies*, 94, 37–53.

Corcos, D., et al. (2021). Cross-taxon congruence between predatory arthropods and plants across Mediterranean agricultural landscapes. *Ecological Indicators*, 123, 107366.

Coyle, D. R., et al. (2017). Soil fauna responses to natural disturbances, invasive species, and global climate change: current state of the science and a call to action. *Soil Biology & Biochemistry*, 110, 116–133.

Cuperus, F., et al. (2023). Effects of field-level strip and mixed cropping on aerial arthropod and arable flora communities. *Agriculture, Ecosystems & Environment*, 354, 108568.

DeLonge, M. S., et al. (2016). Investing in the transition to sustainable agriculture. *Environmental Science & Policy*, 55, 266–273.

Den Herder, M., et al. (2017). Current extent and stratification of agroforestry in the European Union. *Agriculture, Ecosystems & Environment*, 241, 121–132.

Ebeling, A., et al. (2018). Plant diversity induces shifts in the functional structure and diversity across trophic levels. *Oikos*, 127(2), 208–219.

Eo, J., et al. (2023). Composition, abundance and biodiversity of terrestrial arthropods in pastures and their relationship with landscaping plants. *International Journal of Agriculture and Natural Resources*, 50(1), 12–22.

Gallé, R., et al. (2020). Arthropod functional traits shaped by landscape-scale field size, local agri-environment schemes and edge effects. *Basic and Applied Ecology*, 48, 102–111.

Garrett, H. E. G., et al. (2021). Alley cropping practices. *North American Agroforestry*, 163–204. http://dx.doi.org/10.1002/9780891183785.ch7
Gayer, C., et al. (2021). Flowering fields, organic farming and edge habitats promote diversity of plants and arthropods on arable land. *Journal of Applied Ecology*, 58(6), 1155–1166.
Gotelli, N. J. & Colwell, R. K. (2001). Quantifying biodiversity: procedures and pitfalls in the measurement and comparison of species richness. *Ecology Letters*, 4(4), 379–391.
Grimm, V., et al. (2020). The ODD protocol for describing agent-based and other simulation models: a second update to improve clarity, replication, and structural realism. *Journal of Artificial Societies and Social Simulation*, 23(2). https://doi.org/10.18564/jasss.4259
Gunstone, T., et al. (2021). Pesticides and soil invertebrates: a hazard assessment. *Frontiers in Environmental Science*, 9, 122.
Horrigan, L., et al. (2002). How sustainable agriculture can address the environmental and human health harms of industrial agriculture. *Environmental Health Perspectives*, 110(5), 445–456.
Jactel, H., et al. (2021). Tree diversity and forest resistance to insect pests: patterns, mechanisms, and prospects. *Annual Review of Entomology*, 66, 277–296.
Kaiser, L., et al, (2017). The plant as a habitat for entomophagous insects. In *Advances in Botanical Research* (Vol. 81, pp. 179–223). Academic Press.
Kanda, Y. (2013). "Investigation of the freely available easy-to-use software EZR for medical statistics." Bone Marrow Transplantation. https://www.nature.com/articles/bmt2012244.pdf.
Koricheva, J. & Hayes, D. (2018). The relative importance of plant intraspecific diversity in structuring arthropod communities: a meta-analysis. *Functional Ecology*, 32(7), 1704–1717.
Kovács-Hostyánszki, A., Espíndola, A., Vanbergen, A. J., Settele, J., Kremen, C., & Dicks, L. V. (2017). Ecological intensification to mitigate impacts of conventional intensive land use on pollinators and pollination. Ecology letters, 20(5), 673–689.
Kranz, A. J., et al. (2019). Effects of shrub crop interplanting on apple pest ecology in a temperate agroforestry system. *Agroforestry Systems*, 93, 1179–1189.
Kuchenbecker, J., et al. (2022). Community structure of ants (Hymenoptera: Formicidae) in an open habitat: the importance of environmental heterogeneity and interspecific interactions. *Revista Mexicana de Biodiversidad*, 93, e933900.
Le Provost, G., et al. (2021). Contrasting responses of above-and belowground diversity to multiple components of land-use intensity. *Nature Communications*, 12(1), 3918.
Loreau, M., et al. (2021). Biodiversity as insurance: from concept to measurement and application. *Biological Reviews*, 96(5), 2333–2354.
Main, A. R., et al. (2020). Reduced species richness of native bees in field margins associated with neonicotinoid concentrations in non-target soils. *Agriculture, Ecosystems & Environment*, 287, 106693.
Menezes, I. C., Santos, M., Bugalho, L., & Pereira, M. G. (2024). The Effects of Tree Shade on Vineyard Microclimate and Grape Production: A Novel Approach to Sun Radiation Modelling as a Response to Climate Change. Land, 13(11), 1970.

Published by Burleigh Dodds Science Publishing Limited, 2025.

Moreno, G., et al. (2018). Agroforestry systems of high nature and cultural value in Europe: provision of commercial goods and other ecosystem services. *Agroforestry Systems*, 92, 877–891.

Mosquera-Losada, M. R., et al. (2023). Policy challenges for agroforestry implementation in Europe. *Frontiers in Forests and Global Change*, 6, 1127601.

Nair, P. R., et al. (2021). Other ecosystem services of agroforestry. In *An Introduction to Agroforestry: Four Decades of Scientific Developments* (pp. 563–581). Springer.

Nakagawa, S. & Cuthill, I. C. (2007). Effect size, confidence interval and statistical significance: a practical guide for biologists. *Biological Reviews*, 82(4), 591–605.

Nicholls, C., et al. (2016). Agroecology: principles for the conversion and redesign of farming systems. *Journal of Ecosyst Ecography,* S5, 10.

Pardon, P., et al. (2019). Gradients in abundance and diversity of ground dwelling arthropods as a function of distance to tree rows in temperate arable agroforestry systems. *Agriculture, Ecosystems & Environment*, 270, 114–128.

Pinheiro, J., et al. (2023). nlme: linear and nonlinear mixed effects models. R package version 3.1-164, https://CRAN.R-project.org/package=nlme

Prather, R. M., et al. (2020). Abiotic factors and plant biomass, not plant diversity, strongly shape grassland arthropods under drought conditions. *Ecology*, 101(6), e03033.

R Core Team (2020). R: A language and environment for statistical computing. R Foundation for Statistical Computing, Vienna, Austria.URL https://www.R-project.org/.

Rafter, J. A., et al. (2002). Multiple comparison methods for means. *Siam Review*, 44(2), 259–278.

Robinson, S. I., et al. (2018). Soil temperature effects on the structure and diversity of plant and invertebrate communities in a natural warming experiment. *Journal of Animal Ecology*, 87(3), 634–646.

Rumohr, Q., et al. (2023). LandS: Vegetation modeling based on Ellenberg's ecological indicator values. *MethodsX*, 11, 102486.

Sáenz-Romo, M. G., et al. (2019). Ground cover management in a Mediterranean vineyard: impact on insect abundance and diversity. *Agriculture, Ecosystems & Environment*, 283, 106571.

Santos, M., et al. (2013). Converting conventional ecological datasets in dynamic and dynamic spatially explicit simulations: current advances and future applications of the Stochastic Dynamic Methodology (StDM). *Ecological Modelling*, 258, 91–100.

Santos, M., et al. (2018). How good are the ecological assumptions and predictions made in the past? Insights from a dynamic modelling approach applied to changing landscapes. *Ecological Indicators*, 90, 226–230.

Santos, M., et al. (2022a). Contribution of home gardens to sustainable development: perspectives from a supported opinion essay. *International Journal of Environmental Research and Public Health*, 19(20), 13715.

Santos, M., et al. (2022b). Why do agroforestry systems enhance biodiversity? Evidence from habitat amount hypothesis predictions. *Frontiers in Ecology and Evolution*, 9, 630151.

Santos, M., et al. (2023). Side effects of European eco schemes and agri-environment-climate measures on endangered species conservation: clues from a case study in mountain vineyard landscapes. *Ecological Indicators*, 148, 110155.

Schröder, P., et al. (2019). Discussion paper: sustainable increase of crop production through improved technical strategies, breeding and adapted management: a European perspective. *Science of the Total Environment*, 678, 146–161.

Serrão, J. E., et al. (2022). Side-effects of pesticides on non-target insects in agriculture: a mini-review. *The Science of Nature*, 109(2), 17.

Silva, N. R. D., et al. (2023). Diversity of soil surface arthropod is influenced by plant strata in semi-arid conditions. Available at SSRN 4357808.

Stein-Bachinger, K., et al. (2021). To what extent does organic farming promote species richness and abundance in temperate climates? A review. *Organic Agriculture*, 11(1), 1–12.

Struik, P. C. & Kuyper, T. W. (2017). Sustainable intensification in agriculture: the richer shade of green. A review. *Agronomy for Sustainable Development*, 37, 1–15.

Ulanowicz, R. E. (2018). Biodiversity, functional redundancy and system stability: subtle connections. *Journal of the Royal Society Interface*, 15(147), 20180367.

Van Schalkwyk, J., et al. (2020). Environmental filtering and spillover explain multi-species edge responses across agricultural boundaries in a biosphere reserve. *Scientific Reports*, 10(1), 14800.

Vanbergen, A. J., et al. (2020). Transformation of agricultural landscapes in the anthropocene: nature's contributions to people, agriculture and food security. In *Advances in Ecological Research* (Vol. 63, pp. 193–253). Academic Press.

Vasconcellos-Neto, J., et al. (2017). *Spider–plant interactions: an ecological approach. Behaviour and ecology of spiders: contributions from the neotropical region* (pp. 165–214). Springer International Publishing.

Wezel, A., et al. (2020). Agroecological principles and elements and their implications for transitioning to sustainable food systems: a review. *Agronomy for Sustainable Development*, 40, 40.

Wickham, H. (2016). *ggplot2: Elegant Graphics for Data Analysis*. Springer-Verlag, New York. ISBN 978-3-319-24277-4. https://ggplot2.tidyverse.org.

Wilensky, U. (1999). *NetLogo*. Center for Connected Learning and Computer-Based Modeling, Northwestern University, Evanston, IL. http://ccl.northwestern.edu/netlogo/

Wright, A. J., et al. (2021). Biodiversity and ecosystem functioning: have our experiments and indices been underestimating the role of facilitation? *Journal of Ecology*, 109(5), 1962–1968.

Chapter 10

Development in forest farming

Margaret Gove Bloomquist, North Carolina State University, USA

1 Introduction

Growth has dominated the temperate agroforestry world over the last decades (Fig. 1). Forest farming is gaining traction as a useful land practice. There is increased consumer interest and demand for forest-farmed and sustainable wild-crafted, domestic, and local products, and a priority for conservation of plants and forested spaces. Forest farming can be part of a healthy ecosystem for environmental, social, and economic benefits (Chamberlain, 2018; Mudge, 2014).

The term forest farming refers to modern practices of humans stewarding and cultivating crops under a forest canopy and includes non-timber tree materials. Forest farming knowledge has developed from generations of people interacting with the land and diverse, often native, species of trees, shrubs, botanicals, fungi, and ferns. It is part of a 'managed woodlot' in permaculture terminology (Mudge, 2014). Non-timber forest products and forest farming are important to the people for subsistence, livelihood, and cultural purposes,

http://dx.doi.org/10.19103/AS.2024.0141.16

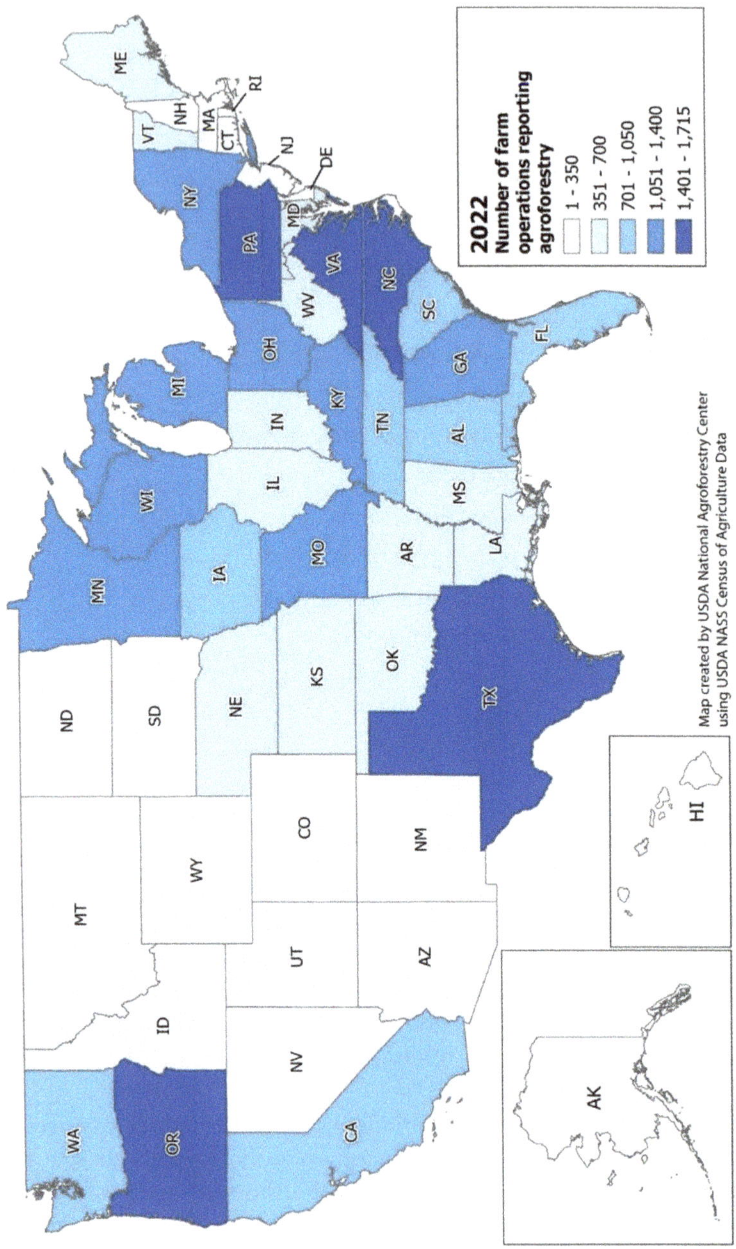

Figure 1 NASS 2022 U.S. Census of Agriculture. Agroforestry Map (Source: United States Department of Agriculture - National Agroforestry Center. 2025).

as they have been for thousands of years. Many cultures have a reciprocal relationship to the forests, past and present. From subsistence relationships with the forest to supplemental enterprises, gatherers, forest stewards, and farmers. Ethnobotanical records and present use showcase humans' relation to the forest and aspects of the modern term forest farming. The Mohican Indian tribe gathers from the woods what are known as 'herbs' for medicine and food (Cook, 2013), as one of countless examples.. Agroforestry management and recognition around the world involve practices that may be considered forest farming. The United Nations FAO Sustainable Development Goals include the sustainable management of forests (FAO UN, 2025).

The United States Department of Agriculture Natural Resource Conservation Service (USDA NRCS) defines forest farming as "Managing or establishing stands of trees or shrubs in coordination with the management and/or cultivation of understory plants or non timber forest products" (USDA NRCS, 2022). Mudge and Gabriel (2014) define forest farming as cultivating crops under a modified or maintained forest canopy to favor growth. Forest farming is a growing sector of agroforestry in the United States (U.S.). The increased adoption of the assigned federal code by state NRCS offices is one indication of this. U.S. federal and tribal NRCS Code 379 outlines the conservation practice standard for forest farming, including the term multi-story cropping (USDA NRCS, 2022). Forest gardening, food forests, wild stewardship, permaculture, and regenerative agriculture are production systems and strategies that contribute to forest farming.

Many variables help define forest farming operations including the location, scale, site parameters, interaction of species, and level of management (Fig. 2). Non-timber forest products are a result of productive forest farmed systems (Chamberlain, 2018; Mudge, 2014). The 2022 U.S. Census of Agriculture shows increases in Agroforestry practices across the country and especially in Eastern states (Kellerman, 2025). Other types of reporting for forest farming may be gleaned from other groups such as questions about NTFPs on the U.S. National Woodland Owners Survey for private landowners (Kellerman, 2025). In the European Union, agroforestry has grown in importance in the past three decades. In Northern Europe, forest farming is a relatively new recognized form of agroforestry, along with forest gardening (Schaffer, 2024).

This chapter focuses on temperate agroforestry and forest farming of woodland botanicals. The focus on the eastern United States and understory plants reflects the author's location and experience. Syrups, nuts, floral and craft materials, mushrooms, and other non-timber forest products (NTFPs) are a large part of current forest farming, while not covered in depth in this chapter. Introductory forest farming information including definitions, species identification, cultivation, and case studies of a variety of forest farming methods and products can be found in the chapter on 'Agroforestry practices:

Figure 2 A wild simulated forest farm in Appalachia, U.S.A (Source: Margaret Bloomquist, 2024).

forest farming' by Kenneth Mudge in the 2019 text *Agroforestry for sustainable agriculture* (Mudge, 2019).

2 Forest farming is multidisciplinary

Forest farming is multidisciplinary in nature with stakeholders from diverse backgrounds and with varying expertise and experience. Common goals are to optimize production and support 'conservation through cultivation', 'under the canopy', two phrases credited to early adopters of and path makers of modern forest farming in Appalachia, USA. It is a management practice to diversify both income and biological diversity (Chamberlain, 2009). The focused cultivation of high value NTFPs is a practice synergic to: managing timber and corresponding shade levels in the canopy, ecosystem needs for restoration or management, or wild harvesting and stewarding populations, in addition to cultivation (Mudge, 2014, 2019).

The wide range of species of trees, plants, fungi, and the products they become presents countless opportunities for interaction (Chamberlain, 2018), like the people and professions that surround it. Agroforestry practices are often managed together and overlap (Schafer, et al. 2024, Mudge, 2014). Forest farming provides unique opportunities for landowners and stewards (Chamberlain, 2009), as well as challenges due to the complexity of the systems. Forest farming brings together the fields of forestry, plant sciences, traditional knowledge, agronomy, and horticulture, along with food safety, regulatory considerations, consumers, and industry.

Forest farming spans world geography and cultures. In the 2010s, organizations, institutions, and forest farmers gathered to collaborate on temperate forest farming challenges and opportunities. In the eastern U.S., the Appalachian Beginning Forest Farmer Coalition was formalized in 2015 with support from the USDA Beginning Farmer and Rancher Development program. The time period marked the start of a flow of federal, state, and philanthropic support for Forest Farming and Agroforestry practice expansion in the U.S. Regional groups in the Midwest, Southeast, Northwest, and Northeast U.S. are expanding surrounding forest farming. The USDA National Agroforestry Center (NAC) supported the need for a national American Forest Farming Association. An example from the U.S., and Costa Rica, illustrates some cooperation in the North and Central Americas; The North American Agroforestry Conference was hosted by the Tropical Agricultural Research and Higher Education Center (CATIE) in Costa Rica in 2024, with the North American Temperate Agroforestry Association (AFTA), providing a window to international and tropic to temperate exchange and collaboration. A track on forest farming was included in the conference (CATIE, 2024).

3 Developing forest farming practices

Agroforestry systems can provide ecosystem benefits along with production potential, more so than agriculture and forestry alone (Nabuurs, 2022; Schaffer, 2024). Forest farming and other Agroforestry practices are recognized as tools to minimize risk associated with climatic changes and extreme weather (Mudge, 2014). Climate Smart Agriculture and Forestry programs of the USDA supported the expansion of agroforestry practices (USDA, 2024). The geographic distribution and range of traditionally harvested crops may change with climatic variability (Chamberlain, 2018), forest farming provides an opportunity for accessibility via cultivation under the canopy.

Saps, syrups, and honey are prominent forest-farmed products, as well as nuts, berries, and more (Davis, 2014; Mudge, 2014; Chamberlain, 2018). Non-timber forest products have not traditionally been valued in forest management plans (Smith-Hall and Chamberlain, 2023). Inclusion of non-timber forest products in land and forest valuations recognizes the investments of landowners and stewards. Incorporating forest farming practices and species into land use management plans is becoming more recognized, and is dependent on the local office providing. Forest farming encourages more time in the forest for land stewards and healthy understory management potential.

Mushroom collection and cultivation is an increasingly popular forest farming practice and can be an enterprise with faster turnaround time for product harvest than perennial species and tree crops. Shitake mushrooms grown outdoors under shade cover on inoculated wood logs have cultivation and best practice guides available from a variety of resources including the University of Vermont (University of Vermont, 2013) and the University of Kentucky (University of Kentucky, 2022). Wild mushroom harvesting and permitting for restaurant sales is another development in the U.S. and is new for many states. Medicinal and specialty mushroom cultivation and incorporation of mushrooms to the forest farm and outdoor ecosystem are practiced. Indoor controlled climate with sanitized media mushroom production remains the prominent source of the majority of mushrooms to the market.

Accessibility to forest farming is considered with diverse and creative land access strategies. Public land access may be restricted to collection and permitting may be required for harvesting, distributing, or selling forest farmed materials. In some areas, indigenous people have rights to collect from and manage populations on public land. For example in North Carolina, U.S., the Eastern Band of Cherokee Indian are permitted bright orange labeled harvest bags to gather traditionally used plants like sochan, *Rudbeckia laciniata,* on public lands in certain areas. An example of the U.S. Department of Agriculture and Department of Interior allowing a limited amount of stewards to participate in resource management. Stewarding populations on leased land, similar to a hunting lease, may provide benefits to the landless and all parties. Forest farmed and wild stewarded populations are unique opportunities for widespread accessibility to forest farming. Demand for regional products offers developing opportunities for forest farming, for example a focus on eco types for branding and regional economic development.

Urban forest farming is a growing practice. Urban Agroforestry and urban forest gardens have been primarily for food production, and can offer additional functions by incorporating cultural, medicinal, diversity, and conservation goals to urban areas, modeling the tropical forest garden (Hemmelgarn, 2021). Food forests are another term assigned to urban agroforestry practices; unique design and management are inherent in urban agroforestry (Favor, 2023).

The growth in recognition of forest farm establishment in the U.S. is in contrast to the continued decline of forested land and increased land use change. Agroforestry systems often operate on small landholdings and on private land, true in Appalachia, U.S., and are at a niche level in Sweden, for example. There is a decline in Agroforestry systems in Europe and an anticipation of continued reduction (Schaffer, 2024). Yet, growth in Agroforestry adoption is highlighted in the latest U.S. Census of Agriculture (COA) data. Kellerman et. Al. (2025) reported 2022 COA results showing growth in the number of farms practicing agroforestry in the U.S., totaling 32 717 or 1.7% of all U.S. reporting farms. An increase in agroforestry practice adoption on farms is attributed to increased understanding of systems, and technical and financial assistance (Kellerman, 2025).

Challenges to implementing agroforestry systems include access to and development of best practice information, gaps in technology preventing the adoption of agroforestry practices, and lack of planting stock (Schaffer, 2024). The supply chain gap of unavailable planting stock and seed of forest farmed species is identified across stakeholders in the U.S. A majority of the planting stock for forest farming operations is gathered from wild populations. Nursery production for wholesale, conservation plantings, and retail are viable opportunities.

Technological opportunities include tissue culture and biological products for crop management. Modern tools available for forest farmers include digital site selection software, plant and tree identification applications, and applying tools from other disciplines to forest farming, for example, from forestry. PlantShoe is an online tool developed at the Center for Geospatial Information Technology (CGIT) at Virginia Tech and Polytechnic School (VA Tech) that assesses site selection for forest-farmed botanicals. The tool evaluates ideal habitat from forests up to 2,000 acres for select woodland species, namely ramps (ramsons), *Allium tricoccum*, black cohosh, *Actaea racemosa*, goldenseal, *Hydrastis Canadensis*, and bloodroot, *Sanguinaria canadensis*. It includes citizen science data (https://plantshoe.frec.vt.edu/) (Fig. 3).

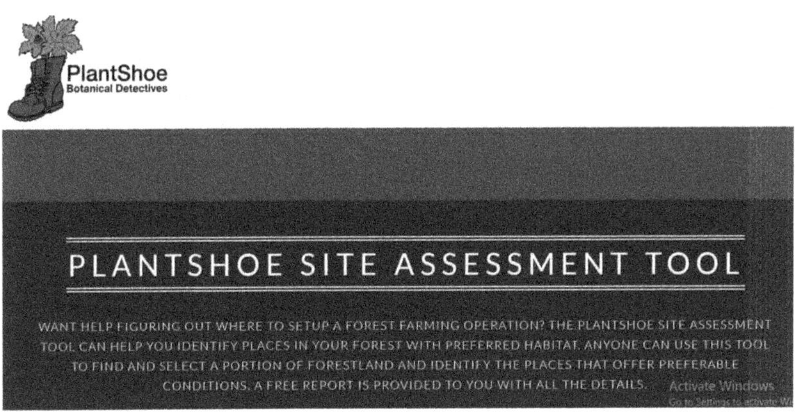

Figure 3 PlantShoe site assessment tool for forest farmed woodland botanicals (Source: Virginia Tech University, 2024).

4 Wild stewardship and wild harvesting

Traditional ecological knowledge forms the basis of many forest farming practices. The ethnobotany of stewarded and cultivated forest botanicals is exemplified the world over. Wild crafting, also known as wild harvesting, gathering, or collecting, is often practiced with forest farming. Mudge (2014) suggests 'assisted wildcrafting' as an inclusive term for the practices. Wild simulated is another term used when blending understory management with natural plant population sizes and practices under the forest canopy (Davis, 2014).

Wild harvesting is responsible for the majority of plant material supplied for raw materials, international markets, and forest farming planting stock, and poses risks to wild populations of forest botanicals (Predny, 2005; Davis, 2014; Chamberlain, 2018). Because many woodland botanicals of conservation status are harvested for their roots (Chamberlain, 2018), the removal of whole plants contributes to the challenge of sustainable harvesting. The quantity and species of wild-harvested NTFPs are reported in a variety of manners, from county and regional-based information to international trade data. The U.S. National Forest Service has completed population dynamic studies that inform permitting of select NTFPs. Permitting program information from state and national forests provides additional data regarding the volume of NTFPs legally removed from these forests. Poaching of NTFPs from state and federal lands, theft from private lands, and illegal and unmonitored trade skew the availability of accurate information.

It is estimated that half of the species of plants traded that are native to North America are from the Southern Appalachians (Greenfield, 2003). American ginseng, *Panax quinquefolius*, is a wild-harvested species and industry of uniqueness and is considered separately from other woodland botanicals and NTFPs. Other herbs, roots, bark, and fungi were traditionally for personal use and small-scale local collections contributed to the trade numbers, today, harvests have expanded to provide for commercial markets (Greenfield, 2003).

For North America, the American Herbal Product Association (AHPA) has produced tonnage reports for select plants since 1999. The latest tonnage survey results represent data from 2011 to 2017 and include quantitative figures for fresh and dried volumes of 38 plant species reported from those that handle raw materials including producers, distributors, and suppliers. The AHPA (2021) wild harvested tonnage reports show *Actaea racemosa,* black cohosh, root harvested in 2011 at 473 731 dry pounds, and 134 302 dry pounds in 2017 with a range of annually harvested recorded of 130 294 - 473 731, and an average of 219 070 dry pounds per year.

A "Collection to Commerce" report outlines select NTFPs and their markets, specific to an area of southern Appalachia U.S. (Greenfield, 2003). It is based on information reported by ginseng (and other root) dealers. Dealer is a term used to designate licensed ginseng aggregators. American (North Carolina Department of Agriculture and C and S, N.D.). At the time of the study in the early 2000s, Greenfield and Davis acknowledged that there was little information on the quantity or economic input of NTFPs being harvested in western North Carolina forests (Greenfield, 2003).

Sustainable wild harvesting, with land permissions through leases or written consent, provides opportunities for private landowners participation in the NTFP supply chain. Where the practices and definitions of stewarding wild populations, wild harvesting, and forest farming overlap is to be determined. Conservation through stewardship and cultivation continues to maintain similar goals for woodland species. Stewarding wild populations in the forest farming context focuses on expanding existing populations and increasing productivity or population health. Monitoring populations and timing successive harvests for plant removal are important. Poaching from public lands and theft of managed populations of woodland botanicals continue to be an issue facing conservation and forest farming goals.

Dr. Jeanine Davis of NC State University states that "there will always be a place for wild-harvesting", and for some species and places that can be sustainable and a source of tradition (2014). Ann Armbrecht notes in the book, *The Business of Botanicals* (2020), that only a small amount of the medicinal plant species traded are cultivated, and that U.S. farmers are beginning to grow more medicinal herbs.

5 Cultivation of woodland botanicals

Healthy forest environments provide the natural needs of forest farmed crops, including shade, water, nutrient cycling, soil development, and a myriad of other services. Information on the cultivation of woodland botanicals is formed from thousands of years and many ways of sharing knowledge of the people who have lived in relationship with forest understory crops.

While modern cultivation information has been available for more than a century for a few nontimber forest products, such as *Hydrastis canadensis*, goldenseal, they are not fast nor popular crops to grow (Predny, 2005). production best practices are often gleaned from wild populations, including site descriptions, plant physiological timing, harvest information, and import and ex Data for port numbers. A view of the quantity of cultivated material traded of wild-harvested forest botanicals is referenced in the American Herbal Product Association's annual tonnage reports; AHPA's tonnage reports derived from stakeholder input and show fluctuating volumes demanded for select woodland botanical species (AHPA, 2021). It was estimated that in 1999, there were 200–300 acres of cultivated goldenseal in North America (Davis, 1999). In the case of American ginseng, quantitative information is better represented by the U.S. Fish and Wildlife Service ginseng export and dealer reports.

Research-based information from the USDA and other technical service providers are available. John Lloyd published "The Cultivation of Hydrastis" in 1912 by the American Pharmaceutical Association (Lloyd, 1912). A USDA goldenseal cultivation bulletin from 1914 mentions the plant's close association with American ginseng in the wild, and describes the commercial history and cultivation of goldenseal (VanFleet, 2014). Shade-grown industries produce the majority of products for top forest farmed botanicals such as American ginseng, but are produced much differently and serve a different market than wild roots. Ramp cultivation best practices researched in the southeast USA (Greenfield, 2001), and in Quebec (Bernatchez, 2013) detail planting and establishment through harvest information for forest farmed ramp cultivation.

Site selection and companion species education are helpful for forest farming. For example, Bob Beyfuss' ginseng site selection tool is available from Cornell Extension (Beyfuss, 2017) along with plant and tree selection tools such as the book *Identifying Ginseng Habitat in the Mid-Atlantic* by Justin Wexler (2019). Additional production information for forest farmed materials can be referenced in Kenneth Mudge's chapter Agroforestry practices: Forest Farming (2019) by this publisher. Other leading cultivation and marketing of forest-farmed botanical resources for the eastern U.S. are found in the more information and references sections of this chapter. Additional resources are educational bulletins from the USDA and Agricultural Extension Services in states and counties; and the books *Growing and Marketing Ginseng, Goldenseal, and other Woodland Medicinals* (Davis, 2014), and the *Forest Farmers Handbook* (Rural Action, 2023), which provide species specific information on the establishment, maintenance, and marketing of forest farmed woodland botanicals. Growing information for forest farmed species is also available from nurseries and plant catalogs, and books. Richo Cech's 2002 text notes that black cohosh, *Actaea racemosa*, can be cultivated in many areas of the U.S. beyond its eastern distribution, and if provided basic requirements, it produces well (Cech, 2002).

Cultivated and certified organic supply of NTFPs is reported in the tonnage surveys from the American Herbal Product Association (2021). For example, goldenseal has both cultivated and wild-harvested supply reported. Other cultivated NTFPs include managed forests for syrup and sap production, and outdoor in situ mushroom cultivation, one book is Organic Mushroom Farming and Mycoremediation: Simple to Advanced and Experimental Techniques for

Indoor and Outdoor cultivation (Cotter, 2014). Food crops including greens, berries, and nuts are common enterprises and companion plants to woodland botanicals in forest farming settings. Ken Mudge and Steve Gabriel's 2024 book *Farming the* Woods details forest farming of fruits, nuts, and mushrooms. Artisan forest farmed materials include bark, hulls, and dye plants such as bloodroot, *Sanguinaria Canadensis*, and yellowroot, *Xanthorhiza simpliscissma*, as well as basket willows, *Salix sp.* Other forest farmed products are mosses and pine straw.

6 Establishment and maintenance of forest farmed crops

Forest farming plots require attention during establishment, especially when new species or outsourced planting material is brought in. It is important to assist new seedlings and transplants to adjust to the new location, and provide support to plants in the first few seasons. Forest farmers should Monitor, weed, and irrigate if needed, as well as consider adjusting conditions such as light levels and other growth factors for optimal production. One can ensure plant survival and establishment are successful by maintaining appropriate airflow, moisture, and shade for the crop.

Tree crops may be mature or they may take time to establish or manage for the production of nuts, fruits, or syrups. Tree and shrub species are established once maturity is reached and populations become reproductive. For nut crops and tree syrup species, this can be 5–10 years from transplanting.

Protection of seedlings and young plants is best done through prevention, physical, and sometimes biological or chemical measures. The maintenance phase of forest farmed plots refers to the time period between established, often two or more-year-old plantings, and when the species is closer to harvest time. Crop populations are considered surviving and hopefully thriving at this stage. Pests, pathogens, and surrounding vegetation and landscape events provide risks to forest farmed botanicals (Fig. 4); plants and fungi must be protected from human and wildlife risks as well (USDA NRCS, 2022).

Appalachia, U.S. has witnessed a bottleneck in the growth of cultivated forest farming botanicals for the medicinal herb industry. With many plants in the ground and populations stewarded, and with encouragement for continued forest-farmed cultivation, the pioneers of modern forest farming techniques experience issues. Risks include seasonal and irregular weather patterns, such as heavy rains and droughts, or changes in neighboring property land uses. Herbivore, mammal, and rodent browse can be exploratory or detrimental to forest-farmed crops. Access to forest-farmed plots is reliant on roads and trails that must be maintained. Consider the distance, slope, and accessibility of sites in forest farm management. Visiting forest farming sites and checking

Figure 4 Pest and disease-affected forest farmed American ginseng plant, *Panax quinquefolius*. (Source: Margaret Bloomquist. 2024).

the status of populations routinely helps mitigate the risks (Davis, 2014; USDA NRCS, 2022). Opening up areas of the forest presents new opportunities for natural and introduced species to thrive and more acreage for the forest farmer to manage.

Incorporating new populations or new species into the forest farm creates opportunities as well as new risks for the plant and ecosystem. Activity in and movement of inputs and tools around the forest farm can introduce or spread pathogens. Preventive management is the best approach to abating risks of pests and diseases on the forest farm (Davis, 2014). Plant tops, leaves, berries, seeds, and bark are eaten by wildlife. Mole, vole, groundhog, raccoon, skunk, turkey, deer, wild hog, and bear are examples of animals that may affect forest-farmed operations. Slugs, other insects, and nematodes also pose risks to forest-farmed crops. At maturity, forest-farmed plantings face additional risks. Mature plants may be vulnerable due to other issues or may be more attractive to insects and herbivores herbivores, or humans.

Pests and diseases affect forest-farmed crops. Wild botanical populations are affected by natural pathogens and various causes for dieback, leaf blights, and other symptoms. Assessments of risk thresholds and management decisions are informed by a small body of knowledge that is specific to woodland botanical species. Forest farming systems that simulate wild population densities and site selection may exhibit similar levels of pathogen pressure to wild populations. Awareness of pathogens affecting forest farming is rising. A list of pathogens affecting woodland botanical species is helpful for identification and management for growers and technical service providers and is found in Appendix II of the book *Growing and Marketing Ginseng, Goldenseal, and other Woodland Medicinals* (Davis, 2014).

Established production and marketing of maple syrup and best production practices for sugaring are available. A specific geographic range is the ideal climate in the North and Northeast U.S. A compendium of information is available from the Tree Sugaring Resource Hub by the U of MO Center for Agroforestry (University of Missouri, 2025). Other tree species are also tapped for saps and syrups including walnut and birch. Opportunities are limited to certain species such as nuts, pawpaws, syrups, and berries (Mudge, 2014).

Current research looks to support forest farming practices and provide tools for management, such as with biological controls and companion botanicals. The interplanting and use of compost teas with goldenseal, for the purposes of fungal pathogen mitigation in American ginseng, was explored. Gao et al. suggest the use of goldenseal extracts as a control product for fungal pathogens to control disease of American ginseng, and have developed extraction methods with performance results (Gao, 2024).

Invasive species management is important for most regions and requires labor and financial resources. Common invasive plants of forest farmed areas in the eastern U.S. include *Ligustrum sinense*, privet, *Rosa multiflora*, multiflora rose, *Celastrus orbiculatus*, bittersweet, *Berberis thunbergii*, barberry, *Microstegium vimineum*, stilt grass, and *Pueraria montana*, kudzu. The U.S. Forest Service provides resources and information on the removal of and effects of invasive species like privet (Hanula, 2009 via USDA).

7 Harvesting forest farmed crops

In temperate areas forest-farmed crops are harvested throughout the year and include products from all seasons. Saps and syrups in the winter, along with dormant bark or wood harvests. Spring brings life to the forest farm with mushrooms, spring ephemerals, fiddleheads, and greens to the table and markets. Seeds, trees, flowers, aerial parts, moss, fungi, berries, leaves, and fruits are managed as raw materials or as a companion to forest-farmed crop production.

The scale and accessibility of a harvest site denote the tools and labor required. Common tools for root digging are pick axes, screw drivers, tree spades, and soil knives for hand harvesting practices. Digging forks, spades, or other tools may be used for larger populations, woods-grown densities, and for select woodland species harvests. Loppers, pruners, and saws are also tools of the trade. It is important to protect the integrity of forest-farmed crops through harvest and post-harvest handling. resources may be explored regarding Good Agricultural and Collection Practices for medicinal herbs (Davis, 2016; American Herbal Products Association, 2023), or food safety information for edible products. The World Health Organization Good Agricultural and Collection Practices guidelines are an additional resource (World Health Organization, 2003).

The European species, *Allium ursinum*, ramsons, are an important wild spring green with traditional medicinal and edible properties. Ramsons are a plant of the well-drained, moist woodland soil that blankets the forest floor in ideal conditions (Sobolewska, 2015); see also the Plants for a Future website (2010-2025). *Allium tricoccum,* ramps, are a North American species and an important non-timber forest product growing demand in recent years with expanded markets, restaurants, and ramp festivals. Ramps have traditionally been and are most commonly wild harvested, and there is 'almost no cultivation happening' (Mudge, 2014). Cultivation and conservation campaigns for the species have emerged as the demand for ramps in high-end dining and local food movements expands. Research from Nilson, Burkhart, et al, of Pennsylvania State University state that harvest timing differences across the season result in different weights of ramps. The yield weight was greatest at 30 days post-emergence, and the ideal harvest time was between 30 and 45 days. These findings promote harvest timing techniques to contribute to the sustainability of wild populations by waiting later in the season to harvest. Opportunities for forest farming include optimal harvest time for greatest weight per lot, and reducing labor at harvest for similar weight of product (Nilson, 2023).

Horticultural and agronomic data from forest farmed systems continue to develop. Work in North Carolina on calcium and pH studies in the 1990s and early 2000s in a shaded system showed specific preferences for optimal root growth of goldenseal (Davis, 2020). A 2013 study from Quebec suggests fertilized ramps exhibited better growth from planted bulbs in establishment stages, including larger bulb size (Bernatchez, 2013). Appendix 5 of Davis and Person's 2014 book is *Jo-Ann McCoy's Disease List* that provides a window in to the occurrence of pests and diseases to consider in the management of forest-farmed botanicals (Davis, 2014).

Root crops such as goldenseal are often harvested in the fall (Predny, 2005), following frost in areas that experience low temperatures and seasonality. Harvest timing is important for population sustainability (Nilson, 2023). To allow for seed dispersal, harvesting goldenseal roots after the berries ripen is recommended (Predny, 2005). Some buyers will offer a premium price for high alkaloid levels. Burkhart et al. analyzed the effects of drying temperature on the alkaloids of goldenseal; drying temperature was found to affect the levels of canadine, but did not alter the levels of hydrastine and berberine in the study (Burkhart, 2024). These studies suggest that forest farming is a viable production system for woodland botanicals.

Accurate harvest yield data is not available for forest-farmed woodland botanicals (Predny, 2005). Enterprise budgets offer a foundation for modeling the economic potential of forest farming enterprises including from Burkhart and Jacobsen (2008), and Davis and Persons (2014). Research is underway for forest farming of these perennial crops. The Forest Farming Calculator tool by the U.S. Forest Service, developed with Appalachian Sustainable Development and support from USDA NRCS, incorporates enterprise budgets and published information to date (USDA Forest Service, 2023).

Mushroom production is a viable short-maturation crop to have on a forest farm. Outdoor mushroom cultivation via wood chip inoculum, cut logs, totems, and other substrates is a part of Agroforestry and on farm research. Harvest protocol and education especially with identification is important and may be required to sell wild collected mushrooms to restaurants in North Carolina (NC), U.S. (NC Department of Health and Human Services, 2017).

Harvesting perishable food crops is subject to guidelines for different markets. Forest farmed crops may be subject to the U.S. Food Safety and Modernization Act (FSMA), Good Manufacturing Practices (GMPs), the U.S. Food and Drug Administration (FDA), or other best practices and regulations. See section 16.10 for more information.

Post-harvest handling of forest farmed botanicals is important. Proper handling, washing for some species, and ensuring there are no contaminants by removing debris and adulterants are necessary. Drying, cutting, further post-harvest processing, and proper storage are required of most producers and collectors of NTFPs (World Health Organization, 2003; Davis, 2014, 2016). A small market for fresh raw material sales exists, often due to consolidators and buyers preferring to manage the post-harvest quality or to processors of fresh materials.

8 Effects of forest farming on ecosystems

Agroforestry systems including forest farming are designed to be multifaceted, garner a variety of ecosystem benefits, and are multifunctional. Forest farming practices continue to evolve and provide social, economic, and environmental services to communities and the environment. Ecosystem services are provided under the managed canopy in regard to preserving water cycles and minimizing stress to the understory (Mudge, 2019). Forest-farmed products provide ecological services as well as livelihood, food security, spiritual and ceremonial, and cultural ecosystem services (Chamberlain, 2018). Other benefits of agroforestry systems include climate mitigation potential and increased biodiversity (Schaffer, 2024). Forest farming provides cultural ecosystem services (CES), which are non-material services to humans such as forest bathing, supporting health, cultural, and personal identity (Forest Research, 2025).

Forest farming a diversity of crops and active management of the understory in natural and managed forests can be a positive tool for resilience to climatic changes and weather extremes, which pose risks to livelihood and the cultural availability of forest farmed products (Chamberlain, 2018; Mudge, 2014). Incorporating multiple agroforestry practices with forest farming is common. Forest botanicals such as ramps (ramsons) are indicator species of healthy forests (McLachlan, 2001; Chamberlain, 2018). Shrub and tree species provide wildlife benefits like food and shelter in non-timber forest production sites (Chamberlain, 2018; Mudge, 2014).

Forest farming practices can contribute to increased diversity of tree and crop species and support their interaction soil health and habitat improvement are other purposes of the conservation practice (USDA NRCS, 2022). Ecosystem services, as defined by the NRCS that are supported by forest farming management systems, also include wildlife habitat (UMCA, 2011; Mudge, 2014; USDA NRCS, 2022). Impacts on forest habitat for other species are in need of research for temperate forest farming systems. Forest farming provides an opportunity for engaged forest management on private lands.

Forest farming systems are designed to preserve or improve ecosystem functions (Chamberlain, 2009). In areas of the world where larger-scale forest farming is growing, such as the Eastern U.S., the impact of practices on forest habitat and NTFP integrity is unknown

9 Developing markets and supply chains

Short-term production of forest-farmed products can grow below the canopy while longer-term crops and timber mature (UMCA, 2011). Forest farming plots are often small-scale and provide supplemental income and value to producers (Davis, 2014). Successful forest farming operations must manage forests and production, and may practice vertical business integration and value-added production. There are a growing number of large-scale woods grown and wild simulated forest farming operations.

The size of the NTFP market is challenging to assess and compare year to year. International and national efforts provide insight for certain species. Many market size estimations are based on extrapolation from formal programs and are known to underreport actual numbers (Frey, 2019). Commercially gathered forest botanicals have been exported to international markets from the Eastern U.S. for at least 300 years (Rural Action, 2023). Chamberlain et al (2018) report that the 2013 U.S. market for wholesale wild-harvested NTFPs was $960 million

U.S., and the overall market size of NTFPs was close to $1 billion U.S. in 2013–2014. International export markets of NTFPs were valued at $4.5 billion U.S. in the period 1999–2013. NTFP imports to the U.S. during the same period were valued at $6.5 billion U.S. (Chamberlain, 2018; Frey, 2019).

The supply and demand of the medicinal herb and wild harvesting industries are variable and fluctuate annually and along with broader economic ups and downs. Volatility and unpredictable markets of supply and demand affect NTFPs (Frey, 2019; Chamberlain, 2018). The latest health trend, the closure of a dealer or aggregator in an area that previously purchased from rural wild harvesters and cultivators, or a natural disaster, can alter these markets very quickly (Davis, 2014). Long-term crops require knowledgeable management and stewardship to anticipate and work with market swings.

Predny and Chamberlain described the historical price range for goldenseal in: Goldenseal, an Annotated Bibliography (2005). Prices in the 1880s were about $0.10 U.S. per dry pound ($0.05 US per Kilo), and $0.58 US per dry pound ($0.26 US per Kilo) in 1890 (Lloyd, 1912; Predny, 2005). The recorded consumption of goldenseal in the late 1800s was 150 000 dry pounds. The export to Europe in 1883 was 15 000 pounds of goldenseal roots. Exported botanicals such as goldenseal are now shipped to Europe, then processed and returned to the U.S. as finished products (Predny, 2005). Fast forward to the 1990s when prices dipped to $10.00 U.S. per pound, and at the turn of the 21st century, prices paid for goldenseal were between $30 and $50 U.S. per pound (Davis, 1999).

The highest volume traded herb commodities from wild and cultivated sources in North America include *Ulmus rubra*, slippery elm, bark, *Prunus serotina*, black cohosh root, black cherry, bark, and goldenseal root. The tons of dry materials per year are illustrated in the AHPA's figures, such as: *Six High-Volume Commodities*, with available data from 1999 to 2017 (2021). The report notes that declines in harvested material in single years usually mark held material in the supply chain and are not necessarily indicative of the market. Additionally, some trends in the North American tonnage reporting were that 2011 yielded the highest reported tonnage in all species over the time period, and 2014 was a low volume year, with the following years increasing in the tonnage reported (Fig. 5). (AHPA, 2021).

Chamberlain (2018) describes five dynamic market segments for commercial non-timber forest products: culinary, medicinal and dietary supplement, decorative, landscape and nursery stock, and fine arts and craft materials. The non-commodity supply chains historically have been unique, informal, and changeable (Chamberlain, 2018). The medicinal herb industry continues to grow as it has over the past decades. The local and regional food market growth and consumer demand for sustainable and natural products are apparent.

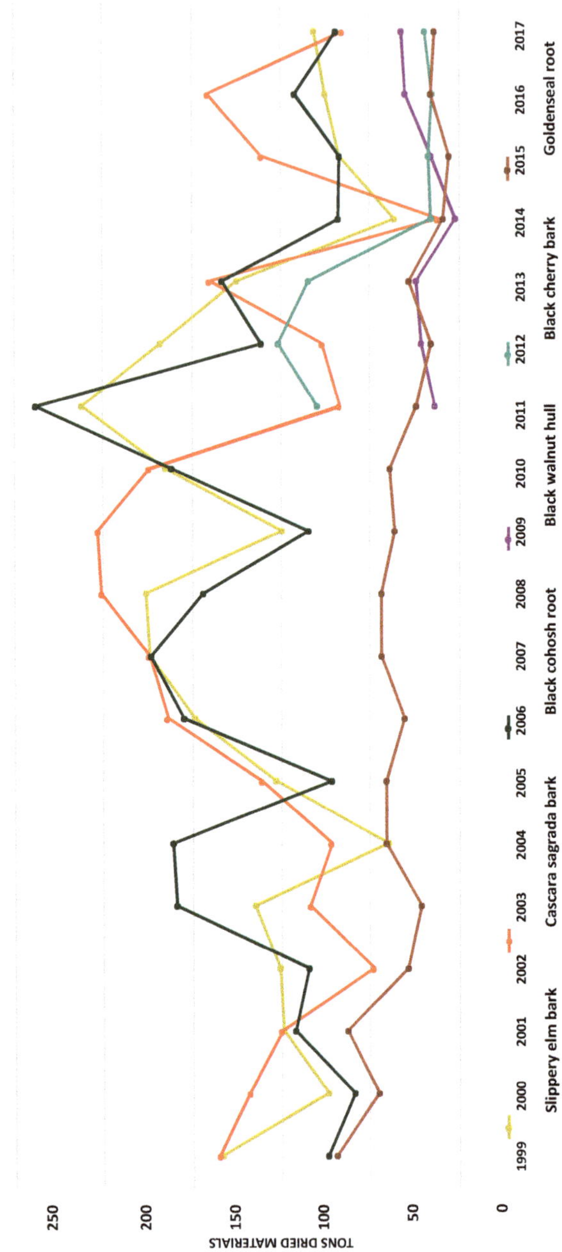

Figure 5 Six high-volume commodities, using available data from 1999–2017 (Source: American Herbal Products Association, 2021).

Greenfield and Davis describe the NTFP industry as complex and 'closed', or based on private networks (Greenfield, 2003). There are many players in the NTFP trade and supply chain; however, little is known about the unique organization of the formal and informal markets (Frey, 2019). Frey et al. reviewed economic and market information and highlighted the lack of information and awareness that contributes to challenging management of resources and economic potential of the diverse markets (2019).

An example of a window into the woodland botanical supply chain of in the 2000's in southern Appalachia is summarized by Greenfield and Dressler (2012) for *Sanguinaria Canadensis*, bloodroot. Bloodroot is a plant used for medicine and is an artisan resource used for dyeing other NTFPs such as white oak bark and honeysuckle vine. The summary describes the demand, price, and volume from a unique report with data from 15 NTFP dealers (Greenfield, 2003). The report also illustrates the distribution channels of select NTFPs. See the black cohosh distribution chart as an example (Fig. 6). A forest farming example is the Eastern Band of Cherokee Indians cultivating, bloodroot to ensure supply for artisans dye.

Cultivated NTFPs are often sold into the wild-harvested industry utilizing existing market channels. Tracking the trade and markets of NTFPs may be done through evaluating existing information through public records, media and literature, interviews, and other methodologies (Greenfield, 2003).

A need for transition from wild to cultivated sources is evidenced by regulations and demand for cultivated material of select species, including goldenseal and ramps. It was noted in the early 2000s that wild ramp populations were being damaged by intensive harvesting, thus encouraging the cultivation (Davis, 2001). Following the restricted harvest of ramps in Quebec in 1995, commercial production was thought to be a viable alternative to wild harvesting and contribute to the conservation of the wild populations (Bernatchez, 2013).

A focus on cultivated woodland botanicals has been prominent, yet, "Transitioning from wild collection to forest cultivation of indigenous medicinal forest plants in eastern North America is constrained by lack of profitability"; this is the title of Burkhart and Jacobson's 2009 publication and captures the stage of development experienced by temperate forest farming stakeholders. The Pennsylvania researchers applied cost analysis and economic modeling to forest farmed woodland botanical species and found that only one or two high-value herbs were profitable with the prices paid in 1990–2005. A verification system and marketing label were suggested as an incentive forward to increase the price point paid to the forest farmer (Burkhart, 2009).

American ginseng has retained an increase in price in the decades prior to 2014 despite yearly fluctuations (Davis, 2014; Frey, 2019). Goldenseal

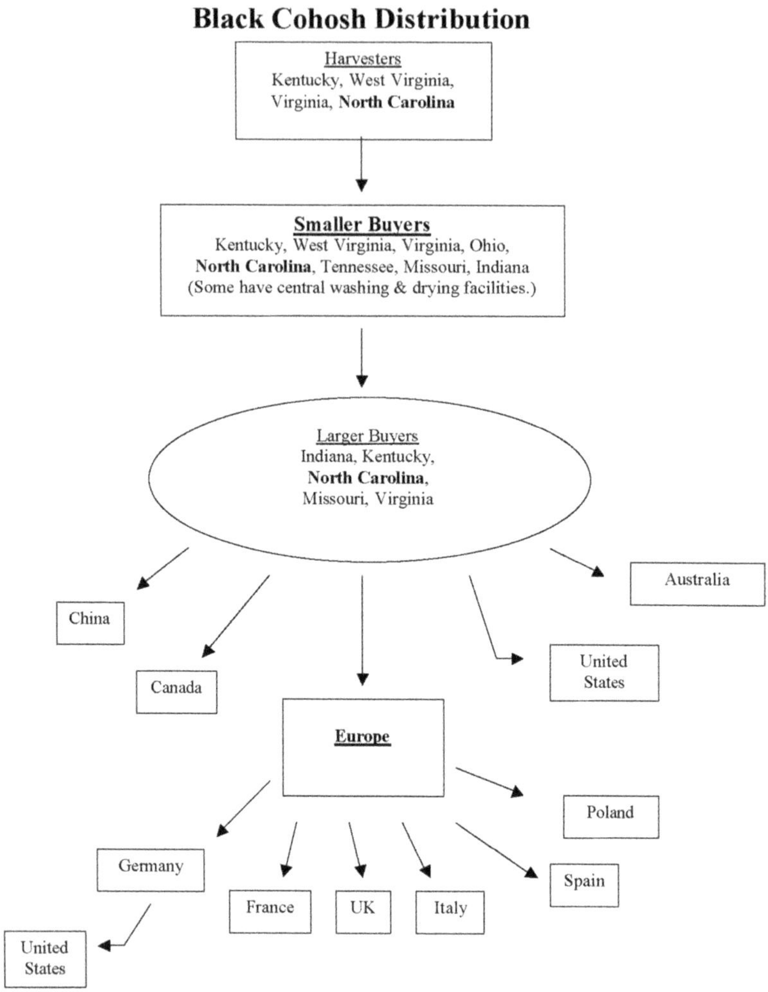

Figure 6 Black cohosh distribution chart (Source: Greenfield and Davis. 2003).

goldenseal, has experienced relative market stability with the inclusion of a cultivated supply, yet despite cultivation efforts the majority of tonnage comes from wild harvesting (AHPA, 2021). The cultivated market for slippery elm is interpreted through aggregated harvest data as well: 45,000 dry pounds of cultivated slippery elm bark were reported in 2013, and reported amounts in the following four years ranged from 1,850 to 12,700 dry cultivated pounds annually. To compare, 424,854 dry pounds of wild harvested slippery elm bark was traded in 2011, with a range of 66,892–335,455 dry pounds between 2011 and 2017 and an average of 207,729 dry pounds annually (AHPA, 2021).

Developing NTFP markets include eco-types with regional and local recognition, branding regarding sustainability and source of origin information, and widening domestic markets. For example, the aforementioned demand for cultivated goldenseal roots increased with wild populations becoming scarce and fueled propagation (Predny, 2005), and cultivation research that provides consistent quality material. A higher price is paid for cultivated goldenseal than wild-harvested (Greenfield, 2003), and premium prices may be available for certified organic and other verifications. Domestic marketing of forest-farmed products is prominent especially for lesser-known and community herbs. Sales to local herbal practitioners and restaurants are viable options (Rural Action, 2023; Davis, 2014). The potential for fresh raw material sales and lower transportation and trade costs are potential benefits of local and domestic marketing.

Certifications, verifications, and branding provide marketing opportunities for forest-farmed botanicals and consumers. Consumer awareness of the transition away from a solely wild collected supply of some products is reflected in anticipated price premiums for cultivated forest botanicals (Rural Action, 2023). The Forest Grown Verification system is a program of the United Plant Savers that certifies the source, practices, and handling of Forest Grown™ products. A label with growing recognition and price premium opportunities, the program lists botanicals currently verifiable (United Plant Savers, 2024). A regional Point of Harvest certification system for wild-harvested material (Appalachian Forest Farmer Coalition, 2022), and the international Fair Wild certification (Fairwild, 2023) are examples of labeling efforts that are bringing the industry and consumers together to support a sustainable woodland botanicals trade (Fig. 7). Third-party verifications, eco-labelling certifications, and recognition of production method are monitored through tonnage surveys by AHPA for (AHPA, 2021).

There is a need for research and education for developing forest farming business support systems (Frey, 2019). Challenges within the forest farmed supply chain present opportunities for new businesses and expanding operations of existing businesses. For example, the lack of available planting stock is a growing demand for forest farming and agroforestry projects across the U.S. There is demand for sustainable planting stock operations (Mudge, 2014; Chamberlain, 2018). Another example is the growing support for a domestic market for American ginseng and other woodland botanicals in the U.S.

A market for fresh raw material is present at low volumes and showing growth in some sectors. The demand for fresh cultivated forest botanicals is interpreted from AHPA's harvest reports for fresh plants; black cohosh root, cultivated, fresh tonnage reports 74 pounds in 2012, 195 fresh pounds in 2014, and 40 fresh cultivated pounds in 2017 (AHPA, 2021). Aggregators may

Figure 7 Forest Grown and Certified Wild labels (Sources: United Plant Savers and Fair Wild. 2025).

provide quality control for post-harvest handling, including washing, drying and consolidated distribution, as well as access to different markets.

Other NTFPs have large and more developed markets, namely pine straw, tree nuts, floral materials, and tree syrups. For example, in the U.S., the maple syrup is tracked nationally and has had a steady growth rate over the past century (Frey, 2019). Firewood collection, charcoal, selecting and planting white oak for basket making and yields, and musical instruments, are other forest farming and stewarding enterprises.

10 Rules and regulations affecting forest farming

Regulations and frameworks affecting forest farming are housed by multiple local, national, and international agencies and entail rules regarding natural resource management, agricultural considerations, land use, and trade (Chamberlain, 2018). Rules and regulations regarding the harvest of native forest botanicals often stem from conservation goals. For example, the establishment of legal harvest seasons for American ginseng in the U.S.; And the Quebec provincial government banning the commercial harvest of ramps in 1995 (Bernatchez, 2013). Rules and regulations also apply to forest-farmed products that enter the food, medicine, nursery, and cosmetic industries.

Forest-farmed or stewarded material destined for the food supply chain may also be subject to rules and regulations. The U.S. Food Safety Modernization act (FSMA) regulates at the federal level. The dietary supplement, cosmetics, and other U.S. Acts guide the industry. In Canada, the Natural Health Products Regulations, the Cosmetic Program of Health Canada, and the Canadian Food Inspection Agency house rules and regulations for NTFP products (Gray, 2011).

Edible forest farmed products entering the retail market and direct-to-consumer sales may be subject to different guidance. On the U.S. state level, cottage laws or other rules apply to small-quantity sales of value-added products such as maple syrup, tea, or soap. Good Manufacturing Practices (GMPs) apply to processed products in the U.S. Nursery sales may be subject to state law jurisdiction or that of cross-border movement of live plant materials domestically, or internationally.

International initiatives that affect the NTFP trade include the Convention on International Trade in Endangered Species (CITES), the International Tropical Timber Agreement, the Montreal Process, the Convention on Biological Diversity, the International Union for the Conservation of Nature, the Ramsar Convention on Wetlands, and the Food and Agriculture (FAO) Organization of the United Nations (Chamberlain, 2018).

In the U.S., three federal laws shape the regulatory framework affecting NTFPs: The Lacey Act, The National Environmental Policy Act, and the Endangered Species Act. The U.S. Forest Service manages NTFPs on public lands, as well as the Bureau of Land Management, the National Park Service, the U.S. Fish and Wildlife Service, and the Department of Defense. U.S. law designates access to NTFPs for indigenous populations. State and local laws and additional regulations at the town or district level help define NTFP management as well. (Chamberlain, 2018).

A U.S. Secretary of Agriculture program was implemented in 2000 to collect fees for the harvest and sale of forest botanical products called the Forest Botanical Products Pilot Program of 2000 (Chamberlain, 2018). County and state-level regulations regarding forest-farmed botanicals include listings on plant conservation lists, requiring a permit to purchase and sell wild material, or restricted access. Regulations developed and enforced for wild-harvested plant material are also in place for cultivated material of the same species.

There are legal harvest seasons for the collection of American ginseng in 19 U.S states that begin in September in an effort to allow mature plants to set seed prior to harvest (United States Fish & Wildlife Service, 2015). International regulations are in place for some forest farmed species as listed in the Appendixes of the CITES, for example American ginseng and goldenseal are listed in Appendix II, necessitating restricted and documented trade for the threatened species (2024). Collection permits, trade licenses, and other registrations are required in many areas for individual species or groups of NTFP plant materials.

In the state of North Carolina (NC) U.S. goldenseal was put back on the NC Plant Conservation Board's list of protected plants in 2020. A permitting program is in place for those handling goldenseal, and designating the proof of origin is an effort to track the trade of the plant in its southern range. "Goldenseal is a North Carolina Protected Plant Species. The collection of protected plants from land you do not own is prohibited. Sale or distribution of wild-harvested plants is prohibited. A Certificate of Origin issued by the NC Department of Agriculture and Consumer Services is required for the sale of propagated Goldenseal." (North Carolina Department of Agriculture & Consumer Services Plant Conservation Program, 2021).

Professional organizations also participate in setting rules and best practices in the industry. The American Herbal Products Association tracks the trade of medicinal plants, many of which are native woodland botanicals harvested for their roots (AHPA, 2021; Chamberlain, 2018). Organizations such as the United Plant Savers (UPS) are internationally recognized, and the Species at Risk list and other services provided influence the trade and consumer connection to the industries associated with forest farming (United Plant Savers, 2022). In the somewhat unregulated medicinal plant trade, the UPS 'at risk list' provides awareness of needs that aren't on the CITES lists. For example UPS added *Triullium sp.*, Trillium, to the "critical list" and suggests alternatives for the plant to the industry (Leopold, 2022). Slippery elm is also a forest-stewarded plant and is on the 'at risk list' of the United Plant Savers (Cech, 2002).

11 Where to look for further information

- Agroforestry Coalition – https://agroforestrycoalition.org/
- American Forest Farming Association (AFFA) - https://www.appalachianforestfarmers.org/affa

- Appalachian Forest Farmer Coalition – www.Appalachianforestfarmers.org
- Association for Temperate Agroforestry (AFTA) - https://www.aftaweb.org/
- Catalyzing Agroforestry- https://www.catalyzingagroforestry.org/
- Consultative Group on International Agricultural Research (CGIAR) - https://www.cgiar.org/research/center/world-agroforestry-centre/
- Northeast Forest Farming Coalition (NFFC) – https://www.northeastforestfarmers.org/
- The Center for International Forestry Research and World Agroforestry (CIFOR-ICRAF) - https://www.cifor-icraf.org/
- United States Department of Agriculture (USDA). National Agroforestry Center (NAC) – https://www.fs.usda.gov/nac/
- University of Missouri Center for Agroforestry – https://centerforagroforestry.org/

12 Acknowledgements

Dedicated to the forest farming mentors who have helped shape and support the way we experience forests and the plants within them. The late Bob Beyfuss, Joe Hollis, Chuck Marsh, Michael McGuffin, Chester Crane, and Johnny Crisson. To my current guides in this work and exploration Jeanie Davis and John Munsell, among others.

Most of all, the chapter is dedicated to my mother, Cindy Gove Bloomquist.

13 References

American Herbal Products Association. 2021. *Tonnage Surveys of Select North American Wild- Harvested Plants, 2011–2017*. Silver Spring, MD: American Herbal Products Association. https://www.ahpa.org/Files/Document%20Library/Tonnage%20Surveys/AHPA%202011-2017%20Tonnage%20Survey.pdf

American Herbal Products Association. 2023. Good Agricultural and Collection Practices and Good Manufacturing Practices for Botanical Materials. May 2023 (Revised). https://www.ahpa.org//Files/GACP-GMP/GACP-GMP-Sections/2024/AHPA_GACP_GMP_Full%20Guidance_May%202023.pdf

Appalachian Forest Farmer Coalition. 2022. Point of Harvest (PoH) Program. https://www.appalachianforestfarmers.org/point-of-harvest

Armbrecht, A. 2020. *The Business of Botanicals: Exploring the Healing Promise of Plant Medicines in a Global Industry*. White River Junction, VT: Chelsea Green Publishing.

Association for Temperate Agroforestry. 2024. North American Agroforestry Conference. https://www.aftaweb.org/157-naac-2023/318-naac-registration-now-open.html

Bernatchez, A., et al. 2013. Testing fertilizer, gypsum, planting season and varieties of wild leek (Allium tricoccum) in forest farming system. *Agroforestry Systems*, 87(5), 10.

Beyfuss, B. 2017. The Practical Guide to Growing Ginseng. Cornell Cooperative Extension Agroforestry Resource Center. https://hdl.handle.net/1813/54771

Burkhart, E. P. and Jacobsen, M. G. 2008. Transitioning from wild collection to forest cultivation of indigenous medicinal plants in eastern North America. *Agroforestry Systems* 76: 437–453.

Cech, R. 2002. *Growing At-Risk Medicinal Herbs: Cultivation, Conservation, and Ecology*. Williams, OR: Horizon Herbs.

Chamberlain, J.L., et al. 2009. Forest Farming Practices. In H.E. Garrett (Ed.), *North American Agroforestry: An Integrated Science and Practice*. 2nd Edition. Dallas, TX: American Society of Agronomy.

Chamberlain, J.L., et al. 2018. *Assessment of Nontimber Forest Products in the United States under Changing Conditions*. General Technical Report SRS-232. Asheville, NC: USDA Forest Service, Southern Research Station. https://www.fs.usda.gov/treesearch/pubs/56484.

Convention on International Trade in Endangered Species. 2024. Convention on International Trade in Endangered Species of Wild Fauna and Flora Appendices I, II and III. Valid from 25 May 2024. https://cites.org/sites/default/files/eng/app/2024/E-Appendices-2024-05-25.pdf

Cook, M. 2013. Medicine Generations: Natural Native American Medicines Traditional to the Stockbridge-Munsee Band of Mohicans Indian Tribe. ISBN-13:978-1482779042.

Cotter, T. 2014. *Organic Mushroom Farming and Mycoremediation: Simple to Advanced and Experimental Techniques for Indoor and Outdoor Cultivation*. White River Junction, VT: Chelsea Green Publishing.

Davis, J.M. 1999. "Forest Production of Goldenseal". Agroforestry Notes (USDA-NAC). 15. https://digitalcommons.unl.edu/agroforestnotes/15

Davis, J.M. and Greenfield, J. 2001. *Cultivation of Ramps (Allium tricoccum and A. burdickii)*. Raleigh, NC: Department of Horticultural Science, North Carolina State University.

Davis, J.M. and Persons, W.S. 2014. *Growing and Marketing Ginseng, Goldenseal and Other Woodland Medicinals*. 2nd Edition. Gabriola Island: New Society Publishers.

Davis, J. 2016. *Suggested Good Agricultural and Collection Practices for North Carolina Medicinal Herbs*. AG-810 Bulletin.

Davis, J. and Dressler, A. 2012. Blood root (*Sanguinaria Canadensis* L.). Mountain Horticultural Crops Research and Extension Center – North Carolina State University, North Carolina, USA.

Davis, J. and McCoy, J.-A. 2020. *Commercial Goldenseal Cultivation Horticulture Information Leaflets*. NC Extension. https://content.ces.ncsu.edu/commercial-goldenseal-cultivation

Fairwild Foundation. 2023. Fairwild Standard. https://www.fairwild.org/

Favor, K. 2023. Urban Agroforestry Tipsheet: Food Forests and Beyond. Published January 2023. National Center for Appropriate Technology IP630, Slot 668, Version 092823.

Food and Agriculture Organization of the United Nations (FAO UN). 2025. Sustainable Forest Management. https://www.fao.org/forestry/sfm/our-work/en

Frey, G. E. 2019. Markets and market values of nontimber forest products in the United States: A review, synthesis, and identification of future research needs. *Journal of Forestry*, 2019, 613–631. https://doi.org/10.1093/jofore/fvz051

Gao, Y., et al. 2024. Goldenseal (Hydrastis canadensis L.) extracts inhibit the growth of fungal isolates associated with American ginseng (Panax quinquefolius L.). *Molecules*, 29(3), 556. https://doi.org/10.3390/molecules29030556

Garrett, H.E.G. Editor. 2009. *North American Agroforestry: An Integrated Science and Practice*. 2nd Edition. Madison, WI: American Society of Agronomy.

Gray, B. 2011. The Boreal Herbal: Wild food and Medicine Plants of the North. Aroma Borealis Press. Whitehorse, Yukon, Canada.

Greenfield, J. and Davis, J.M. 2003. *Collection to Commerce: Western North Carolina Non-Timber Forest Products and Their Markets. 2003 Report*. Raleigh, NC: N.C. State University, Department of Horticultural Science.

Hanula, J.L., et al. 2009 Chinese Privet (Ligustrum sinense) removal and its effect on native plant communities of Riparian Forests. *Invasive Plant Science and Management*, 2, 292–300.

Hemmelgarn, H.L. and Munsell, J.F. 2021. Exploring 'Beyond-Food' Opportunities for Biocultural Conservation in Urban Forest Gardens. *Urban Agriculture & Regional Food Systems*, 6, e220009. https://doi.org/10.1002/uar2.20009

Kays, H. 2024, June 15. Word from the smokies: sochan program grows relationship among tribe, park, plants. *The Citizen Times*. https://www.citizen-times.com/story/news/local/2024/06/15/smokies-sochan-program-grows-relationship-among-tribe-park-plants/74084253007/

Kellerman, T., et al. 2025. Agroforestry across the United States: results of the 2022 census of agriculture. *Agroforestry Systems*, 99, 34. https://doi.org/10.1007/s10457-024-01096-2

Krueger, ET. Al. 2020. Describing medicinal non-timber forest product trade in eastern deciduous forests of the United States. *Forests*, 11(4), 435. https://doi.org/10.3390/f11040435

Leopold, S. 2022. *Ending the Trillium Trade. Journal of Medicinal Plant Conservation*. Rutland, OH: United Plant Savers.

Lloyd, J.U. 1912. *The Cultivation of Hydrastis. The Journal of the American Pharmaceutical Association*. Columbus, OH: The Stoneman Press.

Mclachlan, S. and Bazely, D. 2001. Recovery patterns of understory herbs and their use as indicators of deciduous forest regeneration. *Conservation Biology*, 15, 98–110. https://doi.org/10.1111/j.1523-1739.2001.98145.x

Mudge, K. and Gabriel, S. 2014. *Farming the Woods: An Integrated Permaculture Approach to Growing Food and Medicinals in Temperate Forests*. White River Junction, VT: Chelsea Green Publishing.

Mudge, K. 2019. Agroforestry practices: forest farming. In M.R. Mosquera-Losada and R. Prabhu (Eds.), *Agroforestry for Sustainable Agriculture*. Cambridge, UK: Burleigh Dodds Science Publishing. www.bdspublishing.com

Mudge, K. 2019. Agroforestry practices: forest farming. In Mosquera-Losada, M. R. and Prabhu, R. (eds), Agroforestry for *sustainable agriculture*, Cambridge, UK: Burleigh Dodds Science Publishing,

Nabuurs, G.-J., et al. 2022: Agriculture, Forestry and Other Land Uses (AFOLU). In *IPCC, 2022: Climate Change 2022: Mitigation of Climate Change. Contribution of Working Group III to the Sixth Assessment Report of the Intergovernmental Panel on Climate Change*. Cambridge, UK and New York, NY: Cambridge University Press. https://doi.org/10.1017/9781009157926.009

Nilson, S.E., et al. 2023. Ramp (Allium tricoccum Ait.) weight differs across the harvest season: implications for wild plant stewardship and forest farming. *Agroforestry Systems*, 97(1), 97–107. https://doi.org/10.1007/s10457-022-00790-3

North Carolina Department of Agriculture & Consumer Services. 2021. Goldenseal Certificate Origin Application. Plant Conservation Program. https://www.ncagr.gov/plant-industry/goldenseal-certificate-origin-application/download?attachment

North Carolina Department of Agriculture & Consumer Services. n.d. Current NC Ginseng Dealer List. Plant Industry Division Plant Conservation Program – Ginseng. https://www.ncagr.gov/plant-industry/current-nc-ginseng-dealer-list/download?attachment

North Carolina Department of Health and Human Services. 2017. Wild Foraged Mushrooms in North Carolina Food Establishments. North Carolina Food Code Manual, Section 3-201.16. https://ehs.dph.ncdhhs.gov/docs/position/WildForagedMushroomPositionStatement-FINAL-May12-2017.pdf

O'Brien, L. 2025. Cultural Ecosystem Services, Values and Benefits. Forest Research. https://www.forestresearch.gov.uk/research/cultural-ecosystem-services-values-and-benefits/

Plants for a Future. 2010-2025. Allium ursinum. https://pfaf.org/user/plant.aspx?LatinName=Allium+ursinum.

Predny, M.L. and Chamberlain, J.L. 2005. *Goldenseal (Hydrastis canadensis): An Annotated Bibliography*. General Technical Report SRS-88. Asheville, NC: U.S. Department of Agriculture, Forest Service, Southern Research Station. 67 p. https://www.srs.fs.usda.gov/pubs/gtr/gtr_srs088.pdf

Royse, D.J. 2009. *Cultivation of Shiitake on Natural and Synthetic Logs*. University Park, PA: The Pennsylvania State University College of Agricultural Sciences. Pdf9/09_4194.

Rural Action, United Plant Savers, and Appalachia Sustainable Development. 2023. *The Forest Farmers Handbook: A Beginners Guide to Growing and Marketing At-Risk Forest Herbs*. 2nd Edition.

Schafer, C. et al. 2024. Motivation and perception of farmers on the benefits and challenges of agroforestry in Sweden. Agroforestry Systems 98: 939–958.

Smith-Hall, C. and Chamberlain, J. L. (eds). 2023. The Bioeconomy and Non-timber Forest Products, Earthscan-Routledge, London, UK.

Sobolewska, D., et al. 2015. Allium ursinum: botanical, phytochemical and pharmacological overview. *Phytochemistry Reviews*, 14(1), 81–97. https://doi.org/10.1007/s11101-013-9334-0

Trozzo, K.E., et al. (2021). Forest farming: who wants in? *Journal of Forestry*, *119*(5), 478–492. https://doi.org/10.1093/jofore/fvab023

United Plant Savers, 2022. Species at Risk List. https://unitedplantsavers.org/wp-content/uploads/2022/07/22229-UpS-Species-At-Risk-List-2022-rev-7-22.pdf

United Plant Savers. 2024. Forest Grown Verified Program. https://unitedplantsavers.org/fvg/

United States Department of Agriculture and Natural Resources Conservation Service. 2022. Conservation Practice Standard Forest Farming Code 379. https://www.nrcs.usda.gov/sites/default/files/2022-09/Forest_Farming_379_CPS_NHCP_2020.pdf

United States Department of Agriculture Forest Service. 2023. Tools; Forest Farming Calculator. https://www.fs.usda.gov/nac/resources/tools/forest-farming-calculator.shtml

United States Department of Agriculture (USDA). 2024. Climate-Smart Agriculture and Forestry. https://www.farmers.gov/conservation/climate-smart#your-operation

United States Department of Agriculture (USDA). National Agriculture Statistics Services. 2024. https://www.nass.usda.gov/Surveys/Guide_to_NASS_Surveys/Agroforestry/index.php

United States Fish & Wildlife Service. 2015. Wild American Ginseng Information for Dealers and Exporters. https://www.fws.gov/sites/default/files/documents/factsheet-american-ginseng-harvesters-dealers-exporters.pdf

University of Kentucky. 2022. Kentucky Shitake Production Workbook: Harvesting. University of Kentucky Cooperative Extension Service. https://mushrooms.ca.uky.edu/sites/mushrooms.ca.uky.edu/files/for85.pdf

University of Missouri Center for Agroforestry (UMCA). 2011. *2011 Annual Report: Agroforestry for Ecosystem Services* (M. Gold, I. Cernusca, and C. Cleavelin, Eds.). The Center for Agroforestry at MU.

University of Missouri. 2025. Tree Sugaring Resource Hub. The Center for Agroforestry. https://centerforagroforestry.org/resource/tree-sugaring-resource-hub/

University of Vermont. 2013. Best Management Practices for Log-Based Shiitake Cultivation in the Northeastern United States. University of Vermont Center for Sustainable Agriculture Extension. https://www.uvm.edu/sites/default/files/media/ShiitakeGuide.pdf

VanFleet, W. 2014. *Goldenseal under cultivation. USDA Farmers Bulletin 613*, Washington (DC), USA: United States Department of Agriculture (USDA),

Virginia Polytechnic Institute and State University. 2024. PlantShoe. https://plantshoe.frec.vt.edu/

Wexler, Justin. 2019. *Identifying Ginseng Habitat: The Northeast and Mid-Atlantic*. Treadwell, NY: Wild Hudson Valley.

World Health Organization. 2003. *WHO Guidelines on Good Agricultural and Collection Practices (GACP) for Medicinal Plants*. Geneva, Switzerland: WHO. https://iris.who.int/bitstream/handle/10665/42783/9241546271.pdf?sequence=1

Chapter 11

Developing urban agroforestry in the current crisis

S.M. Newman, BioDiversity International Ltd, UK

1 Introduction

The general aim of this chapter is to contribute to the body of knowledge for developing or designing urban agroforestry in the face of a perfect storm linked to the climate change crisis, the biodiversity emergency, and the aftermath of the coronavirus disease 2019 (COVID-19) pandemic. The key urban agroforestry lessons learned in relation to COVID-19 in the UK included the following:

- Current value chains involving supermarkets are unstable and local food supply is essential for resilience;
- Direct sales from urban agroforests can be more profitable for growers than selling wholesale, e.g. to supermarkets; and
- When governments are pressed in a crisis, they can think 'out of the box', make resources available and solve problems at pace.

http://dx.doi.org/10.19103/AS.2024.0141.17

The approach in this chapter is to find positives in relation to all of these issues and the way forward based on a review of research and agroforestry consultancy experience in over 50 countries. The specific aims of this chapter relate to trying to find answers to the following questions:

- What agroforestry sub-types can be developed in the urban context?
- What are the operationally useful definitions that can help to advance the 'out of the box thinking' that is needed for high-impact research for a rapid response?
- What are the dimensions of resilience?
- How is the research literature organised as a reflection of key issues?
- In the light of the literature and UK experience, (1) what are the unique opportunities in urban agroforestry compared to rural agroforestry? and (2) what are the key research questions in the light of a perfect storm linked to recent experience of climate change, biodiversity declines and pandemics?
- What are the operationally useful measures of effectiveness for the urgent action required?
- What are the key steps in the design process?
- What case studies in the UK could contribute to the body of knowledge?
- What are the recommendations for further research?
- What are the sources of further information?

The sections in this chapter are organised in relation to these questions.

2 Agroforestry sub-types for urban areas

Newman and Gordon (2018) discuss the importance of windbreaks, riparian strips and lines of, e.g. fruit and nut trees as linear agroforestry types. These have great potential in urban systems planted alongside roads, canals, rivers and railways. It is surprising how much land is available if one calculates the potential length of these features and then multiples this by say 10 m to allow a 5 m strip along each side of the feature.

They also describe silvopastoral systems, silvoarable systems, forest farming and orchard systems. These are, again, all possible in the urban environment if there is enough land. Most of the land available, however, is at the urban fringe or in the peri-urban zone. As a rule of thumb, a 10 km ring around the edge of a built area will approximate to at least ten times the green space available within the city/town limits. The only system not described by Newman and Gordon is the silvo-aquatic system, where a strip of trees gives beneficial shade to understorey floating gardens.

3 Operationally useful definitions relating to urban agroforestry

Operationally useful definitions differ from standard definitions in that they empower conceptual thinking on design. The 'original' definition of agroforestry from the World Agroforestry Centre (WAC) (based in Nairobi, Kenya) is:

> a collective name for land use systems and technologies where woody perennials (trees, shrubs, palms, bamboos, etc.) are deliberately used on the same land management unit as agricultural crops and/or animals, either in some form of spatial arrangement or temporal sequence. In agroforestry systems, there are both ecological and economic interactions between the different components.

This definition is fine for classification of systems by professional agronomists or foresters but lacks the notion that urban agroforestry can be at the landscape scale. It also lacks the power to enthuse lay persons or youth. A more suitable definition for this audience could be as follows:

> Urban agroforestry is a form of nature-based solution that can be applied at the whole city or town scale. The design of an agroforestry subsystem embodies the idea that trees or other woody plants can act as a 'mother' to adjacent crops, mushrooms, or animals and in certain circumstances the 'children' can return the favour.

The definition of agriculture by Spedding (1988) is clear in meaning to most people. Agriculture

> is an activity ... carried out primarily to produce food and fibre (and fuel as well as many other materials) by the deliberate and controlled use of mainly terrestrial plants and animals.

The word farmer has an etymology that gives a different understanding of the concept than most people's idea. People imagine that a farmer must have a farm, so the place is key. The etymology of the word farm however is linked to the Latin term *firma* meaning contract. In this sense, anyone who has a 'contract with society' to provide food could be described as a farmer.

The word forest is problematic; however, in that it has two very different definitions and has very strong psychological undertones. The FAO definition of forest is structural:

> Land spanning more than 0.5 hectares with trees higher than 5 meters and a canopy cover of more than 10 percent, or trees able to reach these thresholds in situ. It does not include land that is predominantly under agricultural or urban land use.

This includes single-species plantations and agroforestry (intercropped timber trees used in the crop establishment phase is known as 'Taungya' in some countries). In England, the traditional definition was legal and linked to land designated for use by royalty or nobility for hunting purposes. As well as woodland and forest areas, this land could contain pasture for deer etc., with interspersed or boundary trees or, in extreme cases, might not include any trees at all.

Differing ideas of what a 'forest' means are highlighted in the following statements:

- *'The forest is a dark place full of mystery and at times frightening'.*
- *'A forest is balanced, and its components are in harmony'.*
- *'Cultivating the soil in a forest will destroy the balance and destroy mycorrhizae'.*
- *'Disturbance by macrofauna is the key to maintaining high species diversity in key taxa'.*

The definition of a food forest is evolving fast and there are now over 25 large (over 0.5 hectare) scale food forests in The Netherlands. One definition of a food forest from The Netherlands is as follows:

- a type of agroforestry;
- a productive ecosystem with predominantly woody (edible) species;
- designed on the analogy of a natural forest with;
  - a canopy of tall trees
  - at least three other vegetation layers
  - a robust size
  - a rich undisturbed forest soil
  - a large rapidly increasing biodiversity
- no livestock or annual crops; and
- without the use of heavy machinery manure fertilisers and/or pesticides.

This contrasts with the definition of a forest garden. Based on the discussions with Robert Hart in the 1980s, a forest garden is:

> a garden consisting of at least 3 layers (tree, shrub and ground) with up to 80 species including trees that could produce good quality timber as a by-product of nut production designed primarily for food production, with the potential to support a family of up to 10 people on an area of half a hectare.

Robert's forest garden contained many annual plants. Many forest gardens in the UK contain chickens for egg production and the garden benefits from increased nutrients and weed control.

The definition of urban is simple, it means pertaining to a city or town. Cities and towns throughout the world vary enormously in the current percentage of green space where plants and/or crops are grown. One extreme example is the new city of Milton Keynes in England, which has a green infrastructure of around 76% of its land surface.

4 Dimensions of 'resilience'

For individual 'agriculturalists' to be resilient in a society, ideally, they need to have skills, friends, land, shelter and money. This notion was embodied in the sustainable livelihood approach developed for use in overseas development during the 1990s. This was based on the idea that a person has assets linked to five types of capital:

1 Human capital, e.g. health, nutrition, education, knowledge and skills, capacity to work and capacity to adapt.
2 Social capital (e.g. networks and connections, friends, patronage, neighbourhoods, kinship), relations of trust and mutual understanding and support, formal and informal groups, shared values and behaviours, common rules and sanctions, collective representation, mechanisms for participation in decision-making, leadership.
3 Natural capital, e.g. land and produce, water and aquatic resources, trees and forest products, wildlife, wild foods and fibres and biodiversity, environmental services.
4 Physical capital, e.g. infrastructure (transport, roads, vehicles, secure shelter and buildings, water supply and sanitation, energy, communications), tools and technology (tools and equipment for production, seed, fertiliser, pesticides, traditional technology).
5 Financial capital, e.g. money savings, credit, and debt (formal, informal), remittances, pensions and wages.

This is still considered to be useful today and is embedded in many aspects of approaches used for assessment using the economics of ecosystem and biodiversity.[1] However, it should be noted that the approach underplays the significance of power relationships and the fact that one person's livelihood can undermine those of another.[2]

For the monitoring and evaluation of approaches to climate change resilience and for the prioritisation of actions, the simplest quantitative measures of performance relate to mortality, morbidity and assets. Climate

1 https://teebweb.org
2 The Sustainable Livelihoods Approach (adb.org)

change resilience places more emphasis on bouncing back from a crisis than the older concept of climate change adaptation. It also acknowledges the positive aspects of climate change, such as new crop possibilities and the availability of climate change finance.

5 Current literature on urban agroforestry

A literature search was confined to the use of CAB Abstracts which contains summaries of journal articles, conference proceedings and other literature in the field of applied life sciences, nutrition, food science, human health, environmental science, agriculture and plant sciences, as well as tourism, leisure and the hospitality industries. The period of records was from 1973 to March 2023. Two searches were carried out:

- Agroforestry using the keywords Agroforestry and Urban together in the article title.
- Food Forest using the keywords Urban and Food and Forest together in the article title.

Table 1 shows the results for search 1. Table 2 shows the results for search 2.

Three of the most useful papers in the context of designing urban agroforestry for resilience in the current crisis are the review by Taylor and Taylor Lovell (2021), the modelling approach to predicting ecosystem services provision by the systems by Zanzi et al. (2021) and nutritional yield of peri urban food forests by Nytofte and Henriksen (2019).

Table 1 Literature search: agroforestry

Issue	Number of published papers
Urban environmental restoration	1
Food safety and nutrition	2
Multifunctional design guidelines	1
Biodiversity (species diversity) including agrobiodiversity	3
Urban planning, green infrastructure (including green space) and architecture	2
Title only available (book of abstracts)	6
Traditional systems	2
Governance, food/wood security and common land	3
Total	19

Table 2 Literature search: forest farming

Issue	Number of published papers
Urban environmental restoration	2
Food safety, contaminated land	3
Multifunctional design guidelines or reflections	6
Biodiversity (species diversity) including agrobiodiversity	3
Urban planning, green infrastructure (including green space) and architecture	0
Title only available (book of abstracts)	1
Traditional systems	1
Governance, food/wood security food sovereignty and common land	10
Environmental psychology, e.g. biocultural conservation, gender/ race aspects and social benefits	7
Carbon sequestration and bioenergy	4
Not relevant (mainly papers pre-2017)	102
Emergency, e.g. COVID	1
Total	140

6 What are the unique opportunities in urban agroforestry?

Based on reviewing the literature, the most important and possibly unique opportunities for urban and/or peri-urban agroforestry are:

- capture of 'land betterment value' linked to compulsory purchase of land by local government. This is explained further in case study 1 below;
- increased availability of volunteers, skilled and unskilled labour and funds for employment as part of improving mental health and reducing criminality;
- larger market for produce and ecosystem services (especially water and waste processing);
- larger area of flat roofs and wall surfaces for rooftop and vertical agroforestry;
- larger area of water infrastructure, e.g. reservoirs, balancing lakes (used to reduce flooding from road runoff) and sewage treatment works (also a source of nutrients) for agroforestry with a floating garden understorey possibly combined with aquaculture;
- greater social capital is linked to ethnic diversity (knowledge on growing, cooking and well-being linked to food);
- lower transport costs (financial and carbon) from 'field to fork';

- increased tree health linked to contaminated land; and
- increased availability of organic wastes and composts made from waste for mulching, tree protection and soil fertility improvement.

Based on the review of the literature and over 40 years practical experience, the key research questions are as follows:

- Given the management complexity of the systems, what are the key steps in designing successful urban agroforestry?
- If the aim of the agroforestry system is to feed the maximum number of people per unit area, what are the most useful insights on species combinations?
- What part of the city or town would give the greatest surface area for urban agroforestry?

7 Urban agroforestry design

A key publication in the UK is the *Agroforestry Handbook*[3] published in 2019 by the Soil Association. The key elements of agroforestry design can be summed up in the acronym PAMASAL. This stands for

Purpose: Where do you want to get to?

Advice: Where will you get advice and support?

Measures of success: How will you measure efficiency, effectiveness and impact?

Agroecology: How will you capitalise on agroecology or in other words let nature do some of the work?

Starting points: Where do you want to start from? e.g. pastureland, arable land or an orchard etc.

Adaptive management: As things develop, what tools will help you to carry out adaptive management?

Layout: What varieties/species, spatial arrangement and sequencing will be part of your design?

Ideally, the purpose of the system should be a simple single objective that, when achieved, is verifiable for a lay person. There will always be co-benefits, but management will be simpler if you have one overarching purpose, e.g. to produce food or to reduce some form of pollution etc.

The advice needed to develop agroforestry is quite considerable and the most commercially successful agroforestry ventures involve a partnership. A typical approach would be to use three partners, one focused on the trees, another on the understorey crop or animal and another on adding value to the products, e.g. fermentation/preservation of fruits.

3 the-agroforestry-handbook.pdf (soilassociation.org)

Key agronomic measures of success (if restricted to yield) can be divided into three approaches:

- Main crop yield;
- Feedstock yield; and
- Multicrop yield.

For main crop yield, the focus is on one species and a yield target is stated. Other crops or animals can then be added and, if they do not reduce the yield of the main crop, then the system is acceptable. For feedstock yield, idea is to produce something for conversion, e.g. feed for an animal or biomass for bioenergy. Essentially, the species composition is not as relevant as the specification of the resulting fodder or biomass. Interestingly, food per unit area measured in calories would be an example of feedstock yield. The final measure, multicrop yield, is measured using the land equivalent ratio, which is the sum of the relative yields. If both walnut and wheat are known to give 10 tonnes per ha as monoculture and, when mixed on the same area, gave 5 tonnes, then the LER would be 5/10 plus 5/10 or 1. This means no advantage. In my own work, I have found 7/10 plus 7/10 which gives a value of 1.4. This means that 40% more land would be required to obtain the same yield from monocultures.

The best way to look at the 21st agroforestry agroecology design toolkit is to think of the acronym MSP:

Modulation. This is where one component manages or modulates the physical environment of another component. A tree may reduce the heat loss by livestock through acting as a windbreak, for example.

Synthesis. This is where one component changes the chemical environment of another component. A shrub may fix atmospheric nitrogen and this nitrogen may be made available to an adjacent non-N-fixing crop.

Partitioning. This means sharing. It can happen in both space and time. A deep-rooted tree may get its phosphate requirements from a deep layer of soil below the roots of an adjacent crop and avoid competition.

Any design has to start with some growing space. This will have properties linked to plant nutrients, elevation and exposure that will determine species selection. In most cases, the starting point will be one of the following:

- Pastureland
- Arable land
- Orchards
- Woodland
- 'Edge land'
- Water bodies

It is important to factor in adaptive management as part of any agroforestry design. Trees and woody plants will grow, the effects on adjacent land use will change and new results, opportunities and constraints will appear. It may be useful to design an agroforestry management plan that uses phases such as establishment, mature and over-mature and builds in a plan to adapt to these phases.

For layout there are four variables:

1 Species;
2 Spatial pattern on the ground;
3 Use or occupation of three-dimensional space; and
4 Phasing.

8 Measures of effectiveness: 'the forest garden challenge'

At a virtual international conference on forest gardens held in 2021, I reviewed the low productivity of most forest gardens in the literature and announced a resilience challenge to designers and practitioners to achieve the following:

- 70 adults have adequate food energy per hectare;
- Double the number of friends and useful advisors (social capital) that a person has access to without cost;
- Total income per hectare per adult from the garden minus costs greater than the minimum wage £8.91 per hour, which is £17 143 per year based on 52 weeks of 37 hours;
- 50% Cost reduction on family food purchases; and
- Level of family indebtedness reduced by at least 50%.

9 Case study 1: feeding people using urban agroforestry in the UK

In a paper published in 2019, Nytofte and Henriksen studied the yield and nutritional potential of a 0.8 ha peri-urban food forest (99 species) in Coldstream in Scotland between 2011 and 2017. They found that the average annual yield was 8913 kg dry matter per ha, and this could provide 518 844 kcal per ha per year. This energy is enough to support 0.6 adults per ha per year, assuming that the average adult requires 2500 kcal per day (this is 912 500 kcal per year). They also found that the dry matter of food from a forest garden gave 9868 g protein, 9394 g fat and 85 627 g carbohydrates.

Assuming a carbohydrate-rich diet, where the maximum recommended 60% of energy comes from carbohydrates and the remaining 40% (if divided between 25% fat and 15%) from protein, then 1 ha of food forest with the same species composition as the garden cottage food forest would be able to supply up to seven males or five females with carbohydrate, four males or five females with fat and three males or five females with protein. They suggest that this could be potentially increased by incorporating more protein and fat crops such as legumes and nut trees. In conclusion, the food yield is very low: 0.3 of the 'Robert Hart target' of 20 people per ha.

This represents a major opportunity for improving symbiosis through using agro-ecological principles. How does the food production in a forest garden compare with (1) the most productive monoculture in the UK, (2) a model of optimised agroforestry and (3) the maximum theoretical limit for photosynthesis?

The UK 2018 potato crop averaged 41.7 tonnes per ha. However, some fields yield above 80 tonnes per ha.[4] Assuming 73% moisture, this would give 11 dry tonnes per ha on average, with 22 tonnes per ha in some fields. These figures are based upon the current UK practice of applying fertiliser and sprays which currently have a high carbon footprint and cost.

According to Nix (2020), for a maincrop potato crop with an average expected yield (45 tonnes per ha), the combined cost would be £1087 per ha, which represents 21% of the variable costs of £5213 per ha. The gross margin expected from this financial model would be £1403 per ha. Given the level of care and attention that a good gardener can achieve, 22 tonnes per ha is a realistic target using careful water management and biological approaches to fertilisation. A good gardener should also make a better gross margin given reduced casual labour and agrochemical input costs. The carbon footprint of the operation should be very low and, if designed carefully, could be carbon negative.

About 100 g of fresh potato has a calorific value of 82 kcal, which is equivalent to 820 000 kcal per tonne. Using the above yield data, this equates to 34 194 000 kcal per ha on average or 65 600 000 kcal per ha in some fields. The calorie yield is therefore 126 times better on fields that would represent yields attainable by a good gardener.

Assuming that the average adult requires 2500 kcal per day (i.e. 912 500 kcal per year) enough to support 72 adults per ha per year assuming total digestion (100% food conversion) of raw potato. The suggestion of Nytofte and Henriksen, about nut crops, is very relevant for forest, gardens as they have great potential in the production of high value crops in terms of both income and nutrition. Table 3 shows the potential energy contribution.

4 Are 100t/ha potato crops possible? – *Farmers Weekly* (fwi.co.uk)

Table 3 Potential energy contribution from nuts

Tree	Kcal per 100 g dry kernel	Per tonne	Target yield (UK) tonnes per ha fresh nuts	Total energy per ha kcal assuming zero moisture	Number of persons supported (2500 kcal × 365 days)	Number of persons Fiddle factor 40%
Walnut	654	6540000	5	32700000	35.8	14
Oak	353	3530000	0.5	1765000	1.9	1
Chestnut	369	3690000	2.1	7749000	8.5	3
Hazelnut	628	6280000	3	18840000	20.6	8

It is difficult to find data that shows the percentage dry kernel available from a freshly harvested nut crop. I have assumed that yields for harvested nuts do not include the green husk. I have used a general 'fiddle' factor of 40% dry kernel for all freshly harvested nuts in my table to allow for moisture content and the weight of the discarded shell.

Optimised walnut monoculture could produce enough energy to support 14 adults per year, assuming total digestion (100% food conversion) of dry kernels. This is almost three times that of the forest garden studied.

Assuming that, in an optimised agroforestry system, it is possible to get 100% of the yield of each component, an ideal walnut–potato agroforestry system could support 72 + 14 or 86 adults per hectare per year.

Recent calculations on the maximum theoretical productivity of the planet now give a figure of 200 dry tonnes per ha (DeLucia et al., 2014). Assuming that the calorific value of this biomass was the same as potato of 820 000 kcal per tonne, this gives 164 000 000 kcal which is enough to support 180 adults per hectare per year. Table 4 compares the number of people supported in terms of energy requirements (2500 kcal by 365 days) by the different systems.[5]

In conclusion, the 'productivity' of UK forest gardens is only 3% of the target set by Robert Hart. An optimised forest garden based on potatoes and walnut could produce over four times the productivity of what he envisaged. The challenge is to obtain these targets and at the same time be at least carbon neutral.

How should we reflect on the Hart target if we set 2031 as the attainment date? Given the current biodiversity crisis and the need for land sparing (for biodiversity) in the UK to achieve this, I think 70 adults supported per hectare by forest gardening or agroforestry is an especially useful target that would be a boon to creative thinking. If this is attained, then the average garden would

5 Calories in Potatoes, Raw, Peeled, Nutrition Information | Nutracheck

Table 4 Comparing different systems

System	People supported per hectare	Hart relationship
Forest garden studied	0.6	3%
'Hart target'	20	
Optimised potato monoculture	72	× 3.6
Optimised potato walnut agroforestry	86	× 4.3
Maximum photosynthesis	180	× 9

support (0.02 × 70) 1.4 people or (0.04 × 70) 2.8 people. A key question that follows on from this is

> Should we get the food energy from (1) a tuber, a sallet, a grain, or an oil from an oil seed (e.g. rape) or nut (e.g. walnut or hazel), or (2) a combination of these?

The answer is a combination as this would give better nutrition. In agroecological terms, where the energy comes from in terms of the tree layer, or the field layer is a function of a plants, seasonality, and role in temporal partitioning. In terms of the design of the walnut potato system, the ideal would be a cold tolerant (plant in January?) potato, intercropped under a late leafing walnut. The potato would be harvested say in July with a follow-on crop such as shade-tolerant leguminous sallet crop that could withstand frost and cold in November to December.

10 Case study 2: land and finance available for urban agroforestry in the UK

The greatest opportunity for land in the UK is at the urban fringe or in the peri-urban zone. In UK planning terms, this is conventionally known as the 'Green Belt'. We now need to think of it as a 'green life jacket' with ecosystem services that can save the neighbouring city or town. If even better integrated, the revenue generated from peripheral forest villages or forest garden towns could pay for climate change resilience in a more socially just world.

In 2012, the UK Government issued the National Planning Policy Framework,[6] which stated in Para. 52 stated that

> the supply of new homes can sometimes be best achieved through planning for larger scale development, such as new settlements or extensions to existing villages and towns that follow the principles of Garden Cities.

The UK Town and Country Planning Association has made passing reference to food growing in one of its Garden City Principles[7] – 'Beautifully and imaginatively

6 *National Planning Policy Framework*. Department for Communities and Local Government, Mar. 2012. www.gov.uk/government/publications/national-planning-policy-framework
7 The TCPA's Garden City Principles are set out at www.tcpa.org.uk/garden-city-principles

designed homes with gardens, combining the best of town and country to create healthy communities, and including opportunities to grow food'.

The case study is based on compulsory purchase of agricultural land at agricultural land values of £20 000 per ha. Planning authorities have the statutory power to do this and are at liberty to give further inducements to landowners to sell. The garden city of Milton Keynes was developed using this approach and is a global exemplar of water harvesting and management.

Newman (2018) updated the work of Ebeneezer Howard, whose seminal work provided the vision for combining the virtues of town and country. This led to the legislation and community of practice associated with modern town planning. Newman (2018) developed a replicable spreadsheet model that showed how new agroforestry settlements could be integral to the restoration of the countryside (natural capital) and not an imposition upon it. In other words, in place of Ebenezer Howard's *To-morrow*, we could have the chance today to follow a peaceful path to real rural and community reform in the face of the perfect storm.

The financial case was based on the purchase of 1000 acres with a 30 acres new forest village (housing 2500 people) to give the footprint for urban agroforestry at the landscape scale. Analysis of the 1000-acre model gave a total project cost (land purchase and building) of £38.5 million. A target annual rent from the village of £10.92 million would be met in year 2 of the project. A 20% return on the £38.5 million could be paid to investors after year 5 (the 'vanishing point' in Howardian economics). After the vanishing point, an annual perpetual income (rent-rate) to the 'village development board' would be over £11.5 million – as per the Letchworth Garden City model. The resulting 'nature restoration and community belonging fund' or co-operative commonwealth could provide outcomes including:

- a carbon-negative village;
- improved biodiversity and ecosystem services and enhanced natural capital;
- improved health and a cottage hospital;
- improved education via a forest school-cum-university (forest campus);
- carbon-negative transport;
- shops, markets and places that build on the strengths of ethnic diversity and the variety of learning skills;
- a citizens' wage; and
- other sustainable development goals.

The model assumed no investment costs for rural land improvements and no increase in rents over a 20-year period.

There are now major sources of finance available from pension funds, carbon sequestration linked funds and Impact funds with attractive low rates of interest. Environmentally and socially aware companies have the possibility to use eventual carbon sequestration in a supported new urban agroforestry model, as part of their greenhouse gases mitigation goals as part of a tripartite environmental stewardship contract, leading to a win-win situation to environment and society. The three parties could include the company, government and the forest village/town community. The contract would need a simple target verifiable by a lay person, e.g. earth worm density increase (strongly correlated with soil carbon), a reasonable time frame, e.g. 20 years, a legal protection fund for the weakest partner and a statement of synergy (what each partner brings).

11 Conclusion and future trends

Further research is required so that practitioners can design and develop urban agroforestry at scale. The two driving questions remain:

- What is the most efficient way to produce food from diverse systems in the urban context in terms of number of persons supported per unit area?
- What infrastructure (including carbon-negative housing) could be built into a bankable multi-partner landscape scale urban agroforestry project in a manner that provides a multi-million-pound income for a climate change and biodiversity resilience fund for ever?

12 Where to look for further information

How to be inspired when thinking about:	Reference
Temperate agroforestry	Gordon et al. 2018
Orchard intercropping and LER>2	Newman 1986
Agroecology and earthworms	Darwin 1881
Water in agroforestry	Yeomans 1958
Trees in agroforestry based upon indigenous knowledge, e.g. China/Corsica	Smith 1929
Reform of urban governance and the synergy between agriculture and urban development	Howard 1898
Landscape scale urban agroforestry and the peaceful pathway to reform	Newman 2018
Finding and using indigenous knowledge	Agarwal and Narain 1997
Nut based livelihoods	Squatriti 2013
Growing food in containers as part of food forests or agroforestry or with further development; floating gardens	(304) From Zero to 1st Early Potato Hero: Guide for Growing Determinate Potatoes in Containers – YouTube

13 References

Agrawal, A. and Narain, S. 1997. *Dying wisdom: rise, fall and potential of India's traditional water harvesting systems*. Centre for Science and Environment, New Delhi.

Darwin, C. 1881. *The formation of vegetable mould*. John Murray, London.

DeLucia, E. H., et al. 2014. The theoretical limit to plant productivity. *Environ. Sci Technol.* 48, 9471–9477.

Gordon, A. M., Newman, S. M. and Coleman, B. R. W. 2018. *Temperate agroforestry systems* (2nd ed.). CAB International, Boston, MA, p. 313. ISBN 9781780644851.

Howard, E. 1898. *Tomorrow: A Peaceful Path to Reform* (subsequently republished as: Garden Cities of Tomorrow). Swann Sonnenschein and Co., London.

Newman, S. M. 1986. A pear and vegetable interculture system: land equivalent ratio, light use efficiency and dry matter productivity. *Exp. Agric.* 22, 383–392.

Newman, S. M. 2018. Rural new settlements in 2018: what might Ebenezer Howard Say? *Town Count Plan.* 2018, 301–305.

Newman, S. M. and Gordon, A. M. 2018. Temperate agroforestry: key elements, current limits and opportunities for the future. Chapter 12. In Gordon, A. M., et al. (Eds.), *Temperate agroforestry systems* (2nd ed.). CAB International, Boston, MA, pp. 274–298.

Nix, J. 2020. *(By G Redman) John Nix pocketbook for farm management*. Agro Business Consultants Ltd, Melton Mowbray. ISBN 978-0-9576 939-7-5

Nytofte, J. L. S. and Henriksen, C. B. 2019. *Sustainable food production in a temperate climate - a case study analysis of the nutritional yield in a peri-urban food forest*. Urban Forestry & Urban Greening vol 45 Elsevier GmbH.

Smith, J. R. 1929. *Tree crops: a permanent agriculture*. Harcourt Brace, New York. ISBN 0-933280-44-0.

Squatriti, P. 2013. *Landscape and change in early medieval italy: chestnuts, economy, and culture*. Cambridge University Press, Cambridge. ISBN 978-1-107-03448-8.

Spedding, C. R. W. 1988. *An introduction to agricultural systems* (2nd ed.). Elsevier Applied Science, London.

Yeomans, P. A. 1958. *The challenge of the landscape*. Keyline Publishing, Sydney.

Taylor, J. R. and Taylor Lovell. S. 2021. Designing multifunctional urban agroforestry with people in mind. Urban. *Agriculture and Regional Food Systems*, 6, 20016.

Zanzi, A., et al. 2021. Forecasting agroforestry ecosystem services provision in urban regeneration projects: experiences and perspectives from milan. *Sustainability*, 13, 2434. https://doi.org/10.3390/su13052434

Index

www.ingramcontent.com/pod-product-compliance
Ingram Content Group UK Ltd.
Pitfield, Milton Keynes, MK11 3LW, UK
UKHW021826100926
8556IPUK00003B/178

9 781801 467193